The First Humans

Vertebrate Paleobiology and Paleoanthropology Series

Edited by

Eric Delson
Vertebrate Paleontology, American Museum of Natural History,
New York, NY 10024, USA
delson@amnh.org

Ross D. E. MacPhee
Vertebrate Zoology, American Museum of Natural History,
New York, NY 10024, USA
macphee@amnh.org

Focal topics for volumes in the series will include systematic paleontology of all vertebrates (from agnathans to humans), phylogeny reconstruction, functional morphology, paleolithic archaeology, taphonomy, geochronology, historical biogeography, and biostratigraphy. Other fields (e.g., paleoclimatology, paleoecology, ancient DNA, total organismal community structure) may be considered if the volume theme emphasizes paleobiology (or archaeology). Fields such as modeling of physical processes, genetic methodology, nonvertebrates or neontology are out of our scope.

Volumes in the series may either be monographic treatments (including unpublished but fully revised dissertations) or edited collections, especially those focusing on problem-oriented issues, with multidisciplinary coverage where possible.

Editorial Advisory Board

For other titles published in this series, go to
www.springer.com/series/6978

The First Humans - Origin and Early Evolution of the Genus *Homo*

Contributions from the Third Stony Brook
Human Evolution Symposium and Workshop
October 3 - October 7, 2006

Edited by

Frederick E. Grine

Departments of Anthropology and Anatomical Sciences
Stony Brook University
Stony Brook, NY 11794 USA

John G. Fleagle

Department of Anatomical Sciences
Stony Brook University
Stony Brook, NY 11794 USA

Richard E. Leakey

Department of Anthropology and Turkana Basin Institute
Stony Brook University
Stony Brook, NY 11794 USA

Springer

Editors
Frederick E. Grine
Departments of Anthropology
and Anatomical Sciences
Stony Brook University
Stony Brook
NY 11794
USA

John G. Fleagle
Department of Anatomical Sciences
Stony Brook University
Stony Brook
NY 11794
USA

Richard E. Leakey
Department of Anthropology and
Turkana Basin Institute
Stony Brook University
Stony Brook
NY 11794
USA

ISBN: 978-1-4020-9979-3 e-ISBN: 978-1-4020-9980-9
DOI 10.1007/978-1-4020-9980-9

Library of Congress Control Number: 2009927083

Cover illustration: Photographs of KNM-ER 3733 (*left*) and KNM-ER 1470 (*center*) by Robert Campbell (© and courtesy of Robert Campbell and the Kenya National Museum), photograph of SK 847 (*right*) by Frederick E. Grine (© and courtesy of Frederick E. Grine and the Transvaal Museum). Image design by Luci Betti-Nash.

Printed on acid-free paper

Springer is part of Springer Science+Business Media (www.springer.com)

Preface

There are some issues in human paleontology that seem to be timeless. Most deal with the origin and early evolution of our own genus – something about which we should care. Some of these issues pertain to taxonomy and systematics. How many species of *Homo* were there in the Pliocene and Pleistocene? How do we identify the earliest members the genus *Homo*? If there is more than one Plio-Pleistocene species, how do they relate to one another, and where and when did they evolve? Other issues relate to questions about body size, proportions and the functional adaptations of the locomotor skeleton. When did the human postcranial "Bauplan" evolve, and for what reasons? What behaviors (and what behavioral limitations) can be inferred from the postcranial bones that have been attributed to *Homo habilis* and *Homo erectus*?

Still other issues relate to growth, development and life history strategies, and the biological and archeological evidence for diet and behavior in early *Homo*. It is often argued that dietary change played an important role in the origin and early evolution of our genus, with stone tools opening up scavenging and hunting opportunities that would have added meat protein to the diet of *Homo*. Still other issues relate to the environmental and climatic context in which this genus evolved. Were there global or pan-African climatic events that relate to the appearance and/or extinction of *Homo* species, and if so, can they be tied to the appearance or disappearance of these species in any meaningful way? Did *Homo* species live in environments that differed from those inhabited by earlier hominins, and can any general trends through time be inferred from paleontological and isotopic evidence?

The announcement, over 4 decades ago, of the fossil remains of *Homo habilis* from Olduvai Gorge by Louis Leakey, Phillip Tobias, and John Napier marked a number of major changes in our knowledge and interpretation of human evolution. We have certainly come a long way since the early 1960s towards appreciating the origin of our genus. New finds and analyses have provided some critical information, and have, at the same time, led to new questions. Still, there remain a significant number of unresolved issues. This is the way it should be, and what one expects. Although some of the questions appear to be the same as before, new techniques and interpretations have opened up other avenues of enquiry and have led to new questions for which answers can hopefully be found.

In an effort to update, address and hopefully synthesize our current understanding of this preeminently significant development in human evolution, we organized the Third Stony Brook Human Evolution Symposium and Workshop in 2006. An international group of acknowledged experts in their respective fields assembled for 5 days of discussion and debate on a wide range of topics related to the origin of the genus *Homo*. This volume is the result of those activities. The chapters they have contributed to it represent what we know, and what knowledge we still wish for in the quest to understand the evolution of the first humans.

The workshop was sponsored by Stony Brook University and the Turkana Basin Institute. Many people and organizations made it possible, including the Office of the Provost of Stony Brook University, the Wenner-Gren Foundation, and generous contributions from David and Maureen Acker, Ed and Frances Barlow, Bill and Kathy Cleary, Charles and Ursula Massoud, Allan and Diana Rothstein, Jim and Marilyn Simons, and Kay Harrigan Woods. The success of the symposium and workshop owes much to the tireless efforts of Elizabeth Wilson, Christopher Gilbert, Danielle Royer, Matthew Sisk and Ian Wallace.

The chapters in this volume underwent formal peer-review, and we would like to take this opportunity to thank our colleagues who gave so freely of their time and expert opinion to assist in this process. We also thank Eric Delson, senior co-editor of the Vertebrate Paleontology and Paleoanthropology Series, for his support, assistance and encouragement in bringing this volume to publication. Robert Foley, Marta Lahr and the other faculty, staff and students at the Leverhulme Center for Human Evolutionary Studies, The University of Cambridge, graciously provided Fred Grine with accommodation and support over the 2007–2008 academic year while he was on sabbatical, and during the height of his editorial work on this volume.

Stony Brook University

Frederick E. Grine
John G. Fleagle
Richard E. Leakey

Contents

Contributors

Robert J. Blumenschine
Center for Human Evolutionary Studies, Department of Anthropology, Rutgers University, The State University of New Jersey, 131 George Street, New Brunswick, NJ 08901-1414, USA
rjb@rci.rutgers.edu

René Bobe
Department of Anthropology, University of Georgia, Athens, GA 30602, USA
renebobe@uga.edu

Dennis M. Bramble
Department of Biology, University of Utah, Salt Lake City, UT 84112, USA
Bramble@bioscience.utah.edu

M. Christopher Dean
Department of Cell and Developmental Biology, University College London, Anatomy Building, Gower Street, London WC1E 6BT, UK
ucgacrd@ucl.ac.uk

John G. Fleagle
Department of Anatomical Sciences, Stony Brook University Medical Center, Stony Brook, NY 11794-8081, USA
john.fleagle@stonybrook.edu

Frederick E. Grine
Departments of Anthropology and Anatomical Sciences, Stony Brook University, Stony Brook, NY 11794-4364, USA
frederick.grine@stonybrook.edu

Christopher P. Heesy
Department of Anatomy, Arizona College of Osteopathic Medicine, Midwestern University, 19555 N 59th Avenue, Glendale, AZ 85308, USA
metacephalon@gmail.com

William L. Jungers
Department of Anatomical Sciences, Stony Brook University Medical Center, Stony Brook, NY 11794-8081, USA
william.jungers@sunysb.edu

William H. Kimbel
Institute of Human Origins, School of Human Evolution and Social Change, Arizona State University, P.O. Box 874101, Tempe, AZ 85287-4101, USA
wkimbel.iho@asu.edu

Susan G. Larson
Department of Anatomical Sciences, Stony Brook University Medical Center, Stony Brook, NY 11794-8081, USA
susan.larson@stonybrook.edu

Meave G. Leakey
National Museums of Kenya, Department of Earth Sciences (Palaeontology), Nairobi 00502, Kenya; Turkana Basin Institute, Stony Brook University, Stony Brook, NY 11794-4364, USA
meave.leakey@stonybrook.edu

Richard E. Leakey
Department of Anthropology and Turkana Basin Institute, Stony Brook University, Stony Brook, NY 11794-4364, USA
richard.leakey@stonybrook.edu

Julia A. Lee-Thorp
Division of Archaeological, Geographical and Environmental Sciences, University of Bradford, Bradford BD7 1DP, UK
J.A.Lee-thorp@Bradford.ac.uk

Daniel E. Lieberman
Departments of Anthropology and Organismic & Evolutionary Biology, Harvard University, 11 Divinity Avenue, Cambridge, MA 02138, USA
danlieb@fas.harvard.edu

David Lordkipanidze
Georgian National Museum, 3 Purtseladze Street, 0105 Tbilisi, Republic of Georgia
dlordkipanidze@museum.ge

Mark A. Maslin
Environmental Change Research Centre, Department of Geography, University College London, 26 Bedford Way, Pearson Building, Gower Street, London WC1H 6BT, UK
mmaslin@geog.ucl.ac.uk

David A. Raichlen
Department of Anthropology, University of Arizona, 1009 E. South Campus Drive, Tucson, AZ 85721, USA
raichlen@email.arizona.edu

Kaye E. Reed
School of Human Evolution and Social Change, Institute of Human Origins, Arizona State University, P.O. Box 874101, Tempe, AZ 85287-4101, USA
kreed.iho@asu.edu

G. Philip Rightmire
Department of Anthropology, Peabody Museum, Harvard University, Cambridge, MA 02138, USA
gprightm@fas.harvard.edu

Hélène Roche
Directeur de recherches au CNRS, UMR 7055 du CNRS, Université Paris X – Nanterre, Maison de l'Archéologie et de l'Ethnologie, 21 allée de l'Université, 92023 – Nanterre, France
helene.roche@mae.u-paris10.fr

Samantha M. Russak
School of Human Evolution and Social Change, Institute of Human Origins, Arizona State University, P.O. Box 872402, Tempe, AZ 85287-2402, USA
Samantha.russak@asu.edu

Robert S. Scott
Department of Anthropology, Rutgers University, The State University of New Jersey, 131 George Street, New Brunswick, NJ 08901-1414, USA
robertsc@rci.rutgers.edu

John J. Shea
Department of Anthropology and Turkana Basin Institute, Stony Brook University, Stony Brook, NY 11794-4364, USA
john.shea@stonybrook.edu

B. Holly Smith
Museum of Anthropology, The University of Michigan, Ann Arbor, MI 48109-1107, USA
bhsmith@umich.edu

Emma J. Smith
Pharmaceutical Product Development, Inc., 929 North Front Street, Wilmington, NC 28401, USA
Emma.Smith@wilm.ppdi.com

Heather F. Smith
School of Human Evolution and Social Change, Arizona State University, P.O. Box 872402, Tempe, AZ 85287-2402, USA
Heather.F.Smith@asu.edu

Matt Sponheimer
Department of Anthropology, University of Colorado at Boulder, Boulder, CO 80309, USA
msponheimer@gmail.com

Phillip V. Tobias
School of Anatomical Sciences, Institute of Human Evolution, University of the Witwatersrand Medical School, 7 York Road, Parktown 2193, South Africa
Phillip.Tobias@wits.ac.za

Martin H. Trauth
Institut für Geowissenschaften, Universität Potsdam, POB 601553, D-14415 Potsdam, Germany
trauth@geo.uni-potsdam.de

Peter S. Ungar
Department of Anthropology, University of Arkansas, Old Main 330, Fayetteville, AR 72701, USA
pungar@uark.edu

Bernard A. Wood
Center for the Advanced Study of Hominid Paleobiology, Department of Anthropology, George Washington University, 2110 G Street NW, Washington DC 20052, USA
bwood@gwu.edu

Part I
Retrospectives and Theoretical Perspectives

Chapter 1
Early Humans: Of Whom Do We Speak?

Richard E. Leakey

Keywords *Homo habilis* • *Homo erectus* • *Homo rudolfensis* • Koobi Fora • Turkana Basin • Olduvai Gorge • archeology • Oldowan • history of discovery

I would like to begin by saying that the manuscript in this volume by my dear friend and mentor, Professor Phillip V. Tobias, is particularly enjoyable to me because it allows me to reflect, and remember some of the events that he describes. I think my father probably thought that he *had* successfully browbeaten Phillip to accept the designation of *Homo habilis* as a new species of our genus. It is, I am told, a trait that I have inherited from Louis: we think we have achieved something, or persuaded someone and proceed on that basis when, in fact, we have made nowhere near the point we were trying to make. So I think that Phillip and my father were *both* correct in their beliefs: Philip was not browbeaten, but Louis certainly would have tried, although he would have denied all along that he was trying! I particularly like Phillip's reference to a characteristic that Louis had (and which I am told I also have). He would pick something up and say, "Isn't that *obvious*?" to which people would respond "No!" This is exactly the response I get, and the response that my father got from Phillip. In this case, at least, Phillip has had the good grace to say that Louis *had* been right, and that it *had* been obvious – Phillip just hadn't noticed the obviousness. So, for the family history, let me just give my Dad a little support, which I was loathe to give when he was alive, because I now find myself more inclined to appreciate his qualities.

I totally agree with the validity of the taxon *Homo habilis,* and I think that Phillip Tobias, along with John Napier and Louis Leakey, was perfectly correct in launching that species (Leakey et al., 1964). The type specimen, OH 7, which Phillip has alluded to as "Jonny's child," and whose mandible and parietals formed the basis of the new designation, is *Homo habilis*. And, I think it is correctly a species of the genus *Homo* (cf. Wood and Collard, 1999; Collard and Wood, 2007; Wood, 2009). By the same token, however, I think some of the specimens from Olduvai that Phillip (Tobias, 1991) and others (Johanson et al., 1987) have included in *H. habilis* still have a question mark next to that allocation.

Whilst I was active at Lake Turkana in northern Kenya at Koobi Fora, we found a great many fossils and we published them fairly rapidly through an initial announcement in the journal *Nature* and subsequently with more detailed descriptions in the *American Journal of Physical Anthropology*. It was a policy in the early days that the group working on the early hominids – myself, Alan Walker, Bernard Wood and Michael Day – would not engage in the naming game. We intended to find additional material, and needed time to study specimens in detail. Subsequently, this reserved group dispersed slightly, and we all went on to do different things, and some of us started talking about species. Specimens such as KNM-ER 1470 that were first identified simply as representing the genus *Homo* in the Upper Burgi Member of the Koobi Fora Formation (Leakey, 1973) were shortly thereafter allocated to a novel species *Homo rudolfensis* (Aleexev, 1986). Others found it appropriate to provide taxonomic names for other specimens we had discovered and described but not named, including a mandible from the Okote Member of the Koobi Fora Formation (KNM-ER 992), which was attributed to a species called *H. ergaster* (Groves and Mazak, 1975). The relationship between *Homo habilis* and *Homo ergaster* is not clear to me and I am of the opinion that there is a conundrum which arises in part from the incompleteness of material upon which early work was done. No one would question the quality of the work that was being done in the 1960s, but I think one could argue about the quality of the material and its real geochronological age. I do feel that the mandible, the parietals and the hand and foot bones of OH 7 and OH 8, which formed the basis of the original study, and to which were later added OH 16 and OH 24 (Leakey and Leakey, 1964; Leakey et al., 1971) are specimens that are so fragmentary that we don't really know that much about the anatomy of the type of *Homo habilis*.

Additionally, I am concerned that, although the stratigraphy and the chronology between the specimens at Olduvai is

R.E. Leakey (✉)
Department of Anthropology and Turkana Basin Institute, Stony Brook University, Stony Brook, NY 11794-4364, USA
e-mail: richard.leakey@stonybrook.edu

F.E. Grine et al. (eds.), *The First Humans: Origin and Early Evolution of the Genus* Homo, Vertebrate Paleobiology and Paleoanthropology, superb, there clearly is a problem insofar as there are at least 150,000 and possibly 280,000 years of time separating "Jonny's Child," OH 7, the type specimen of *H. habilis*, and OH 13 (Walter et al., 1991; Tamrat et al., 1995). We know from hindsight that at the upper end of that span, at around 1.65 million years, there are specimens known both in Africa and beyond that are the same age as or somewhat older than OH 13 and which are clearly *not* what OH 7 seems to be (Antón, 2003; de Lumley et al., 2006; Lordkipanidze et al., 2006; Rightmire et al., 2006; Spoor et al., 2007). They are more refined, and you might call them *Homo erectus* or, as some call some of them, *Homo ergaster*. Had we known that *H. ergaster/erectus* made its appearance by 1.6 Ma, a fragmentary skull of that age from the middle of Bed II at Olduvai might have been assigned to that taxon rather than to *H. habilis*. We now have quite a complicated set of finds in well-documented contexts (Brown et al., 2006; Gathogo and Brown, 2006a, b; McDougall and Brown, 2006; Spoor et al., 2007). With the advantage of hindsight, we will need to revisit some of the taxonomic assignments.

I also feel – and I will come back to this for the purposes of this discussion – that the efforts made to determine the cranial capacity of *Homo habilis* were faultless at the time those estimations were made (Leakey et al., 1964; Tobias, 1964; Holloway, 1965). However, when an estimation of cranial capacity is being attempted only from the parietals there is room for error (Pilbeam and Simons, 1965; Wolpoff, 1981). With the benefit of hindsight, and knowing that large-headed individuals such as KNM-ER 1470 and KNM-ER 1590 were about at the time of *H. habilis* (i.e., OH 7), a different guesstimate or calculation is possible. Does the real *H. habilis* have a cranial capacity of less than 750 cc is still a valid question today.

I now return to my opening remarks: I think that there is probably no other issue more important to the general public than the evidence for the first appearance (as it is usually referred to) of our own genus. We are special, we are different, and we are concerned. We face a very complicated future – the world is riven by all sorts of fundamentalist and other concepts – and if we can really pin down this issue of who we are and where we are from once and for all, it would be to everyone's greater good.

For that reason, I do not think it trivial to revisit the earliest species of the genus *Homo*, because the first publication in 1964 was based on specimens from Olduvai that at that time did not have the benefit of comparison with many other specimens that have since been found. If we can step back slightly, I think this workshop will allow us to consider this conundrum: if the sequence of discovery had been different, would Louis Leakey, Phillip Tobias, and John Napier (Leakey et al., 1964) necessarily have come to the conclusions that they came to? This is not to say that they made a mistake. We know, however, that hindsight is often clearer than foresight. My task, in part, is to provoke a discussion around these issues.

I have always been and remain impressed with what obviously impressed Louis – and led him, I think, to make his early mistakes – and that is the relationship between the manufacture of artifacts out of stone, the anatomical changes that would be necessary to manipulate stone as a cutting instrument, and the brain that would be necessary to think about breaking a stone in a certain way in order to do that. It is clear that at the time *Homo habilis* was launched, the oldest (Oldowan of Leakey, 1936) artifacts were the same age as the oldest fossils attributed to the genus *Homo*. That, of course, was in the lowest sequence at Olduvai, which is dated to some 1.87–1.78 Ma (Walter et al., 1991; Tamrat et al., 1995; Blumenschine et al., 2003). Since then, and new discoveries are constantly being made, the threshold for evidence of making stone tools has been pushed back to about 2.5 or 2.6 million years, with four sites in the Hadar Formation that possess lithic artefacts (Harris, 1983; Semaw et al., 1997, 2003; Stout et al., 2005; Roche et al., 2009) and one in the Hata Member of the Bouri Formation that bears evidence of bones both cut and percussed by stones (de Heinzelin et al., 1999). Certainly by 2.4–2.3 Ma ago in Kenya and Ethiopia, there are a number of sites that reveal artifacts that are very clearly quite sophisticated in the way that they were being made (Kimbel et al., 1996; Hovers et al., 2002; de la Torre, 2004; Kibunjia et al., 1992; Kibunjia, 1994; Plummer, 2004; Roche et al., 1999, 2009).

So, somebody was around who was fashioning a cutting edge out of stone that was quite complicated. Were there several contemporaneous species making cutting-edged stone tools? Or do the cutting edges that are represented by the archeological collection represent a single species but populations of it operating with different raw materials and doing things in their own way?

When we talk about a species – and I think there is a real issue here – are we talking about more than one species of *Homo* between 1.8–1.6 Ma, of which only one was ancestral to *Homo ergaster/erectus*? Were there several species – and one or two disappeared for various reasons – but that all had the same general adaptation that had shifted to become much more omnivorous because they could access meat by cutting with a stone through flesh? These are important biological and adaptive questions that we need to understand in terms of ourselves.

I think a more plausible suggestion would be that the adaptation of a slightly more complex, if not larger, brain and the development of a slightly more manipulative ability (*Homo habilis*) may only have happened once. This then was expressed and manifested further in different parts of Africa in different ways, depending on different circumstances, rather than having it happen twice or thrice in one geographical area at the same time. That would be my simplistic view, but the thing about this issue is that we do not know because we do not find only one hominid.

When my mother, Mary, discovered the first hominid at Olduvai, which was the skull of *Australopithecus (Zinjanthropus) boisei*, Louis was convinced that she had found the maker of the tools. The reason was that he had been looking for 30 years with Mary at Olduvai for just that. He knew there were tools, and he knew that those tools had been made by somebody, so when somebody was found, he assumed that it had made the tools. It was therefore a shock to him a year and a half later, in 1960, when my older brother Jonathan came up with something that clearly was not *Zinjanthropus* because of obvious differences in morphology. Louis therefore had to do a slight change of pace and say, well, maybe this second one was a toolmaker and this second one ate the first one. And so we got into a slight complication. The same problem arises today: we find hominids, and we find tools and, depending on our points of view, it is very easy to say, "this one must have been the toolmaker," or that "these were *all* toolmakers." We have to figure out how they relate to one another even though the tools may be different. Are they very different? Even though we use computers, are we different from those people who are still using stone flakes today? What is that difference? We are all the same species today. I believe that species designation as it pertains to our understanding of the evolution of the genus *Homo* is obviously an important and sensitive issue that we need to think about.

I also believe that the relationship between tools and hominins – that is, the relationship between the first appearance of what we call *Homo* and the link between the development of tools and human behavior (perhaps in terms of campsites) – is very critical (Davidson and McGrew, 2005). Hindsight might require a modification of ordering. I know that workers who champion the existence of "*Homo*" *rudolfensis* and support a complex species composition in the Late Pliocene – Early Pleistocene (e.g., Wood, 1992, 1993; Grine et al., 1996, 2009; Grine, 2001), regardless of whether forms such as *H. habilis* and *H. rudolfensis* should actually be considered as members of our genus (Wood and Collard, 1999; Collard and Wood, 2007; Wood, 2009), will possibly have a different standpoint. But I think the purpose of this workshop is to put away our previous convictions and put away our prejudices and say, "let us go back to basics." We need to think not about the sequence of things in which they were *found*, but about the sequence in which these individuals *lived* and about the completeness of the fossils. I believe the archeological record is particularly critical. The evidence for the use of stone is long and complex, and I think will be a very important insight into what was going on in the Late Pliocene and Early Pleistocene. Too often we deal with archeology and with the fossils in isolation, and we fail to talk together about what they may mean in context of one another.

My view is that the *Homo habilis* story is indeed a very complex story. I think specimens have been included in *H. habilis* that may not be *H. habilis*, that may belong to a species or lineage that is later than *H. habilis*, and I think the sequential relationship between these fossils is therefore very critical. Unfortunately, although we have spent an awful lot of time, and an awful lot of money obtaining dates for many of these important archeological and paleontological sites, there are still question marks about the temporal relationship between some of the fossils. We thought we had got it all tidied up at Lake Turkana, but as more people look at things, new questions inevitably arise. It is not possible at this time to say what is the right story. There are a lot of question marks, and we must not assume that we know the answers.

References

Aleexev, V.P., 1986. The Origin of the Human Race. Progress Publishers, Moscow.

Antón, S.C., 2003. Natural history of *Homo erectus*. Yearbook of Physical Anthropology 46, 126–169.

Blumenschine, R.J., Peters, C.R., Masao, F.T. Clarke, R.J., Deino, A.L., Hay, R.L., Swisher, C.C., Stanistreet, I.G., Ashley, G.M., McHenry, L.J., Sikes, N.E., van der Merwe, N.J., Tactikos, J.C., Cushing, A.E., Deocampo, D.M., Njau, J.K., Ebert, J.I., 2003. Late Pliocene *Homo* and hominid land use from western Olduvai Gorge, Tanzania. Science 299, 1217–1221.

Brown, F.H., Haileab, B., McDougall, I. 2006. Sequence of tuffs between the KBS Tuff and the Chari Tuff in the Turkana Basin, Kenya and Ethiopia. Journal of the Geological Society 163, 185–204.

Collard, M., Wood, B.A., 2007. Defining the genus *Homo*. In: Henke, W., Tattersall, I. (Eds), Handbook of Paleoanthropology, Volume 3: Phylogeny of Hominids. Springer, Berlin, pp. 1575–1610.

Davidson, I., McGrew, W.C., 2005. Stone tools and the uniqueness of human culture. Journal of the Royal Anthropological Institute 11, 793–817.

de Heinzelin, J., Clark, D., White, T., Hart, W., Renne, P., WoldeGabriel, G., Beyene, Y.,Vrba, E., 1999. Environment and behavior of 2.5-million-year-old Bouri hominids. Science 284, 625–629.

de la Torre, I., 2004.Omo revisited – evaluating the technological skills of Pliocene hominids. Current Anthropology 45, 439–465.

de Lumley, M.A., Gabounia, L., Vekua, A., Lordkipanidze, D., 2006. Les restes humains du Pliocène final et du début du Pléistocène inférieur de Dmanissi, Géorgie (1991–2000). I – Les crânes, D 2280, D 2282, D 2700. Anthropologie 110, 1–110.

Gathogo, P.N., Brown, F.H., 2006a. Revised stratigraphy of Area 123, Koobi Fora, Kenya, and new age estimates of its fossil mammals, including hominins. Journal of Human Evolution 51, 471–479.

Gathogo, P.N., Brown, F.H., 2006b. Stratigraphy of the Koobi Fora Formation (Pliocene and Pleistocene) in the Ileret region of northern Kenya. Journal of African Earth Sciences 45, 369–390.

Grine, F.E., 2001. Implications of morphological diversity in early *Homo* crania from eastern and southern Africa. In: Tobias, P.V., Raath, M.A, Moggi-Cecchi, J., Doyle, G.A. (Eds), Humanity from African Naissance to Coming Millennia – Colloquia in Human Biology and Palaeoanthropology, Firenze University Press, Firenze, pp. 107–115.

Grine, F.E., Jungers, W.L., Schultz, J., 1996. Phenetic affinities among early *Homo* crania from East and South Africa. Journal of Human Evolution 30, 189–225.

Grine, F.E., Smith, H.F., Heesy, C.P., Smith, E.J., 2009. Affinities of *Homo* fossils from South Africa: phenetic evidence from molar cusp proportions. In: Grine, F.E., Fleagle, J.G., Leakey, R.E. (Eds), The First Humans: Origin and Early Evolution of the Genus *Homo*. Springer, Dordrecht, pp. 49–62.

Groves, C.P., Mazak, V., 1975. An approach to the taxonomy of the Hominidae: gracile Villafranchian hominids of Africa. Časpis Pro Mineralogii a Geologii 20, 225–247.

Harris, J.W.K., 1983. Cultural beginnings: Plio-Pleistocene archaeological occurrences from the Afar, Ethiopia. African Archaeological Review 1, 3–31.

Holloway, R.L., 1965. Cranial capacity of the hominine from Olduvai Bed I. Nature 208, 205–206.

Hovers, E., Schollmeyer, K., Goldman, T., Eck, G., Reed, K., Johanson, D., Kimbel, W., 2002. Late Pliocene archaeological sites in Hadar, Ethiopia. Journal of Human Evolution 42, 1–17.

Johanson, D.C., Masao, F.T., Eck, G.G., White, T.D., Walter, R.C., Kimbel, W.H., Asfaw, B., Manega, P., 1987. New partial skeleton of *Homo habilis* from Olduvai Gorge, Tanzania. Nature 327, 205–209.

Kibunjia, M., 1994. Pliocene archaeological occurrences in the Lake Turkana Basin. Journal of Human Evolution 27, 159–171.

Kibunjia, M., Roche, H., Brown, F.H., Leakey, R.E.F., 1992. Pliocene and Pleistocene archaeological sites west of Lake Turkana, Kenya. Journal of Human Evolution 23, 431–438.

Kimbel, W.H., Walter, R.C., Johanson, D.C., Reed, K.E., Aronson, J.L., Assefa, Z., Marean, C.W., Eck, G.G., Robe, R., Hovers, E., Rak, Y., Vondra, C., Yemane, T., York, D., Chen, Y., Evensen, N.M., Smith, P.E., 1996. Late Pliocene *Homo* and Oldowan tools from the Hadar Formation (Kada Hadar Member), Ethiopia. Journal of Human Evolution 31, 549–561.

Leakey, L.S.B., 1936. Stone Age Africa. Oxford University Press, London.

Leakey, L.S.B., Leakey, M.D., 1964. Recent discoveries of fossil hominids in Tanganyika: at Olduvai Gorge and near Lake Natron. Nature 202, 5–7.

Leakey, L.S.B., Tobias, P.V., Napier, J.R., 1964. A new species of the genus *Homo* from Olduvai Gorge. Nature 202, 7–9.

Leakey, M.D., Clarke, R.J., Leakey, L.S.B., 1971. New hominid skull from Bed I, Olduvai Gorge, Tanzania. Nature 232, 308–312.

Leakey, R.E.F., 1973. Evidence for an advanced Plio-Pleistocene hominid from East Rudolf, Kenya. Nature 242, 447–450.

Lordkipanidze, D., Vekua, A. Ferring, R., Rightmire, G.P., Zollikofer, C.P.E., de Leon, M.S.P., Agusti, J., Kiladze, G., Mouskhelishvili, A., Nioradze, M., Tappen, M., 2006. A fourth hominin skull from Dmanisi, Georgia. Anatomical Record 228A, 1146–1157.

McDougall, I., Brown, F.H., 2006. Precise Ar-40/Ar-39 geochronology for the upper Koobi Fora Formation, Turkana Basin, northern Kenya. Journal of the Geological Society 163, 205–220.

Pilbeam, D.R., Simons, E.L., 1965. Some problems with hominid classification. American Scientist 53, 237–257.

Plummer, T., 2004. Flaked stones and old bones: biological and cultural evolution at the dawn of technology. Yearbook of Physical Anthropology 47, 118–164.

Rightmire, G.P., Lordkipanidze, D, Vekua, A., 2006. Anatomical descriptions, comparative studies and evolutionary significance of the hominin skulls from Dmanisi, Republic of Georgia. Journal of Human Evolution 50, 115–141.

Roche, H, Blumenschine, R.J, Shea, J.J., 2009. Origins and adaptations of early genus *Homo*: what archeology tells us. In: Grine, F.E., Fleagle, J.G., Leakey, R.E. (Eds), The First Humans: Origin and Early Evolution of the Genus *Homo*. Springer, Dordrecht, pp. 135–147.

Roche, H., Delagnes, A., Brugal, J.-P., Feibel, C.S., Kibunja, M., Mourre, V., Texier, P.-J., 1999. Early hominid stone tool production and technical skill 2.34 Myrr ago, in West Turkana, Kenya. Nature 399, 57–90.

Semaw, S., Renne, P., Harris, J.W.K., Feibel, C.S., Bernor, R.L., Fesseha, N., Mowbray, K., 1997. 2.5-million-year-old stone tools from Gona, Ethiopia. Science 284, 625–629.

Semaw, S., Rogers, M.J., Quade, J., Renne, P.R., Butler, R.F., Dominguez-Rodrigo, M., Stout, D., Hart, W.S., Pickering, T., Simpson, S.W., 2003. 2.6-million-year-old stone tools and associated bones from OGS-6 and OGS-7, Gona, Afar, Ethiopia. Journal of Human Evolution 45, 169–177.

Spoor, F., Leakey, M.G., Gathogo, P.N., Brown, F.H., Anton, S.C., McDougall, I., Kiarie, C., Manthi, F.K., Leakey, L.N., 2007. Implications of new early *Homo* fossils from Ileret, east of Lake Turkana, Kenya. Nature 448, 688–691.

Stout, D., Quade, J., Semaw, S., Rogers, M.J., Levin, N.E., 2005. Raw material selectivity of the earliest stone toolmakers at Gona, Afar, Ethiopia. Journal of Human Evolution 48, 365–380.

Tamrat, E., Tohveny, N., Taieb, M., Opdyke, N., 1995. Revised magnetostratigraphy of Plio-Pleistocene sedimentary sequence of the Olduvai formation (Tanzania). Paleogeography, Paleoclimatolology and Paleoecology 114, 273–283.

Tobias, P.V., 1964. The Olduvai Bed I hominin with special reference to its cranial capacity. Nature 202, 3–4.

Tobias, P.V., 1991. The Skulls, Endocasts and Teeth of *Homo habilis*. Olduvai Gorge, Volume 4. Cambridge University Press, New York.

Tobias, P.V., 2009. *Homo habilis* – a premature discovery: remembered by one of its founding fathers, 42 years later. In: Grine, F.E., Fleagle, J.G., Leakey, R.E. (Eds), The First Humans: Origin and Early Evolution of the Genus *Homo*. Springer, Dordrecht, pp. 7–15.

Walter, R.C., Manega, P.C., Hay, R.L., Drake, R.E., Curtis, G.H., 1991. Laser-fusion 40Ar/39Ar dating of Bed I, Olduvai Gorge, Tanzania. Nature 354, 145–149.

Wolpoff, M.H., 1981. Cranial capacity estimates for Olduvai Hominid 7. American Journal of Physical Anthropology 56, 297–304.

Wood, B.A., 1992. Origin and evolution of the genus *Homo*. Nature 355, 783–790.

Wood, B.A., 1993. Early *Homo*: how many species? In: Kimbel, W.H., Martin, L.B. (Eds), Species, Species Concepts, and Primate Evolution. Plenum, New York, pp. 485–522.

Wood, B.A., 2009. Where does the genus *Homo* begin, and how would we know? In: Grine, F.E., Fleagle, J.G., Leakey, R.E. (Eds), The First Humans: Origin and Early Evolution of the Genus *Homo*. Springer, Dordrecht, pp. 17– 28.

Wood, B.A., Collard, M.C., 1999. The human genus. Science 284, 65–71.

Chapter 2
Homo habilis – A Premature Discovery: Remembered by One of Its Founding Fathers, 42 Years Later

Phillip V. Tobias

Keywords *Homo habilis* • *Australopithecus* • Olduvai Gorge • archeology • history of discovery

Introduction

One of the most dramatic contributions to human evolutionary studies made by the Leakey family was their discovery from 1959 onwards of the hominin fossils to which Louis Leakey, Phillip Tobias and John Napier in 1964 gave the name *Homo habilis*. The first specimens were found in the lower part of Bed I of the superlative Olduvai Gorge sequence in Tanzania. The then new dating method, the potassium-argon technique, shocked everyone by revealing an age of 1.75 million years for Olduvai Bed I (Leakey et al., 1961). This high antiquity was one of the factors conducive to the shocked rejection of *Homo habilis* by almost all our peers.

The delicate parietal bones and modest teeth, which seemed to differ from those of South Africa's *Australopithecus africanus*, were thus of high antiquity. Moreover, they were contemporaneous with an excessively large-toothed australopithecine whose remains were found in the same Bed I (Leakey, 1959a, b, c; Tobias, 1959, 1967). The history of early *Homo* contemporary with robust australopithecines has been recounted elsewhere (Tobias, 1978). The Olduvai discoveries of 1959–1961 were not the first evidence of an apparent *Homo* contemporary and sympatric with an australopithecine to emerge from Africa's soil. A decade earlier, fossils showing every justification to be included in the genus *Homo* had been found in the Swartkrans cave, in South Africa. Swartkrans had already proved to be an immensely rich source of fossils of *Australopithecus (Paranthropus) robustus*. In September 1949, a jaw fragment, SK 45, came to light in Member 1 of the Swartkrans Formation. This and other Swartkrans specimens were originally named by Robert Broom and John Robinson *Telanthropus capensis*. However, in 1957, Simonetta proposed to re-designate it *Homo erectus*, and Robinson (1961) agreed with this. By 1959, when Olduvai began to yield early *Homo* fossils, the existence of a *Homo* contemporary and sympatric with large-toothed robust australopithecines had been well-established in the paleoanthropological record for 10 years. When the Olduvai hominins were being studied from 1961 to 1963, the prevailing view was that *H. erectus* was contemporary with the robust australopithecines.

The Leakeys at Olduvai

Louis and Mary Leakey had searched assiduously for early humans at Olduvai Gorge from 1931 onwards. By 1955, the only hominin skeletal remains available were Hans Reck's 1913 human skeleton (Olduvai Hominid 1), and two thick fragments of cranial vault (OH 2) which Mary Leakey had found at the site of MNK. I studied these two fragments at the Natural History Museum in London in June 1964. They were recorded as *cf. H. erectus* in Mary Leakey's 1971 book. In 1955, a "giant molar tooth" and a canine (OH 3) were found in Upper Bed II at site BK. Much excitement was aroused at the time by the gigantic dimensions of the molar, especially as it showed the hallmarks of a deciduous tooth. It was served up as an hors-d'oeuvre for the sumptuous repast provided by the type specimen of *Australopithecus boisei* 4 years later.

That was all. After 28 years of slog at Olduvai (1931–1958), it could not but seem a poor return. Admittedly the dogged perseverance of Louis and Mary Leakey had been rewarded by thousands of comparative faunal remains and hoards of stone artefacts, but the stone tool-maker, that Louis had believed in perfervidly and sought for decades, remained elusive.

That was the position in 1957 when I paid my first visit to Olduvai, having been invited by Louis Leakey to join a safari he had organized for Richard Foster Flint of Yale University. Apart from some diverting encounters with large felines, one

P.V. Tobias (✉)
School of Anatomical Sciences, Institute of Human Evolution, University of the Witwatersrand Medical School, 7 York Road, Parktown 2193, South Africa
e-mail: phillip.tobias@wits.ac.za

of which came close to terminating the excursion and Dr. Flint's active life, the only fruits of our visit were sundry mammalian fossils, and serious, animated and often heated questioning by Flint of Leakey's paleoclimatic interpretation of the sequence of events at Olduvai. During these skirmishes I often found myself, willy-nilly, acting as referee!

A turning point came in 1959.

1959 – *Annus Mirabilis* at Olduvai

At the end of June or the beginning of July 1959, the Leakeys' longtime senior assistant Heselon Mukiri found a molar tooth in a lump of calcified tuff at MK in Bed I. It proved to be socketed in a small fragment of mandible. Slight as this evidence was, the specimen, OH 4, along with a premolar, represented the first remnant of *H. habilis* to be recovered. Only some weeks later came Mary Leakey's spectacular discovery, on 17th July 1959, of the cranium which became the type specimen of *Australopithecus (Zinjanthropus) boisei*. I called it colloquially a "hyper-robust" australopithecine. Needless to say, this magnificent specimen ousted all other considerations. The shy little jaw and teeth of the first *Homo habilis* (OH 4) were put on one side, and published only in 1964 as one of the designated paratypes of *H. habilis*.

Immediately after the *A. boisei* cranium was found, they brought it down to Johannesburg from Nairobi, the "Dear Boy" sitting on Mary's lap throughout the flight. Thus, it came about that Dart and I were the first two people to see and touch the fine specimen. Incidentally, OH 5 was the first australopithecine to be found away from South Africa, save for Kohl-Larsen's maxilla with two premolars from Eyassi not far from Olduvai in Tanzania. The Eyassi remains had been found shortly before the outbreak of the Second World War, and the news did not become generally available until after the war.

Shortly after the Leakeys' visit to Johannesburg in 1959, we flew to Kinshasa (then called Léopoldville), to participate in the IV Pan-African Congress on Prehistory. There Louis Leakey publicly announced the discovery of what he proposed to call *Zinjanthropus boisei*. In my monograph on this hyper-robust cranium, I proposed to sink the genus *Zinjanthropus* in *Australopithecus*, relegating the former to sub-generic status. Thus its revised nomen became *Australopithecus (Zinjanthropus) boisei*.

A personally momentous incident at the Kinshasa Congress was that Louis and Mary invited me to undertake the definitive study of the "Zinj" cranium. Thereafter all of the hominin fossils that Louis and Mary recovered at Olduvai and elsewhere were handed to me to analyse and describe. It was a major turning point in my life. It came at a juncture when I was inclining to devote my career to the study of the living peoples of Africa. That thrust had grown out of my researches on Bushmen or San in Botswana, and the Tonga people of Zambia. Thus, it came about that the Leakeys rather than my mentor Raymond Dart ushered me into a life among the fossils. It must be admitted that my earlier work on human genetics and physical anthropology of the living flavoured my subsequent interpretations of the fossil hominins.

Another career-fashioning event of 1959 occurred on the home front: at the beginning of the year I succeeded Dart as the Professor and Head of the Department of Anatomy, placing me as the youngest professor at Wits University in charge of a large and active department. It was a position I was to hold for 32 years.

By late 1961 the amount of East African skull material, teeth and postcranial bones had become voluminous. I suggested to Louis Leakey that John Napier of the Anatomy Department at the Royal Free Hospital Medical School in London, and a specialist in the functional morphology of the upper limb and of locomotion, be invited to study the postcranial bones. I would handle the crania, including jaws and teeth, and artificial endocranial casts. Napier accepted and, in turn, involved his colleagues, Peter Davis and Michael Day.

1960 – Jonny's Child

1960 was the first year in which the Leakeys' Olduvai project received major financial aid from the National Geographic Society. Excavation went ahead on an unprecedented scale. A few teeth and calvarial fragments (OH 6), and a tibia and fibula (OH 35) came to light. For a while these disparate remains were left in a suspense account. Louis Leakey was busy claiming that Nutcracker Man (a name I had lightheartedly suggested for "Zinj" at the Kinshasa Conference) was the long-sought Olduvai tool-maker. His view seemed reasonable in the absence of any more advanced hominin at the same time-level. But *was* there no more advanced contemporary? For the moment it seems that *Telanthropus* of Swartkrans was forgotten or overlooked, and so were the puny scraps of OH 4 and OH 6.

At this critical juncture, Jonathan Leakey, the eldest son of Louis and Mary, made a very important discovery on or close to his 20th birthday (4 November, 1960). At FLK NN, in the lower part of middle Bed I, he dug out a juvenile mandible, the greater part of an immature left parietal bone, fragments of a matching right parietal, some other cranial pieces, and 21 wrist-, hand- and finger-bones. Urgently I was called from Johannesburg to examine these new and perplexing specimens designated OH 7. The parietal bones were outsized. I brought with me measurements of the parietal bones of the South African australopithecines. It immediately emerged that the parietals from FLK NN were greater in their measurable dimensions (except thickness), than any known parietals of crania of *A. africanus* and *A. robustus* from South Africa and of *A. boisei* from Olduvai. I was aware, from my earlier studies, how variable were the contributions of the three main vault-bones, frontal, parietals and occipital, to the size of the brain-case as a whole. Nevertheless, one could not escape a tingling feeling that a pair of parietals that were not only extensive from front to back, but laterally widespread, must have covered an endocranial cavity larger than those of any early hominids then available. Just how much larger was the first vexing question. Moreover, even if I could form an estimate of the endocranial capacity of OH 7, this fact on its own would not, of course, exclude OH 7 from having been a large-brained australopithecine.

For a start, I made an arch between the left and right parietals: there were sufficient landmarks to make this possible. The first arch was made on Monday, 5th June 1961, at the Centre for Prehistory and Palaeontology behind the Kenya National Museums. I was helped by Louis and Mary Leakey, Shirley Coryndon (later Shirley Savage), and the distinguished American palaeontologist, George Gaylord Simpson, on a visit from Harvard University, while his wife Anne Roe wrote down the measurements as I took them. We were all conscious of an air of ill-suppressed excitement as I manoeuvred the restored left and right parietals in relation to each other. If I moved the lower margins of the two bones closer together, a strange angulation became apparent along the mid-line (sagittal margins). On the other hand, if I distracted the two lower (squamosal) margins from each other, a negatively-angled hollow appeared along the line of the sagittal suture, flanked by an elevation to each side. Both extreme positions of the mid-line hinging gave us a bizarre and deformed-looking partial calvaria. Somewhere between these two extremes the correct alignment of the two parietals must have lain. Eventually we narrowed down the likely options to a very small range of widths between the lower margins (biasterionic widths) which appeared anatomically reasonable. We were awed by the size of the resulting biparietal arch, substantially greater than the corresponding measurements of the australopithecine parietal bones or biparietal arches.

I brought the specimens to my laboratory in the Wits Anatomy Department. There, Alun Hughes and I made another reconstruction of the arch, slightly different from the Nairobi arch. The difference in biasterionic breadth between the two reconstructions was scarcely 2.0 mm. Next, Alun and I made two partial endocasts that fitted snugly within the two biparietal arches. The mean volumes of these partial endocasts were almost identical (363.6 and 363.4 cm^3). At that time, the smallest total endocranial capacity recorded for an *A. africanus* cranium was 428 cm^3.

I was now confronted with the task of converting the capacity of the part-endocast of OH 7 to an estimated total capacity. Using a series of hominid analogues, I determined a capacity of 680 cm^3 (uncorrected for age – OH 7 was a juvenile). My latest revised estimate for OH 7 is 647 cm^3 (uncorrected for age) and 674 cm^3 (as an estimated adult capacity). Jonny's Child turned out to have an endocranial capacity nearly 50% greater than the average for half-a-dozen specimens of *A. africanus* from South Africa. Indeed, its value was well above the observed range and 95% population limits of the australopithecine capacities. This was the major morphological surprise posed by Jonny's Child who lived 1.75 Ma ago. But, it was not the only unusual feature.

The teeth of the type jaw of OH 7 impressed those of us who had studied the teeth of australopithecines in their relative slenderness. This buccolingual (BL) narrowing connoted the absence of the australopithecine BL broadening which I had called the "australopithecine bulge." In the Olduvai teeth the narrowing applied especially to the third and fourth premolars and the first molar. At the same time, the mesiodistal (MD) diameters of the teeth of OH 7, as well as of OH 4 and OH 6, were somewhat increased. When I compared the two crown diameters to obtain a crown-shape index, its values for the Olduvai little strangers fell outside the entire ranges of index values for the large sample of comparable teeth of *A. africanus*. Were these odd values of the tooth indices for the Olduvai sample the mark of individual variants, whose teeth extended the observed ranges of values for *A. africanus*? Or did they characterise a hominin population whose cheek–teeth lacked the australopithecine bulge?

Stratigraphically speaking, these enigmatic specimens from Olduvai came from a horizon well below the OH 5 or "Zinj" stratum within Bed I. For a time, therefore, Leakey called OH 7 "pre-*Zinjanthropus*", a nickname until we could make up our minds what manner of person this was. Some of our colleagues (such as Nesturkh, 1967) mistook Leakey's intention and used the term as the name of another genus!

With its strange teeth and large endocranial capacity, the youngster posed a real challenge to us. Not surprisingly, we did not come to any precipitate conclusion. For at least 4 years, from 1959 to late 1963, we subjected the new specimens to observation, reconstruction, mensuration and comparisons. From the beginning, Louis was convinced that Jonathan had found an early specimen of the genus *Homo*. There were times when I, too, thought we must have in our hands a representative of something new. More frequently, I held to the conservative view that the new juvenile bones might be regarded as simply widening our concept of the variability of *A. africanus*. The concept of how much *A. africanus* varied was based on the type specimen from Taung, and good samples from Sterkfontein and Makapansgat.

It was not until January 1964 that we finally decided upon the creation of a new species, and *Homo habilis* made its debut through the pages of *Nature* in April 1964. I do not think that we who had cogitated for 4 years had been over-hasty (although that was a criticism levelled against us – in this case by Sir Wilfrid LeGros Clark, 1964). Louis Leakey's mind seems to have been made up some time before, almost intuitively. What produced the change of mind – the conversion – of John Napier and myself?

Conversion in Science: The Case of *Homo habilis*

A myth grew up around my "conversion" to the view that OH 7 belonged to a hitherto unrecognised species of the genus *Homo*. The essence of the myth is that it was Louis Leakey's enormous powers of persuasion and forceful personality that eventually wore down my resistance – and John Napier's too, so that, on this version of the history, we were virtually bludgeoned into supporting Leakey's view!

As far as I can trace, the first published source of this unlikely story was Sonia Cole's (1975) biography of Louis, *Leakey's Luck*. She states (p. 256, emphasis mine):

> Tobias had always been a little unhappy about creating a new name for it, but he had been swayed by Napier's conclusions on the manipulative ability of the hand: 'man the toolmaker' was an accepted definition of man, *habilis* was apparently a tool-maker and therefore he must be a man, yet he did not fit into any of the known species of *Homo*. This argument, *plus the strong pressure brought to bear by Louis*, had persuaded Tobias to associate his name with the creation of the new species; as one colleague put it, *habilis* had been launched mainly by the power of Louis's personality.

In fact, it was neither Louis's persuasive skills nor Napier's views on the hand that converted me, as was clearly stated in contemporary publications. The same point recurred in *Lucy* by Don Johanson and M. Edey (1981: 288). The authors put this question into the mouth of Tim White: "It's not like Louis Leakey hammering and hammering on Phillip Tobias about *Homo habilis* until he had him beaten down?" Their source seems to have been Sonia Cole's book, according to personal communications I received from Don and Tim.

The third reference is by Mary Leakey in her autobiography, *Disclosing the Past*:

> Louis was predictably delighted by the new finds. In 1959 he had regarded Zinj as the maker of the artefacts at FLK, but here was a far better candidate in terms of both brain and hand. This was right in line with his own theories of *Homo* evolving during the earlier Pleistocene. This had to be *Homo*. *He directed his considerable powers of eloquent persuasion towards Phillip Tobias and John Napier….* (M.D. Leakey, 1984, pp. 127–128; emphasis mine)

I set the record straight on the factors leading to my conversion in my monograph on "*The Skulls, Endocasts and Teeth of* Homo habilis" (1991). In any case, anyone who knows me and my personality well will be able to confirm that my individualism extends not only to my 40-years long fight against apartheid and the inroads against academic freedom by the apartheid government of South Africa, but also and overwhelmingly to my scientific studies and interpretations. It is inconceivable that I would be brow-beaten to a certain standpoint even by such fortissimo personalities as Raymond Dart, LeGros Clark and even Louis Leakey. Apart from these personality traits, contemporary letters and records testify to a different motivation in my conversion.

In the interests of historical accuracy, let me here correct the record as to my position in the saga of *Homo habilis*. It is true that I was hesitant to recognise the single specimen OH 7 as representing a new species. Although I pointed out its departures from typical australopithecine morphology, I felt that I needed more evidence before I could exclude the possibility that these were simply the features of an individual, perhaps slightly aberrant member of an extended *A. africanus* hypodigm.

In August 1961, I replied to a query from Sir Wilfrid LeGros Clark. After a long account of my findings on the teeth and endocranial capacity, I summed up as follows:

> Whether this puts it outside the range of the Australopithecinae (which Leakey believed – and which is in my view quite unlikely) or whether it gives us a better idea of what the australopithecine range really was (which is more likely – especially when the evidence of the parietals is coupled with that of the teeth and mandible) remains to be determined; but at the moment, until I can get down to my more detailed study of all these remains later this year, *my view is that it is an early and rather large-brained member of the Australopithecinae*. (Letter, Tobias to LeGros Clark, 1 August 1961)

Two years later, I had not changed this view. In reply to a written enquiry from Sherwood L. Washburn, I summarised the state of my interim conclusions on "Zinj" and "pre-Zinj." My reply was quoted in full by Washburn as a footnote to his chapter in *Classification and Human Evolution* (1963: 196). At that point, in late 1963, I could tell Washburn: "I do not think I have yet seen any features which, individually or collectively, place it outside the probable range for *Australopithecus sensu lato*." (Tobias, 1963). Right up to December 1963 – over 4 years after Jonathan Leakey had found OH 7 – I believed that I could explain the features of the OH 7 juvenile simply as those of an extreme variant of *Australopithecus sensu lato*. For 4 years, "the force of Louis Leakey's persuasive powers" had failed to bully me into accepting that OH 7 should be seen as representing a new early hominin species! On the contrary, my own hesitation may well have given pause to Leakey, so that he did not name the suspected new species, but waited until more and better specimens were discovered. Louis and Mary made this point clear in 1964.

It was only when additional facts and further appraisals were cast into the scale-pan that I felt forced *by the new evidence* to abandon the null hypothesis and to conclude that the new Olduvai specimens represented something different from *Australopithecus*.

New Evidence of a Second Kind of Hominin

By 1963, teeth representing five individuals from three Olduvai localities were available. All of them showed the absence of the "australopithecine bulge." They were smaller teeth. In these respects they differed from those of *A. africanus*. Their traits pointed to a distinctive population, whatever its systematic position, that showed features approaching those of *Homo*.

Independently of my studies on the teeth, Napier, Davis and Day had been analyzing the hand- and foot-bones from FLK NN I. They concluded that they were close in their morphology to those of *Homo sapiens*. They could not, however, show that their structure was different from those of *A. africanus* or other australopithecines, because of the lack of adequate remains of the latter. It could be said that Bed I, Olduvai Gorge, included teeth that showed *Homo* departures and hand- and foot-bones that were remarkably like those of later forms of *Homo*. If the teeth and the limb-bones had belonged to members of the same hominin population, it was one characterised by *Homo*-like features not in a single, but in three major structural–functional complexes. It was now beginning to look as though we were dealing with a population with three character-complexes all pointing consistently in the same direction.

As a fourth and critical line of evidence, in 1963 I turned my attention to the parietal bones and the endocranial capacity. Working in Johannesburg, Alun Hughes and I made a new reconstruction of the biparietal arch of OH 7 and a partial endocast that fitted snugly within this arch. As analogues we made a series of part-endocasts of other specimens for which there were fairly reliable estimates of the total endocranial capacity and for which there were clear impressions of the outlines of the parietal bones. From these I devised a method by which to compute the estimated total endocranial capacity of OH 7. By the end of 1963 the first result was obtained. I was amazed at its high value, 675–680 cm^3. At that time, the mean capacity for *A. africanus* was 504 cm^3, though later studies by Holloway cropped this average to 441–442 cm^3. The mean for *Homo erectus* at that stage was 974 cm^3. Thus, the estimated value for OH 7 lay between the means for *A. africanus* and *H. erectus*. The OH 7 value also lay between the highest value in the *A. africanus* observed range and the smallest value in the *H. erectus* observed range. The value for OH 7 was 40–50% greater than the *A. africanus* mean. We needed more specimens with similarly large capacities. Without them, we could not, however, be sure if this large value, considered alone, was the trait of an isolated and perhaps freakish specimen – with, as some suggested, hydrocephalus! – or the hallmark of a population.

Therefore no fewer than four character-complexes – brain, teeth, hands and feet – were displayed by OH 7 and, at least in respect of the teeth, by the remains of four other individuals stemming from MK, FLK NN, and FLK. If all of those remains represented members of the same species, we were surely dealing not with an isolated extreme variant of the australopithecines, but with a population four of whose major character-complexes approached those of later members of the genus *Homo*. It was clearly a population that had moved away from the *A. africanus* pattern towards the *Homo* pattern of structure in all of the anatomical regions available for study and comparison.

At this crucial juncture late in December, 1963, Napier and I felt that we were able to recognize a new species at Olduvai. We were preparing to publish with Louis Leakey the evidence for this conclusion, when dramatic tidings reached us from East Africa.

Most important to the proving of the point came a shower of new hominin fossils from Olduvai Bed II in October and November 1963. These comprised OH 13, a delicately constructed, probably female hominin specimen from MNK II; OH 14 and OH 15; and OH 16, consisting of the teeth and hundreds of fragments of a probable male skull. This fresh haul of hominin specimens, showing some features in common with the Bed I specimens, was the main factor in clinching my realisation that we now had to deal with a population, not a few isolated sports. It was a population that seemed to be represented from the lowest part of Bed I into the middle of Bed II, some two metres below Tuff IIB: this time range was about a quarter of a million years.

The first news of the exciting 1963 finds was brought to Johannesburg by Louis and Mary Leakey late in December 1963. In January 1964, I left for Nairobi to finalize my volume on *A. boisei* and to study the newly discovered Olduvai specimens. Tentative reconstructions of the calvariae of both OH 13 and OH 16 enabled Ronald J. Clarke, then working at the Kenya National Museum, to make partial endocasts of both specimens. Using the part-endocast method with analogues that I had introduced, we obtained estimates of the total capacity of OH 13 and OH 16 and each was over 600 cm^3. It became plain that this Olduvai group of early Pleistocene fossil hominins had a mean capacity that was nearly 50% greater than the mean value for *A. africanus*. OH 7 was no hydrocephalic pathotype – but the type specimen of a population.

When Louis Leakey showed me the maxilla and mandible of OH 13, with their small, slender but long teeth, Ron Clarke who was standing by, recalls vividly how my eyes widened and shone: I turned to Leakey and said, "Louis, this is *Homo*". Clarke remembers how at that moment Louis gave vent to his characteristic, heaving, panting chuckle of sheer delight!

If there was a quintessential instant of conversion, that was it: no force of argument, coercion or power of personality, but the irresistible weight of evidence – and five or six lines of evidence at that! These were the factors that led me to have a change of mind, and to be converted to the view that we had a new and different species at Olduvai.

Conversion by the Hard Evidence

Just as the sheer weight of hard evidence led to my conversion, so too was it conversion of this sort that led a sceptical world eventually to accept Dart's claims for the Taung skull. The hard evidence also dictated Leakey's change of mind about the authorship of the Oldowan stone artefacts, and hard evidence converted Le Gros Clark to his acceptance, after his earlier rejection, of the hominin status of *Australopithecus*. These four examples show that conversions may be governed by rational factors, in a more reasonable, less subjective way than Thomas Kuhn (1962) would have us believe. Ernst Mayr (1972), likewise, has shown that the adoption of Darwinism, the change of paradigm epitomised in the phrase "the Darwinian Revolution," was a less sudden and a more rational set of events in the history of science than Kuhn's analysis proffered. So was it also with the Dartian revolution and the habiline revolution.

What Was the New Species?

Once we had determined that these Olduvai fossils represented a new species, we next had to decide to which of the existing hominid genera of the day the Olduvai fossils should be assigned. Contrary to John Robinson's (1965a) assertion that "insufficient morphological distance exists between (*A. africanus* and *H. erectus*) to justify the insertion of another species," I was able to show, from an analysis of several metrical features, that there was indeed an appreciable gap between the ranges of the two species. In 1964, Napier and I agreed that the new Olduvai fossil taxon "neatly closes the gap between the most advanced *Australopithecus* and the lowliest *Homo erectus*" (Tobias, 1964). Our careful analysis of morphological features revealed a number of apomorphic features of *Homo*.

So by February–March 1964, Napier and I had become convinced that the Olduvai pygmoids should be classified as a species within the genus *Homo*. Louis Leakey had had this conviction for over 5 years. One might say they were two different kinds of conviction – Leakey's being intuitive, prescient, inspirational, ours being dogged, statistical, functional anatomical – and perspirational!

Two things remained to be done: to find a name for the new species and to determine whether its inclusion in *Homo* necessitated a re-definition of the genus *Homo*.

It was chiefly in respect of the cranial capacity that we found it necessary to re-define *Homo*. It was a relatively small change. Yet some of our colleagues excoriated us for defiling the definition of *Homo*, as though it were sacrosanct like the laws of the Medes and Persians. They overlooked the fact that as respectable a scholar as LeGros Clark had changed the definition of *Homo* between the first and second editions of his seminal work, *The Fossil Evidence for Human Evolution* (1955, 1964). Robinson was to change the definition of *Homo* when in 1965 he proposed to lump *A. africanus* into *Homo,* while Bernard Campbell in 1978 re-worded our 1964 definition.

No-one need be coy or shame-faced about altering a generic diagnosis, if the growing and ever-changing state of knowledge and of taxonomy at any time reveals inadequacies in the previous diagnosis. Yet as recently as 1996, Walker and Shipman (1996: 90) quaintly described what we had done as "shifting the ground rules." In the words of Simpson (1963: 8): "The category genus is necessarily more arbitrary and less precise in definition than the species…There is no absolute criterion for the degree of difference to be called generic." Ernst Mayr, in the same year, wrote, "There is no non-arbitrary yardstick available for the genus as reproductive isolation is for the species." (Mayr, 1963: 340–341).

To find a name for the new species, I went to my predecessor and mentor, Raymond Dart, whose linguistic versatility had been one of his more remarkable skills. This man had invented such exorbitantly sesquipedalian words as *Australopithecus* and Osteodontokeratic. I hoped against hope he would abandon this style in favour of something short and simple. I told Dart what we had and that we had reason to believe that members of the new species were so manually proficient as to have been the probable fabricators of the earliest stone tools made to a set and consistent pattern. When the week-end had passed Dart offered the simplest, shortest and sweetest name he had ever invented – *Homo habilis*. The Latin word *habilis* means "able, handy, mentally skilful" and it is the etymon of the English words habile, able and ability. This was a functionally sound as well as mellifluous name, and I and my co-authors readily adopted it.

The name and species definition were announced in our paper, "A new species of the genus *Homo* from Olduvai Gorge" that appeared in *Nature* on 4th April 1964. Our article revised the diagnosis of *Homo*, created a new species *Homo habilis*, offered a diagnosis of it, and gave its geological horizon. A type specimen and paratypes were identified and briefly described. The essential points of difference from *A. africanus* were that *H. habilis* had smaller teeth that lacked the marked buccolingual bulge of the tooth crowns characteristic of australopithecine cheek–teeth; reduced third molar teeth in relation to the sizes of the first and second molars; relatively elongated cheek–teeth; endocranial capacities, the mean of which was greater by 45% or 10.2 standard deviations than the mean for *A. africanus*. To these morphological traits I was later (1975, 1987) to add the selective development of the impressions of certain brain areas detectable on the surface of the endocasts, notably asymmetrical parietal lobes and the caps over the motor speech cortices of Broca and Wernicke. Among dozens of other traits that serve to differentiate *H. habilis* from *A. africanus* are the shortening of the cranial base with the alignment of the petrous pyramid nearly in the coronal plane (as in modern humans and in robust and hyper-robust australopithecines, but quite different from its sagittal lie in *A. africanus*) (Tobias, 1967, 1991; Dean and Wood, 1981; Wood and Dean, 1981), and the presence in *H. habilis* of a helicoidal occlusal plane (Tobias, 1980a). More recently, Kegley and Hemingway (2005) have argued that fluctuating odontometric asymmetry differentiates *H. habilis* from australopithecine taxa.

The Assault on *Homo habilis*

Even before our article appeared in *Nature*, the pressures had begun to mount. Thus, LeGros Clark wrote to me begging me not to allow Leakey to make a new species for, he asserted, OH 7 was clearly an australopithecine, and this before he had even seen the specimens.

Immediately after the paper was published, our diagnosis and claims were repudiated by such formidable figures as LeGros Clark, Kenneth Oakley, Bernard Campbell, David Pilbeam, Elwyn Simons, Clark Howell, Loring Brace and John Robinson. The opposition continued for nearly 20 years. Was *H. habilis* ahead of its time? The phenomenon of delayed appreciation has been not uncommon. The eminent Finnish palaeontologist, Björn Kurtén, went so far as to declare in 1972 "…practically all of the epoch-making discoveries of new kinds of fossil hominids have been received with doubts and opposition from most contemporary anthropologists."

We live in an age when scientific knowledge is said to double every 10 years (although I have never been quite sure how that figure is arrived at). Yet delays in the acceptance of a selection of scientific discoveries ranged in the 20th century from 8–30 years. Such delays are extraordinary phenomena and not a general feature of scientific discoveries. I had to search long and hard to find even ten examples from many fields of science. Günther Stent (1972) sought to understand the belated acceptance that DNA was the basic hereditary substance. He needed a criterion of a premature discovery, other than its failure to make an impact at the time – which would have been simply a tautology. He proposed that "a discovery is premature if its implications cannot be connected by a series of simple logical steps to canonical, or generally accepted knowledge" (Stent, 1972: 84). I took his analysis further in a presentation to the American Philosophical Society in November 1994. My ten case histories included two of Stent's examples and eight others, two of which were drawn from palaeoanthropology.

At the causal level, I drew attention to the fact that the delayed acceptance of *Australopithecus africanus* and of *Homo habilis* were both good examples of premature discoveries in the strict sense of Stent. In each instance I examined the tenets of the prevailing paradigms, in 1925 in the case of *A. africanus*, and in the early 1960s for the proposed new species *Homo habilis*. Both discoveries proved to be premature *sensu* Stenti: their implications could not be reconciled with the pre-existing mind-set by simple, logical steps. To substitute a new paradigm for the pre-existing one, it was necessary for each of these tenets to be replaced or modified, before the proposal to substitute a new paradigm could gain general acceptance. It took 28 years, from 1925 to 1953 for *A. africanus* to gain general acceptance. With *H. habilis*, the measure of prematurity was about 20 years (from 1964 to 1984). The major tenets of the old paradigm fell away one by one. The tide began to turn late in the 1970s. From Omo at the northern end of Lake Turkana, isolated teeth had suggested to Yves Coppens and Clark Howell the presence of *H. habilis*. In 1977, a fragmentary cranium, L. 894-1, came to light and was published by Clark Howell and Noel Boaz as a member of *H. habilis* or *Homo modjokertensis*. They added (1977: 106), "…without further and more definite data on provenance, particularly from Indonesia, any taxonomic assignation to this early *Homo* taxon would be premature." However, when Vincent Maglio's and Basil Cooke's great work, *Evolution of African Mammals*, appeared in the following year, 1978, Clark Howell's compendious chapter on the Hominidae positively included the species *Homo habilis* and among the remains attributed to it was L. 894-1 from the Shungura. *H. habilis* was acquiring a semblance of respectability!

What I have always considered the quiddity of the early *Homo* saga came in 1980, when a formerly sceptical critic, Milford Wolpoff, in his book on *Paleoanthropology*, declared roundly, "*Homo habilis* is a taxon whose time has come" (1980: 182). I had been obliged for much of that time to carry on the fight on behalf of *Homo habilis* single-handedly, as Louis Leakey had moved on to pastures new at Fort Ternan, whilst John Napier devoted more and more of his time to the popularizing of science in the media, especially the B.B.C. There was an aloneness during much of those 21 years (from 1959 to 1980) as I strove to establish that the taxon that we three had named *Homo habilis* represented a good species of hominid. The task was not lightened over the next 10 years when the international anti-apartheid campaign generated an academic boycott. This affected academics who had chosen to stay on in South Africa and even those of us who were indomitably campaigning against apartheid from within the country. Although the notion that *H. habilis* was a good species is now widely accepted, there were years of intellectual isolation and antagonism before the point gained general acceptance. I make so bold as to liken this time of conceptual solitude, the habiline pre-revolution, to the terrible and unutterable solitariness that Charles Darwin faced after he published *The Origin of Species* (1859), or that Raymond Dart confronted when, for decades after the discovery of the Taung skull, he was virtually alone in his belief that *Australopithecus africanus* had stood upon the threshold of humanity.

Yet it was, too, a time of hope.

Late Thoughts

The twenty-first century brings new problems in the *H. habilis* story. So it should, for knowledge does not stand still. The kinds of questions that challenge scientists today include the following:

Is *H. habilis* still considered a good species? I think most scholars would say, "Yes". The hypodigm of *H. habilis* was initially enumerated by Leakey et al. (1964), modified by me in subsequent writings, by Clark Howell in his 1978 chapter, and by Bernard Wood (1991) in his monograph on the hominin cranial remains of Koobi Fora. Which of these hypodigmata, if any, is still tenable? A number of scholars have seen fit to remove some of the specimens in the earlier hypodigm of *H. habilis* and to place them in one or two more recently-created species such as *Homo ergaster* and *H. rudolfensis*. Such removals still leave a substantial rump of specimens in *H. habilis*, and this residuum is sometimes called *Homo habilis sensu stricto*, it being implied that *H. habilis* correctly applies only or mainly to the original Olduvai sample. There is no consensus as to which specimens, if any, need to be removed from the earlier hypodigm of *H. habilis* and placed in some other species – and if so, in which species.

The expansion of the fossil samples from South and East Africa and other parts of the continent, and the adding and refinement of dating methods, have brought palaeoanthropology to a juncture where the place of *H. habilis* in hominin phylogeny can be approached with greater assurance. The problem must be canvassed in the light of several competing philosophies of biological evolution.

Functional aspects of *H. habilis* were tentatively broached in the Leakey, Tobias and Napier article in 1964. Napier and I took the matter a little further, whilst I carried it much further especially since my claim in 1980 that *Homo habilis* was capable of spoken language. Again I stood alone in making this revolutionary claim, just at a time, incidentally, when some quite serious and competent colleagues were questioning whether the Neandertals were capable of spoken language! My sin in extending the capacity for spoken language back by 2 million years seemed to be unforgivable for some and not worthy of taking note of for others. Yet a trickle of supporters emerged over the dying decades of the twentieth century, notably Dean Falk, Sir John Eccles, Peter Andrews and Chris Stringer, Wilkins and Wakefield, and Terrence Deacon. Was this another premature discovery? At this stage it is too early to say. It is at most a potentially premature discovery! This raises the broader issue: is one justified in importing behavioural traits into the definition of species? Apparently the ornithologists do this frequently and get away with it.

My last point is to ask: were we correct in placing the species *habilis* in *Homo*? From the beginning, this was questioned by some colleagues. Amazingly, within 15 months of the publication of our joint article, no fewer than five nomina had been proposed to accommodate these Olduvai hominins, including *Australopithecus africanus* (subspecies unspecified), *A. africanus habilis, A. habilis, Homo erectus habilis and H. erectus* (subspecies unspecified). A quaint variation was *Homo Paleoanthropus Habilinensis* proposed by Enver Bostanci. In an unforgettable footnote, he wrote: "Because there is a famous story in history Habil and Kabil, we should not have mixed them with fossil man" (Bostanci, 1974: 566). I understand that Kabil and Habil refer to Cain and Abel!

After this initial orgy of naming, the world seemed to have settled on *Homo habilis*. Of late, a few investigators have elected to put *habilis* into *Australopithecus*, thus reverting to one of the suggested nomina of 40 years ago. The only nomen that has enjoyed some longevity and anything like a consensus has been *Homo habilis*. Are there good reasons to change this? I do not know of any, but the ball is in your court!

References

Boaz, N.T., Howell, F.C., 1977. A gracile hominid from Upper Member G of the Shungura Formation, Ethiopia. American Journal of Physical Anthropology 46, 93–108.

Bostanci, E., 1974. Anadolu'da insan kronolojisi ve genus *Homo* icin yeni bir siniflandirma (Chronology of man in Anatolia and a new classification for genus *Homo*). Dil ve Tarih-Cografya Fakültesi Yayin 239, 540–591.

Broom, R., Robinson, J.T., 1949. A new type of fossil man. Nature 164, 322–323.

Campbell, B.G., 1978. Some problems in hominid classification and nomenclature. In: Jolly, C.J. (Ed.), Early Hominids of Africa. Duckworth, London, pp. 567–581.

Clark, W.E. LeGros, 1955. The Fossil Evidence for Human Evolution: An Introduction to the Study of Palaeoanthropology. University of Chicago Press, Chicago, IL.

Clark, W.E. LeGros, 1964. The Fossil Evidence for Human Evolution: An Introduction to the Study of Paleoanthropology. Second Edition. University of Chicago Press, Chicago, IL.

Cole, S., 1975. Leakey's Luck: The Life of Louis Seymour Bazett Leakey 1903–1971. Collins, London.

Darwin, C., 1859. The Origin of Species. John Murray, London.

Dean, M.C., Wood, B.A., 1981. Basicranial anatomy of Plio-Pleistocene hominids from East and South Africa. American Journal of Physical Anthropology 59, 157–174.

Howell, F.C., 1978. Hominidae. In: Maglio, V.J., Cooke, H.B.S. (Eds), Evolution of African Mammals. Harvard University Press, Cambridge, MA, pp. 154–248.

Johanson, D.C., Edey, M., 1981. Lucy – the Beginnings of Humankind. Simon & Schuster, New York.

Kegley, A.D.T., Hemingway, J., 2005. Surveying stress and developmental stability in *Homo habilis* via fluctuating odontometric asymmetry: assessing generic patterns and australopithecine differentiation. In: Strkalj, G., Pather, N., Kramer, B. (Ed.), Voyages in Science. Content Solutions, Pretoria, pp. 35–49.

Kuhn, T.S., 1962. The Structure of Scientific Revolutions. University of Chicago Press, Chicago, IL.

Kurtén, B., 1972. Not from the Apes: Man's Origins and Evolution. Victor Gollancz, London.

Leakey, L.S.B., 1959a. A new fossil skull from Olduvai. Nature 184, 491–493.

Leakey, L.S.B., 1959b. The newly discovered skull from Olduvai: first photographs of the complete skull. Illustrated London News 235, 288 19th September; 217–219, 12th September.

Leakey, L.S.B., Evernden, J.F., Curtis, G.H., 1961. Age of Bed I, Olduvai Gorge, Tanganyika. Nature 191, 478–479.

Leakey, L.S.B., Tobias, P.V., Napier, J.R., 1964. A new species of the genus *Homo* from Olduvai Gorge. Nature 202, 7–9.

Leakey, M.D., 1971. Olduvai Gorge: Excavations in Beds I and II, 1960-1963. Vol. III. Cambridge University Press, Cambridge.

Leakey, M.D., 1984. Disclosing the Past: An Autobiography. Weidenfeld & Nicolson, London.

Mayr, E., 1963. The taxonomic evaluation of fossil hominids. In: Washburn, S.L. (Ed.), Classification and Human Evolution. Viking Fund Publications in Anthropology, Chicago, IL, pp. 332–346.

Mayr, E., 1972. The nature of the Darwinian revolution. Science 176, 981–989.

Nesturkh, M., 1967. The Origin of Man. Progress Publishers, Moscow.

Robinson, J.T, 1965a. *Homo habilis* and the australopithecines. Nature 205, 121–124.

Robinson, J.T., 1965b. Comments on new discoveries in Tanganyika: their bearing on hominid evolution (by P.V. Tobias). Current Anthropology 6, 403–406.

Simpson, G.G., 1963. The meaning of taxonomic statements. In: Washburn, S.L. (Ed.), Classification and Human Evolution. Viking Fund Publications in Anthropology, Chicago, IL, pp. 1–31.

Simonetta, A., 1957. Catalogo sinonimia annotata degli ominoidi fossil de attuali (1758-1955). Atti di Sociedade Toscano Scienze Naturale B, 64, 53–112.

Stent, G.S., 1972. Prematurity and uniqueness in scientific discovery. Scientific American 227, 84–93.

Tobias, P.V., 1959. Studies on the occipital bone in Africa: part IV. Components and correlations of occipital curvature in relation to cranial growth. Human Biology 31, 138–161.

Tobias, P.V., 1961. Letter to W.E. LeGros Clark, 1st August 1961.

Tobias, P.V., 1963. Cranial capacity of *Zinjanthropus* and other australopithecines. Nature 197, 743–746.

Tobias, P.V., 1964. The Olduvai Bed I hominine with special reference to its cranial capacity. Nature 202, 3–4.

Tobias, P.V., 1967. Olduvai Gorge, Vol. 2. The Cranium and Maxillary Dentition of *Australopithecus (Zinjanthropus) boisei*. Cambridge University Press, Cambridge.

Tobias, P.V., 1975. Brain evolution in the Hominoidea. In: Tuttle, R. (Ed.), Primate Functional Morphology and Evolution. Mouton, The Hague, pp. 353–392.

Tobias, P.V, 1978. The earliest Transvaal members of the genus *Homo* with another look at some of the problems of hominid taxonomy and systematics. Zeitschrift für Morphologie und Anthropologie 69, 225–265.

Tobias, P.V., 1980a. The natural history of the helicoidal occlusal plane and its evolution in early *Homo*. American Journal of Physical Anthropology 53, 13–187.

Tobias, P.V., 1980b. L'évolution du cerveau humain. La Recherche 11, 282–292.

Tobias, P.V., 1987. The brain of *Homo habilis*: a new level of organisation in cerebral evolution. Journal of Human Evolution 16, 741–761.

Tobias, P.V., 1991. Olduvai Gorge, Vol. 4. The Skulls, Endocasts and Teeth of *Homo habilis*. Cambridge University Press, Cambridge.

Walker, A., Shipman, P., 1996. The Wisdom of Bones. Weidenfeld and Nicolson, London.

Washburn, S.L., 1963. Behavior and human evolution. In: Washburn, S.L. (Ed.), Classification and Human Evolution. Aldine, Chicago, IL, pp. 190–203.

Wolpoff, M.H., 1980. Paleoanthropology. Knopf, New York.

Wood, B.A., 1991. Koobi Fora Research Project, Vol. 4: Hominid Cranial Remains. Clarendon Press, Oxford.

Wood, B.A., Dean, M.C., 1981. Comparative basicranial morphology of Plio-Pleistocene hominids: background studies and fossil evidence. International Union for Prehistoric and Palaeontological Sciences, X Congrés, Colloquium of Commission VIII, pp. 1–46.

Chapter 3
Where Does the Genus *Homo* Begin, and How Would We Know?

Bernard A. Wood

Keywords Genus category • taxonomy • adaptive grade • phylogenetic relationships • *Australopithecus* • *Homo habilis* • *Homo erectus*

Defining the Genus Category

Genus definitions use four criteria, two lines of evidence (phenotypic and genetic) and fall into two categories. The four criteria are the relationships among taxa, information about their adaptive grade, estimates of the genetic distance that separates them, and the estimated time of divergence. At least one researcher (Dubois, 1988) has suggested that a fifth criterion, evidence of hybridization between species, should be the primary criterion for grouping species into genera. However, even if this suggestion had any merit, it is not clear how evidence of hybridization could be obtained from the fossil record of extinct taxa.

One, or more, of the criteria has been used, or combined, to generate genus definitions that belong to two categories. The first category draws upon phenotypic evidence to make inferences about adaptive grade, and uses both phenotypic and molecular evidence (if the latter is available) to make inferences about relationships. Definitions in the second category rely solely on genetic evidence for either estimates of genetic distance, or for generating estimates about the timing of evolutionary events, which are then converted into criteria for recognizing genera.

Four genus concepts in the first category are reviewed. The first of these, the most widely used genus concept, is the one suggested by Mayr (1950). Mayr proposed that "a genus consists of one species, or a group of species of common ancestry, which differ in a pronounced manner from other groups of species and are separated from them by a decided morphological gap" (Mayr, 1950: 110), and he went on to state that the species united in a such a genus must "occupy an ecological situation which is different from that occupied by the species of another genus, or, to use the terminology of Sewall Wright, they occupy a different adaptive plateau" (Mayr, 1950: 110). Thus, according to Mayr, a genus is a group of species of common ancestry (which for Mayr includes both monophyletic and paraphyletic groups) that is adaptively both homogeneous and distinctive. Mayr et al. (1953: 50) suggest that if "unrelated species acquire a superficial similarity owing to parallel adaptations to similar environments" then evidence about relatedness should be given precedence over evidence about adaptation.

The second of the four concepts in the first category was proposed by Willi Hennig in his 1966 volume "Phylogenetic Systematics." Hennig (1966) suggested that only monophyletic groups (i.e., all – no more and no less – of the species descended from a recent common ancestor) should be accepted as valid genera, and he promoted the view that, within reason, taxa arising at the same time should be assigned the same taxonomic rank (as will be evident, the use of time as a criterion has been advocated by other researchers since Hennig).

The third of the genus concepts in the first category was proposed by Wood and Collard (1999). These authors suggested that a genus should be a monophylum whose members occupy a single adaptive zone. This definition differs from Mayr's (1950) in that it excludes paraphyletic taxa, but *contra* the interpretation of Leakey et al. (2001) and Cela-Conde and Altaba (2002), it does not require the adaptive zone of a genus to be unique. Wood and Collard (1999) suggested that two, or more, genera based on different monophyletic groups could occupy a similar adaptive zone.

The final genus concept in this first category was outlined by Cela-Conde and Altaba (2002). It is similar to the Wood and Collard (1999) definition in that it requires the component species to be monophyletic, but it differs in that it allows one species in a genus, the *species germinalis*, to be in a different adaptive zone from the other species in the genus.

The two genus concepts in the second category draw solely on genetic evidence. The first concept focuses on

B.A. Wood (✉)
Center for the Advanced Study of Hominid Paleobiology,
Department of Anthropology, 2110 G Street NW,
George Washington University, Washington, DC 20052, USA
e-mail: bernardawood@gmail.com

F.E. Grine et al. (eds.), *The First Humans: Origin and Early Evolution of the Genus* Homo,
Vertebrate Paleobiology and Paleoanthropology,

evidence about genetic distance. It suggests that species should be grouped into genera if the genetic distance between them is the same, or less than, the typical genetic distance between pairs of congeneric species in other animal groups (Watson et al., 2001).

The second genus concept in the genetic evidence-only category combines genetic distance and time. With respect to primates it suggests that the species in a primate monophylum should be included in the same genus if they originated between 11 and 7 Ma BP (Goodman et al., 1998). Wildman et al. (2003) justified the 11–7 Ma BP date because the majority of genera in other mammalian orders arose in that time period. As Morris Goodman and his colleagues have noted (Goodman et al., 1998, 2001; Wildman et al., 2003), the implication of their interpretation is that the genus *Homo* would not only include all hominin taxa, but also chimpanzees and bonobos. Subsequently, Colin Groves (one of the authors of the Goodman et al., 1998 study) used a mix of paleontological and genetic evidence to propose an origin time of between 7 and 4 Ma BP as the criterion for delineating extant mammalian genera (Groves, 2001). This led Groves to retain *Homo* and *Pan* as separate genera. However, even more recently Groves and a colleague have proposed that primate monophyletic groups should be recognized as genera if they originated between 6 and 4 Ma BP (Cameron and Groves, 2004). If this criterion were to be adopted, the genus *Homo* would embrace most, if not all, currently accepted hominin taxa.

So, what can we learn from these thumbnail reviews of genus concepts? First, that the sensible strategy is to adopt the adage that "all genera should be clades, but not all clades are genera." Second, most of the researchers who focus their research on the hominin fossil record implicitly, if not explicitly, subscribe to a genus definition that blends information about clades and grades, and this is certainly the case for the way most of these researchers interpret the genus *Homo*. It is widely assumed, but rarely articulated, that the species included in *Homo* form a monophyletic group, and it is also widely assumed, but also rarely articulated, that the taxa all share functional characteristics or competencies. In the case of *Homo* these functional characteristics or competencies have included the ability to use complex language, the ability to make the only type of tool, stone tools, that can be reliably detected in the early archeological record, and the ability to hunt.

History of *Homo* (1758–1964)

The history of the interpretation of the genus *Homo* from its introduction in 1758 by Carolus Linnaeus in the tenth edition of his *Systema Naturae* to the inclusion in 1964 of *Homo habilis*, has been one of episodic relaxation of the criteria used for including taxa in *Homo*. Each of these episodes has resulted in one, or more, species being added to *Homo*.

As originally conceived by Linnaeus, the genus *Homo* incorporated two species, *Homo sapiens* and *Homo sylvestris* (also called *Homo troglodytes* or *Homo nocturnus*). The latter apparently referred to a mythical nocturnal cave-dwelling form from Java, but the discovery of *Homo floresiensis* (Brown et al., 2004; Morwood et al., 2005) may mean that *Homo sylvestris* might not have been mythical after all (Collard and Wood, 2006).

William King was the first to suggest that an extinct hominin species, namely *Homo neanderthalensis*, should be included within *Homo* (King, 1864). Researchers still argue about the taxonomic significance of the differences between *H. sapiens* and *H. neanderthalensis*, but there is no gainsaying that the inclusion of the latter within *Homo* meant expanding the definition of *Homo* so that it included archaic and derived morphology (e.g., discrete and rounded supraorbital margins, midline facial projection, a distinctive parietal and occipital morphology, and robust limb bones with relatively large joint surfaces) that is not seen at all, or is not seen in this combination, in *H. sapiens*.

The next modification of the interpretation of the genus *Homo* took place nearly half a century later, when in 1908, the sole item of fossil evidence, the Mauer mandible, for the taxon *H. heidelbergensis* (Schoetensack, 1908), was added to the genus. This meant that *Homo* now embraced at least one individual with a mandible that combined the absence of a true chin with evidence a robust mandibular corpus. Then came the addition of *Homo rhodesiensis* from the site now known as Kabwe (Woodward, 1921), and subsequently the addition of *Homo soloensis* from Ngandong (initially included in the subgenus *Javanthropus* [Oppenoorth, 1932], but later included in the genus *Homo* [Oppenoorth, 1937]). The addition of the Ngandong fossils, meant that crania substantially more robust than those of modern humans and *H. neanderthalensis* were now included in *Homo*. Nonetheless, the endocranial volumes of all the crania associated with *H. neanderthalensis*, *H. rhodesiensis* and *H. soloensis* are still within, or close to, the modern human range (de Sousa and Wood, 2006).

When Franz Weidenreich formally proposed in 1940 that two existing extinct hominin hypodigms, *Pithecanthropus erectus* and *Sinanthropus pekinensis*, should be merged into a single species and transferred to *Homo* as *Homo erectus* (Weidenreich, 1940) the addition of these taxa resulted in *Homo* subsuming an even wider range of morphology. Subsequently the hypodigms of *Meganthropus* (Mayr, 1944), *Telanthropus* (Robinson, 1961), and *Atlanthropus* (Le Gros Clark, 1964) were also transferred to *H. erectus*. The addition of the *H. erectus* hypodigm at this time, even though it was well before the discovery of the small *H. erectus* crania from East Africa (Antón, 2004: Leakey et al.,

2003; Potts et al., 2004; Spoor et al., 2007) and Dmanisi (Rightmire et al., 2006), substantially increased the range of endocranial volume within the genus *Homo*. Compared with most of the pre-1940 hypodigm of *Homo*, fossils attributed to *H. erectus* have a smaller neurocranium, a lower vault, a broader base relative to the vault, and more complex premolar roots. They also have a substantial, shelf-like torus above the orbits, and there are often both sagittal and angular tori, although the expression of some, or all, of this morphology may be size-related (Antón et al., 2007). The occipital sagittal profile is sharply angulated in *H. erectus*, with a well-marked supratoral sulcus, and the inner and outer tables of the vault are thickened. The cortical bone of the postcranial skeleton is generally thick. The long bones are robust, and the shafts of the femur and the tibia are flattened from front to back relative to those of other *Homo* species. However, all the postcranial elements of *H. erectus* are consistent with a habitually upright posture and obligate long-range bipedalism.

In 1964, Louis Leakey, Phillip Tobias and John Napier announced the discovery at Olduvai Gorge of specimens they assigned to a new species of *Homo*, called *Homo habilis* (Leakey et al., 1964). The type specimen of *H. habilis*, the sub-adult OH 7, recovered in 1960, preserves most of both parietal bones, a partial mandible and several hand bones of a juvenile skeleton. Between 1960 and 1963 further evidence of a "non-robust" fossil hominin was unearthed in Bed I (OH 8) and Bed II (OH 13, OH 14, OH 16) of Olduvai Gorge. The inclusion of this group of specimens in *Homo* substantially widened the range of morphology within the genus, and meant that the Le Gros Clark (1955) diagnosis of *Homo* needed to be amended. In particular, to accommodate *H. habilis* in the genus *Homo* Leakey et al. (1964) were forced to reduce the lower end of the range of brain size of fossils attributed to *Homo* to 600 cm^3. Leakey and his colleagues claimed that other criteria, such as dexterity, an erect posture and a bipedal gait, did not need to be changed because their interpretation of the functional capabilities of the type specimen and the paratypes of *H. habilis* was consistent with these functional criteria (Leakey et al., 1964). Ultimately, however, fresh evidence, and fresh interpretations of existing evidence, has led others to offer rather different functional assessments of the same material (see below).

History of *Homo* Since 1964

Since 1964 various categories of evidence have provided information germane to generating hypotheses about the nature of the genus *Homo*. These include new fossil evidence and additional information about the relationships and adaptive grade of potential early *Homo* taxa. Space constraints prevent other than a selective review of all the three categories of evidence.

New Fossil Evidence

After the announcement of *H. habilis* in 1964, the next significant addition to the *Homo* hypodigm was the recovery of KNM-ER 1470 from the Upper Burgi Member of the Koobi Fora Formation (Leakey, 1973). Morphologically, it presented an apparently unique mixture of a relatively large, *Homo*-like neurocranium, and a broad *Paranthropus*-like face, but it lacked the distinctive combination of small anterior and large postcanine teeth that is typical of *Paranthropus*, especially *Paranthropus boisei*. However, the initial assessments of KNM-ER 1470 paid more attention to its large neurocranium than to its *Paranthropus*-like face, and most researchers allied the new specimen with *Homo* (e.g., Leakey, 1973; Rak, 1987; Bilsborough and Wood, 1988). As a consequence, from 1972 onwards, the genus *Homo* subsumed a substantially wider range of facial morphology (Wood, 1991) than it did prior to the discovery of KNM-ER 1470. Walker (1976) was alone among the early commentators to caution that KNM-ER 1470 may sample a large-brained australopith taxon and may not belong to *Homo* after all. The KNM-ER 1470 cranium was initially not allocated to a species, but instead was referred to an informal category called 'early *Homo*'.

In due course, other cranial specimens from Koobi Fora (e.g., KNM-ER 1590, 1802, 1805, 1813, 3732) (Wood, 1991) and Olduvai (e.g., OH 62) (Johanson et al., 1987) were added to the early *Homo* hypodigm, as were fossils from Members G and H of the Shungura Formation (Howell et al., 1976; Boaz and Howell, 1977; Coppens, 1980), Members 4 and 5 at Sterkfontein (Hughes and Tobias, 1977; Clarke, 1985; Kimbel and Rak, 1993), Member 1 at Swartkrans (Clarke and Howell, 1972; Grine and Strait, 1994; Grine et al., 1993, 1996), the Chemeron Formation (Hill et al., 1992), Uraha (Bromage, et al., 1995), Hadar (Kimbel et al., 1996), the Nachukui Formation in West Turkana (Prat et al., 2005), and Dmanisi in Georgia (Gabunia and Vekua, 1995; Gabunia et al., 2000; Vekua et al., 2002; Lordkipanidze et al., 2005, 2007).

These additions to the hypodigm of "early *Homo*" subsumed a wide range of cranial and dental morphology. For example, the endocranial volumes of the specimens range from just less than 500 cm^3 to around 850 cm^3. The mandibles also vary in size, but all have relatively robust bodies (Wood and Aiello, 1998) and premolar teeth with relatively complex crowns and roots (e.g., Bromage et al., 1995).

The discovery and subsequent analysis of the OH 62 associated skeleton was particularly significant. Although the preservation of this specimen is poor, enough craniodental morphology is preserved to be confident that it does not belong to *P. boisei*. Thus, unless it is the first evidence from Bed I of Olduvai Gorge of a novel taxon, then OH 62 must belong to *H. habilis*. Although several isolated postcranial specimens from Bed I at Olduvai Gorge found before OH 62 had been attributed to *H. habilis* (Leakey et al., 1964), it was subsequently pointed out that it is at least equally likely that this postcranial evidence belongs to *P. boisei* (Wood, 1974; Wood and Constantino, 2007), and Gebo and Schwartz (2006) have recently lent their support to the suggestion that OH 8 belongs to *P. boisei*. Thus, parsimony suggests that the discovery of OH 62 provided the first unequivocal postcranial evidence for *H. habilis*, so it is significant that OH 62 has been interpreted as having limb proportions that are at least as ape-like as those of individuals attributed to *Australopithecus afarensis* (Johanson et al., 1987; Hartwig-Scherer and Martin, 1991; Richmond et al., 2002). It is also likely that the even more fragmentary KNM-ER 3735 associated skeleton also belongs to *H. habilis* (Wood, 1991).

The most recent species to be added to *Homo*, *Homo floresiensis* (Brown et al., 2004) broadened the morphological scope of *Homo* even further. The specimens initially attributed to this species were recovered from deposits in the Liang Bua cave on the Indonesian island of Flores, and are dated to between approximately 74,000 and 18,000 years ago (Brown et al., 2004; Morwood et al., 2004). They include a well-preserved skull and partial skeleton of an adult female as well as several more fragmentary specimens (Brown et al., 2004). Since then additional upper limb fossils belonging to the type specimen, LB 1, have been recovered, together with evidence of a second adult mandible (LB 6), as well as postcranial remains belonging to other individuals (LB 4, 5, 7, 8–9) (Morwood et al., 2005).

Its small brain size makes *Homo floresiensis* a particularly significant addition to *Homo*. The endocranial volume of the partial associated female skeleton, LB1, was initially reported to be 380 cm^3 (Brown et al., 2004), but Falk et al. (2005) increased this figure to 417 cm^3. Even at 417 cm^3, the endocranial volume of *H. floresiensis* is considerably smaller than those of the other species assigned to *Homo*. Adult endocranial volume in *H. habilis* ranges between 509 and 687 cm^3 (de Sousa and Wood, 2006), thus the assignment of the late Pleistocene Liang Bua specimens to *Homo* (Gordon et al., 2008) greatly increases the range of brain size in the genus. The stature estimates of 106 cm for LB 1 (Brown et al., 2004) and 109 cm for LB 8 (Morwood et al., 2005) are only slightly smaller than McHenry's (1991) stature estimate of 118 cm for the *H. habilis* OH 62 partial skeleton. An analysis of calvarial shape as defined by six linear variables, concluded that the LB 1 calvaria is almost certainly not a normal, dwarfed, modern human, but instead it is closest in shape to early African *H. erectus* (Gordon et al., 2008)

New Evidence About Adaptive Grade

In this brief review it is not possible to review all of the research that has been carried out since 1964 relevant to determining the grade of potential early *Homo* taxa. Instead, I will consider some of the research relevant to the reconstruction of the diet, locomotion and life history of these early hominins.

Diet – Postcanine Microwear

The increasing sophistication of dental microwear research has brought both "good news" and "bad news." The "good news" is that researchers are becoming much more discriminating about the specimens they judge to contain evidence about microwear due to contact between the teeth and food, as opposed to the microscopic damage caused by taphonomic factors such as fluvial transport that affects teeth found on the surface. The "bad news" is that when Ungar et al. (2006) and Ungar and Scott (2009) applied these more stringent criteria to their initial sample of 83 early *Homo* specimens from three southern African and seven East Africa sites, it resulted in the sample size being whittled down to 18.

The results of the Ungar et al. (2006) study, when combined with an assessment of the dietary significance of differences in the occlusal slope of worn postcanine teeth (Ungar, 2004), suggest that any differences in microwear between the teeth of specimens assigned to *H. habilis* and those assigned to an archaic hominin such as *Au. afarensis*, may be due more to differences in the physical properties of "fall back" foods rather than to differences in the physical properties of their preferred diets (Ungar and Scott, 2009). The more striking result of the study was the difference between the microwear seen in teeth assigned to *H. habilis* and those assigned to *H. erectus*. Ungar et al. (2006) suggest that "*H. erectus* and individuals from Swartkrans Member 1 ate, at least occasionally, more tough or brittle foods than did *H. habilis* and individuals from Sterkfontein Member 5C" (Ungar et al., 2006: 91).

Diet – Scaling of Teeth and Jaws

The size of the teeth and jaws of an individual should reflect the diet of that individual, and likewise, the sample parameters of the size of the teeth and jaws should broadly reflect the nature of the preferred and fall back foods of the species to which that individual belongs.

McHenry (1988) developed the megadontia quotient (MQ) as a way of comparing the size of the postcanine teeth of hominins with different overall body sizes. The most recent computations of this index are given in McHenry and Coffing (2000). The ends of the range for hominin taxa are 0.9 for *H. sapiens* and 2.7 for *P. boisei*, so, give or take the inevitably substantial error associated with these types of estimates, per unit of body mass the postcanine teeth of *P. boisei* are approximately three times the size they are for *H. sapiens*. The body mass estimates they use for *H. rudolfensis* are almost certainly too large and thus invalidate the MQ for that taxon, but 1.9, the MQ estimate for *H. habilis*, is likely to be closer to the mark. It compares with the MQs of 1.7 and 2.0 for *Au. afarensis* and *Au. africanus*, respectively, so *H. habilis* shows no reduction in relative postcanine tooth crown area compared with the two archaic hominins with the largest hypodigms, and its MQ is only a little smaller than the MQ (2.2) for *P. robustus*. Interestingly, the MQ of early African *H. erectus* (0.9) is the same at that for *H. sapiens*. Thus, in terms of both absolute and relative size, the postcanine teeth of early African *H. erectus* are well within the modern human range.

Bailey (2004) has also shown morphological differences within *Homo* in the relative size of the cusp components of postcanine tooth crowns. A recent study of the relative size of the main cusps of the first permanent maxillary molar (M^1) (Quam et al., in press) has shown that, with respect to relative cusp size, early *Homo* (i.e., *H. habilis sensu lato*) shares the presumed primitive condition seen in *Australopithecus* and *Paranthropus* (i.e., a relatively large metacone and a relatively small paracone), whereas the single early African *H. erectus* specimen shows cusp relative size relationships that are similar to those seen in later *Homo* species.

Wood and Aiello (1998) carried out a comparable exercise for the linear dimensions of the mandibular corpus at M_1. They generated two comparative regressions based on extant taxa for the relationship between actual body mass and the linear dimensions (but mainly the height) of the mandibular corpus at M_1. The first, the simian regression, was based on 23 taxa, and the second, the hominoid regression, was based on the subset of six hominoid taxa that were included in the simian regression. Wood and Aiello (1998) showed that per unit body mass the "average" simian has a larger mandibular corpus than the "average" hominoid. The authors used the height of the corpus and the two comparative regressions to predict body mass for extinct hominin taxa including those included in early *Homo* (*H. habilis sensu stricto* [N = 5]; *H. rudolfensis* [N = 6]; early African *H. erectus* [N = 7]), and then compared those mandible-based body mass predictions with the body masses estimated using either postcranial evidence, or non-mandibular cranial evidence (e.g., orbital height).

The results mirrored McHenry's results for postcanine crown area. The "hominoid mandibular-based" body mass predictions for *P. boisei* were more than three times larger than the estimates of body mass based on non-mandibular evidence, whereas the equivalent body mass predictions for early African *H. erectus* were similar to the estimates of body mass based on non-mandibular evidence. The equivalent "hominoid mandibular-based" body mass predictions for *H. habilis sensu stricto* and for *H. rudolfensis* were, respectively, 75% and 100% larger than the estimates of body mass based on non-mandibular evidence. These discrepancies between mandibular-based body mass predictions and estimates of body mass based on non-mandibular evidence are similar to those seen in *Au. afarensis* and *Au. africanus* (Wood and Aiello, 1998, Figs. 3 and 5) and they suggest that reduction in the size of the masticatory apparatus within the hominin clade did not occur until the emergence at c. 1.9 Ma of early African *H. erectus*.

Locomotion – Limb Proportions

Richmond et al. (2002) examined the significance of the differences in limb proportion between the early hominin associated skeletons assigned to *H. habilis sensu stricto* (i.e., OH 62) and *Au. afarensis* (i.e., A.L. 288–1). They found that the limb proportions of OH 62 are not statistically significantly different from those of A.L. 288-1. More recently Reno et al. (2005) have argued that the humerofemoral index of OH 62 cannot be calculated for OH 62 because the portion of femur it retains – the proximal femur – is a poor predictor of maximum femur length. This claim is supported by an analysis of the relationship between proximal and maximum femur length in extant hominoids, which suggests that the two lengths are uncorrelated. At the least, Reno et al. (2005) have demonstrated that current estimates of the length of the OH 62 femur, and thus of the OH 62 humerofemoral index, must be treated with caution. As a result, the claim that the limb proportions of OH 62 are more primitive than those of archaic hominins (Hartwig-Scherer and Martin, 1991) is weakened. Haeusler and McHenry (2004, 2007) have also investigated the limb proportions of early *Homo* by looking at OH 62 and a second associated skeleton, KNM-ER 3735, from Koobi Fora, Kenya, and they conclude that the limb proportions of both of these skeletons are more modern human-like than chimpanzee-like. However, Haeusler and McHenry's (2004) use of OH 34 to derive the limb proportions of OH 62 is controversial, as are their conclusions with respect to KNM-ER 3735. At the very least there is still doubt about the limb proportions of the two individuals represented by OH 62 and KNM-ER 3735, and it is fair to say that most informed observers would subscribe to the view that the limb proportions of OH 62 are more similar to those of archaic hominins than to those of the limb proportions of modern humans and pre-modern *Homo* taxa. Jungers (2009) has provided convincing evidence that the hominin locomotor skeleton was fundamentally reorganized by 1.8–1.6 million years ago in early *Homo erectus/ergaster*, and elongated hind limbs are clearly part of this new package

Locomotion – Semicircular Canal Morphology

In 1994 Fred Spoor presented the preliminary conclusions of a long-term research project to survey the morphology of the semicircular canals of early hominins (Spoor et al., 1994). Among the early hominin specimens considered in that study were Stw 53, assigned by some to *H. habilis*, and SK 847. The semicircular canal morphology of the former was so different from that of modern humans that Spoor and his colleagues suggested that "Stw 53 relied less on bipedal behavior than the australopithecines" (Spoor et al., 1994: 648). They also suggested that "the extreme differences in labyrinthine morphology between SK 847 and Stw 53 make attribution of both specimens to the same species, on this evidence alone, highly unlikely" (Spoor et al., 1994: 648).

In the same year Spoor (1994) presented the results of his analysis of the labyrinthine morphology of Sangiran 2 and 4, OH 9 and SK 847. He suggested that "the dimensions of the semicircular canals (of these taxa) are similar to those in modern humans" (Spoor, 1994: 254).

Whatever taxon Stw 53 belongs to, be it *H. habilis* or a different early *Homo* taxon (e.g., Grine et al., 1996; Curnoe and Tobias, 2006), these results suggest that the locomotor repertoire of Stw 53 was very different from that of *H. erectus sensu stricto* and modern humans.

Dexterity – Evidence from Carpal Bones

Tocheri (2007) and Tocheri et al. (2007) report the results of a study of the 3D shape of the carpal bones of fossil hominins and the extant great apes; the parts of their study that are relevant to this discussion of the genus *Homo* relate to the carpal bones of LB1 and OH 7, the type specimens of *H. fbr esiensis* and *H. habilis*, respectively. Tocheri (2007) stresses the caveat that the sample of carpal bones within the hominin clade is very small and for some taxa (e.g., *P. boisei* and *H. erectus*) non-existent, but he demonstrates that the derived carpal morphology seen in *H. sapiens* (which is also shared with *H. neanderthalensis*) is not seen in either *H. floresiensis* or *H. habilis*. Instead, Tocheri et al. (2007) make a convincing case that the carpal bones of these two taxa closely resemble the carpal morphology seen in archaic hominins such as *Au. afarensis*. Although they make the point that this primitive wrist morphology did not necessarily preclude its owners from using and making stone tools, the retention of such a primitive carpal morphology in the type specimen of *H. habilis* certainly does not strengthen any claim that the latter taxon should be in the same adaptive zone as modern humans, at least in terms of dexterity.

Life History

Life history is the term used to describe the relative rate at which members of a species proceed through developmental, maturational and reproductive milestones. Life histories reflect the ways taxa have adapted to their ecological context by dividing the energy of individuals between maturation, the maintenance of the individual and its reproduction, with the latter component being further subdivided between the production of offspring and the subsequent maintenance of those offspring. Direct measures of life history (LHVs) include length of gestation, age at first molar eruption, age at weaning, age at sexual maturity, ages at first and last birth, interbirth interval, mean lifespan, and length of post-reproductive lifespan. In modern humans the total life span is relatively long, as are the intervals between developmental and reproductive milestones. Modern humans are exceptional because they wean their infants early, their age at first birth is later than would be expected for a great ape of the same body mass, and they have an absolutely long life span. They also decouple female fertility and mortality so that females have a long post-reproductive lifespan (see Robson and Wood, 2008 for a review of hominin life histories).

With the exception of the age at weaning (Aiello et al., 1991; Skinner, 1997), we cannot yet make direct observations about life history variables on an extinct taxon, but this may change as new methods are devised and applied to the fossil record. But researchers can glean qualitative or quantitative information from the hominin fossil record about ontogeny, and about variables such as body mass and brain size that have been shown empirically within primates to either influence life history, or at the least to be correlated with LHVs (e.g., Godfrey et al., 2003; Hofman, 1984; Martin, 1981, 1983; Sacher, 1975; Smith, 1989, 1992; Smith and Tompkins, 1995; Smith et al., 1995, and see Robson and Wood, 2008 for a discussion of how these correlations differ within broad [i.e., all primates] and narrow [i.e., great apes only] allometric contexts). To distinguish them from life history variables, Skinner and Wood (2006) suggest that indirect measures of life history be referred to as "life history-related variables" (LHRVs).

An important contribution to the debate about the boundary of *Homo* was the Dean et al. (2001) analysis of enamel formation times in the incisors and canines of early hominins. Dean et al. (2001) counted long-period cross striations, used an empirically-derived modal periodicity of 9 days to calculate enamel formation times, and then plotted the latter against enamel thickness. These analyses show that archaic hominins take on average 100 days less than modern humans to reach an enamel thickness of 1,000 μm. The authors conclude that "none of the trajectories of enamel growth in apes, australopiths or fossils attributed to *Homo habilis*, *Homo rudolfensis* ... falls within that of the sample from modern humans"

(Dean et al., 2001: 629). Similarly, in his analysis of root formation time in OH 16 (a specimen assigned to *H. habilis*) Dean (1995) identified a non-modern human-like pattern.

Dean et al. (2001) included a specimen assigned to *H. erectus* in their study. On the basis of their observations of Sangiran S7-37a they concluded that *H. erectus* reached maturity relatively rapidly. In a study reported by Coqueugniot et al. (2004) in which an infant *H. erectus* specimen from Java, Perning 1, was compared with a sample of modern humans and chimpanzees, *H. erectus* was found to have had a pattern of brain growth that was more ape-like than modern human-like (but see Robson and Wood, 2008 for a discussion of the basis for these conclusions). A number of recent studies have examined development in *H. heidelbergensis* (Bermúdez de Castro and Rosas, 2001; Ramirez Rozzi and Bermúdez de Castro, 2004; Bermúdez de Castro et al., 2003; Macchiarelli et al., 2006), and these studies suggest that the enamel formation rates of the anterior teeth of Neanderthals were faster than those in *H. sapiens*, but subsequent studies of the development of Neanderthal postcanine teeth suggests that the developmental timing of *H. neanderthalenis* was modern human-like (Guatelli-Steinberg et al., 2005; Dean et al., 2001). Clearly, larger samples are needed and researchers need to cross-validate their methods, but Smith et al. (2007) have recently provided evidence that the distinctively slow modern human dental development can be traced back to at least 160 ka.

Even if the developmental schedule of *H. heidelbergensis* was not like that of modern humans, it was more similar to the developmental schedule of *H. sapiens* than to those of chimpanzees and gorillas. In contrast, preliminary results suggest that the developmental schedules of *H. erectus sensu stricto*, early African *H. erectus*, *H. habilis*, *H. rudolfensis* were more like those of chimpanzees and gorillas (Robson and Wood, 2008; see also Dean and Smith, 2009).

New Evidence about the Relationships of Potential Homo Species

The pre-1999 phylogenetic analyses that had tested the hypothesis of *Homo* monophyly (Chamberlain, 1987; Chamberlain and Wood, 1987; Wood, 1991, 1992; Lieberman et al., 1996; Strait et al., 1997) were reviewed in Wood and Collard (1999). Three phylogenetic analyses carried out since 1999, (Curnoe, 2001; Cameron and Groves, 2004; Strait and Grine, 2004) are relevant to the ongoing debate about the phylogenetic relationships of potential early *Homo* taxa.

Curnoe (2001) focused on the phylogenetic relationships of three specimens from southern Africa, SK 847, SK15 and Stw 53, all of which are usually considered to represent one or another species of early *Homo*. The extinct hominin OTUs in the Curnoe (2001) study included *Au. afarensis*, *Au. africanus*, *H. erectus sensu lato*, *H. habilis sensu stricto*, *H. rudolfensis*, *P. aethiopicus*, *P. robustus* and *P. boisei*: *Pan troglodytes* was used as the outgroup. The results of Curnoe's analyses do not support the hypothesis that *Homo* is a monophylum. The most parsimonious and consensus cladograms suggest that *H. habilis sensu stricto*, *H. erectus. s.l.* and SK 847, SK 15 and Stw 53 form a clade to the exclusion of the other taxa in the sample, and *H. rudolfensis* is not linked exclusively to other *Homo* taxa.

Cameron and Groves (2004) examined the phylogenetic relationships of 14 hominin species, including early African *H. erectus*, *H. habilis sensu stricto*, *H. rudolfensis* and *H. sapiens*. One of their analyses used 92 characters, the other a subset of 52 characters. The former analysis resulted in eight equally parsimonious trees. In the consensus trees early African *H. erectus*, *H. habilis* and *H. sapiens* were clustered in a clade to the exclusion of the other taxa, and grouped *H. rudolfensis* in a clade with *K. platyops*. A bootstrap analysis of the 92 characters supported a sister group relationship between early African *H. erectus* and *H. sapiens*, but was unable to resolve the relationships of the other early *Homo* taxa at the 70% level of support that is commonly used to identify statistically significant clades in such analyses (Hillis and Bull, 1993). The results of the 52 character analyses were comparable. Twenty cladograms were equally parsimonious. They grouped early African *H. erectus*, *H. habilis* and *H. sapiens* in a clade, and clustered *H. rudolfensis* in a separate clade with *K. platyops*. A bootstrap analysis supported a sister group relationship between early African *H. erectus* and *H. sapiens*, but was unable to resolve the relationships of the other *Homo* taxa at the 70% level. Thus, although both analyses suggested that early African *H. erectus* is more closely related to *H. sapiens* than to any other hominin species, they were unable to resolve the relationships of *H. habilis sensu stricto* and *H. rudolfensis*.

Strait and Grine (2004) carried out a series of maximum parsimony and bootstrap analyses to examine the relationships of several hominin species including early African *H. erectus*, *H. habilis sensu stricto*, *H. rudolfensis* and *H. sapiens*. Their dataset comprised 109 qualitative craniodental characters and 89 craniometric characters recorded on 14 hominin species, plus seven extant non-human primate taxa. The consensus of the most parsimonious cladograms obtained by Strait and Grine (2004) suggests that early African *H. erectus*, *H. habilis sensu stricto*, *H. rudolfensis* and *H. sapiens* form a clade to the exclusion of the other species in the sample, within which early African *H. erectus* and *H. sapiens* form a clade to the exclusion of *H. habilis sensu stricto* and *H. rudolfensis*. The relationships among the (early African *H. erectus*, *H. sapiens*) clade, *H. habilis sensu stricto* and *H. rudolfensis* were unresolved. The results of Strait and Grine's (2004) bootstrap analyses showed that when all the characters and taxa were analyzed together, a (early African

H. erectus, *H. sapiens*) clade was supported by 86% of the bootstrap replicates, but the relationships of the other *Homo* taxa were not resolved at the 70% level. When *K. platyops* was dropped from the all-characters analysis, two clades (early African *H. erectus*, *H. sapiens*) and (early African *H. erectus*, *H. habilis sensu stricto*, *H. rudolfensis*, *H. sapiens*) were supported by more than 70% of the replicates. Similar results were obtained when only the 109 qualitative characters were included. Thus, while Strait and Grine's (2004) parsimony analyses support the hypothesis that *Homo* is a monophylum, this hypothesis is not consistently supported by their bootstrap analyses.

The results of the three latest cladistic analyses thus differ with respect to the phylogenetic relationships of *H. rudolfensis* and *H. habilis sensu stricto*. Although *H. rudolfensis* clustered exclusively with the other *Homo* species in Strait and Grine's (2004) parsimony analyses and in some of their bootstrap analyses, it did not form a clade with the other *Homo* species in the parsimony and bootstrap analyses reported in Curnoe (2001) and Cameron and Groves (2004). The results of the parsimony analyses carried out by Curnoe (2001), Cameron and Groves (2004), and Strait and Grine (2004) offer support for the hypothesis that *H. habilis sensu stricto* is a member of the *Homo* clade.

Nonetheless, none of the bootstrap analyses carried out by Cameron and Groves (2004) and some of the bootstrap analyses conducted by Strait and Grine (2004) failed to support a link between *H. habilis sensu stricto* and later *Homo* species at the 70% level.

This reduces the confidence that can be placed in the hypothesis that *H. habilis sensu stricto* is unambiguously part of the crown group that includes modern humans. It also has to be remembered that the data used in these analyses does not sample all of the craniodental evidence, let alone the postcranial skeleton. For example, it does not include any characters states based on endocranial morphology, and it is noteworthy that the endocranial morphology of KNM-ER 1805, a specimen attributed to *H. habilis sensu stricto*, has been interpreted by one of the researchers most familiar with this evidence as being ape-like (Falk, 1983).

New Interpretations of the Genus Homo

Genetics-Based Interpretations

Morris Goodman and his colleagues have presented their interpretation of how hominin taxonomy should be adapted to reflect the recent molecular and genetic evidence suggesting a particularly close relationship between modern humans and chimpanzees (Goodman et al., 1998, 2001; Wildman et al., 2003). The only paleoanthropologists who have responded to these new data have been Groves (2001), Cameron and Groves (2004) (both reviewed above) and Curnoe and Thorne (2003). The conclusions of the latter study can be summarized by their statement that "the genetic proximity of our analysis of genetic distances between humans and chimpanzees has been used to suggest these species are congeneric. Our analysis of genetic distances between them is consistent with this proposal. It is time that chimpanzees, living humans and all fossil humans be classified in *Homo*" (Curnoe and Thorne, 2003: 201).

It is clear that taxonomic hypotheses generated by paleoanthropologists should take cognizance of comparative molecular evidence, but it makes no sense to generate such hypotheses as if morphological evidence and functional inference have ceased to exist, or as if they had no relevance for such decisions. Molecular biologists have some excuse for being blind to such evidence; paleoanthropologists should know better. Unless we abandon any consideration of the grade of an organism when considering what genus to assign them to, then it makes no sense to include chimpanzees, bonobos and modern humans in the same genus.

Fossil Evidence-Based Interpretations

Although the last 2 decades have seen substantial heat generated by debates about how many species should be recognized among the fossil evidence for early *Homo*, surprisingly little attention has been paid to the definition and identification of the genus *Homo*.

There are two options for putting two of the principles of genus identification (i.e., monophyly and adaptive coherence) into practice. You can either start in the present, or in the past. If you start in the present, and adopt the "top down" option, you start with the type species. In the case of the genus *Homo* you take stock of the derived morphology and behavior of *H. sapiens*, decide on the cardinal features and behaviors you will use to determine the adaptive zone of *H. sapiens*, and then determine the characters you will use to generate phylogenetic hypotheses. Then you work backwards into the past, and apply the same two tests to each hominin taxon you encounter. Is there reliable (i.e., the same cladogram is generated even if you change details of the OTUs and the method) evidence that the taxon is in the same clade as *H. sapiens*? Is there reliable (i.e., reliable quantitative proxies of important behaviors) evidence that the taxon is in the same adaptive zone as *H. sapiens*?

If you adopt the "bottom up" approach, you have to make a subjective judgment about whereabouts in the past you should start to pick up the trail leading to *Homo*. You then work towards the present applying the tests set out above to the hominin taxa you encounter. The difference between this approach and the "top down" option is that the evidence is sketchier, and thus the likelihood that one can satisfy the "reliability" criterion of the two tests, monophyly and adaptive similarity, is diminished.

Ironically, there have been very few attempts to formally assess the relationships of modern humans with respect to *H. neanderthalensis* and *H. erectus*. Eldredge and Tattersall (1975) included all three taxa in the cladogram (see Eldredge and Tattersall, 1975, Fig. 4) presented in their seminal paper that pioneered the application of cladistic methods to hominin relationships. However, the authors did not carry out a formal analysis of the relationships among the taxa, nor did they refer to any specific characters when considering the merits of different cladograms for expressing the relationships among the pre-modern *Homo* taxa within the hominin clade.

Presumably most researchers since then considered the hypothesis of monophyly of later *Homo* (i.e., *H. sapiens*, *H. neanderthalensis*, *Homo heidelbergensis*, *H. erectus s.l.*) to be so obvious that it did not require formal investigation. Although there are grounds for adding *H. habilis sensu stricto* and *H. rudolfensis* to the (*H. sapiens*, *H. neanderthalensis*, *H. heidelbergensis*, *H. erectus s.l.*) clade, I think even the supporters of such an interpretation would accept that the evidence is not as strong as the evidence for including *H. neanderthalensis* and *H. erectus s.l.* within the crown group that includes modern humans.

Thus, as far as relationships are concerned there seem to be two options. You either draw the lower boundary of *Homo* so that it includes *H. habilis sensu stricto* and *H. rudolfensis* (Fig. 3.1, arrow A), or you draw it beneath early African *H. erectus* so that it excludes *H. habilis sensu stricto* and *H. rudolfensis* (Fig. 3.1, arrow B).

As far as adaptive grade is concerned, the problem is more complicated. If the criteria are restricted to what can be deduced about the adaptive grade of a taxon from its morphology, then it could be argued that if the combination of a modern human-sized brain and obligate long range bipedalism are the criteria, then the boundary of *Homo* would be set so that it includes *H. heidelbergensis,* but not *H. erectus s.l.* or *H. floresiensis*. If a modern human body shape and obligate bipedalism are deemed to be the criteria, then the boundary would be set so that *Homo* would include early African *H. erectus*, but not *H. habilis sensu stricto* and *H. rudolfensis* (but see Haeusler and McHenry, 2004, 2007 for an alternative interpretation). But even that solution results in a hominin genus that embraces a substantial range of ontogenies and life histories (Robson and Wood, 2008). If *H. habilis sensu stricto* and *H. rudolfensis* are included in *Homo* for relationship reasons, this poses problems for any genus definition that insists on adaptive coherence for the same genus would include taxa with a range of cranial and postcranial morphology (including very different semicircular canals) that imply different dietary and locomotor adaptations. Furthermore, the adaptive strategies of *H. habilis sensu stricto* and *H. rudolfensis* are probably closer to the adaptive strategy of the type species of the genus *Australopithecus* (i.e., *Au. africanus*) than they are to *H. sapiens*, the type species of *Homo*.

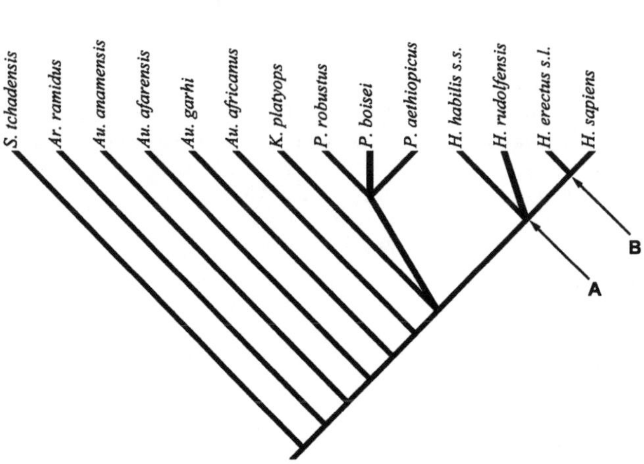

Fig. 3.1 A cladogram setting out one hypothesis of relationships among early hominins. The nodes A and B represent two hypotheses for the *Homo* clade. If *Homo* were to include node A, it would embrace the species presently included in early *Homo* (i.e., *H. habilis sensu stricto* and *H. rudolfensis*). If *Homo* was defined so as to exclude node A, and include just node B, then it would be confined to early African *H. erectus* and temporally later, more derived *Homo* species.

Conclusions

Seven years after the publication of Wood and Collard (1999), and approaching this problem afresh, there is still ambiguity about where to draw the lower boundary of the genus *Homo*. However, I am still of the opinion that the combination of evidence about relationships and adaptive grade is in favor, albeit narrowly, of excluding *H. habilis sensu stricto* and *H. rudolfensis* from the genus *Homo*. What Kimbel (2009) suggests with respect to the origin of the genus *Homo* logically also applies to attempts to define *Homo*, namely "we have come as far as we can with the evidence at hand."

Acknowledgments I am pleased to acknowledge the support of George Washington University's Academic Excellence Initiative, the VPAA at GW, Don Lehman and the NSF. I am also grateful to Mark Collard for allowing me to draw on ideas we developed in a series of jointly authored papers.

References

Antón, S.C., 2004. The face of Olduvai Hominid 12. Journal of Human Evolution 46, 337–347.

Antón, S.C., Spoor, F., Fellmann, C.D., Swisher III, C.C., 2007. Defining *Homo erectus*: size considered. In: Henke, W., Tattersall, I. (Eds.),

Handbook of Paleoanthropology, Volume 3. Springer, New York, pp. 1655–1693.

Aiello, L.C., Montgomery, C., Dean, M.C., 1991. The natural history of deciduous tooth attrition in hominoids. Journal of Human Evolution 21, 397–412.

Bailey, S., 2004. A morphometric analysis of maxillary molar crowns of Middle-Late Pleistocene hominins. Journal of Human Evolution 47, 183–198.

Bermúdez de Castro, J.M., Rosas, A., 2001. Dental development in Lower and Middle Pleistocene hominids from Atapuerca (Spain). American Journal of Physical Anthropology. Supplement 32, 39.

Bermúdez de Castro, J.M., Ramirez Rozzi, F., Martinon-Torres, M., Sarmiento Perez, S., Rosas, A., 2003. Patterns of dental development in Lower and Middle Pleistocene hominins from Atapuerca (Spain). In: Thompson, J.L., Krovitz, G.E., Nelson, A.J. (Eds.), Patterns of Growth and Development in the Genus *Homo*. Cambridge University Press, Cambridge, pp. 246–270.

Bilsborough, A., Wood, B.A., 1988. Cranial morphometry of early hominids: facial region. American Journal of Physical Anthropology 76, 61–86.

Boaz, N.T., Howell, F.C., 1977. A gracile hominid cranium from Upper Member G of the Shungura Formation, Ethiopia. American Journal of Physical Anthropology 46, 93–108.

Bromage, T.G., Schrenk, F., Zonneveld, F.W., 1995. Palaeoanthropology of the Malawi Rift: an early hominid mandible from the Chiwondo Beds, northern Malawi. Journal of Human Evolution 28, 71–108.

Brown, P., Sutikna, T., Morwood, M.J., Soejono, R.P., Jatmiko, Saptomo, E.W., Due, R.A., 2004. A new small-bodied hominin from the Late Pleistocene of Flores, Indonesia. Nature 431, 1055–1061.

Cameron, D.W., Groves, C.P., 2004. Bones, Stones and Molecules. Elsevier, Amsterdam.

Cela-Conde, C.J., Altaba, C.R., 2002. Multiplying genera versus moving species: a new taxonomic proposal for the family Hominidae. South African Journal of Science 98, 229–232.

Chamberlain, A.T., 1987. A taxonomic review and phylogenetic analysis of *Homo habilis*. Ph.D. Dissertation, The University of Liverpool, Liverpool.

Chamberlain, A.T., Wood, B.A., 1987. Early hominid phylogeny. Journal of Human Evolution 16, 119–133.

Clarke, R.J., 1985. *Australopithecus* and early *Homo* in South Africa. In: Delson, E. (Ed.), Ancestors: The Hard Evidence. Liss, New York, pp. 171–177.

Clarke, R.J., Howell, F.C., 1972. Affinities of the SK 847 cranium. American Journal of Physical Anthropology 37, 319–335.

Collard, M., Wood, B.A., 2006. Defining the genus *Homo*. In: Henke, Tattersall, I. (Eds.), Handbook of Paleoanthropology, Volume 3. Springer, New York, pp. 1575–1610.

Coppens, Y., 1980. The differences between *Australopithecus* and *Homo*; preliminary conclusions from the Omo Research Expedition's studies. Pergamon, Oxford.

Coqueugniot, H., Hublin, J.-J., Veillon, F., Houët, F., Jacob, T., 2004. Early brain growth in *Homo erectus* and implications for cognitive ability. Nature 431, 299–302.

Curnoe, D., 2001. Cranial variability in East African "robust" hominins. Human Evolution 16, 169–198.

Curnoe, D., Thorne, A., 2003. Number of ancestral human species: a molecular perspective. Homo 53, 201–224.

Curnoe, D., Tobias, P.V., 2006. Description, new reconstruction, comparative anatomy, and classification of the Sterkfontein Stw 53 cranium, with discussions about the taxonomy of other southern African early *Homo* remains. Journal of Human Evolution 50, 36–77.

de Sousa, A., Wood, B., 2006. The hominin fossil record and the emergence of the modern human nervous system. In: Kaas, J.H. (Ed.), Evolution of Nervous Systems, Volume 4. Elsevier, Amsterdam, pp. 291–336.

Dean, M.C., 1995. The nature and periodicity of incremental lines in primate dentine and their relationship to periradicular bands in OH 16 (*Homo habilis*). In: Moggi-Cecchi, J. (Ed.), Aspects of Dental Biology: Palaeontology, Anthropology and Evolution. Florence, International Institute for the Study of Man, pp. 239–265.

Dean, M.C., Leakey, M.G., Reid, D., Schrenk, F., Schwartz, G.T., Stringer, C., Walker, A., 2001. Growth processes in teeth distinguish modern humans from *Homo erectus* and earlier hominins. Nature 414, 628–631.

Dean, M.C., Smith, B.H., 2009. Growth and development of the Nariokotome youth, KNM-WT 15000. In: Grine, F.E., Fleagle, J.G., Leakey, R.E. (Eds.), The First Humans: Origin and Early Evolution of the Genus *Homo*. Springer, Dordrecht, pp. 101–120.

Dubois, A., 1988. The genus in zoology: a contribution to the theory of evolutionary sysytematics. Mémoirs du Muséum Nationale D'Histoire Naturelle, Séries A 140, 1–122.

Eldredge, N., Tattersall, I., 1975. Evolutionary models, phylogenetic reconstruction, and another look at hominid phylogeny. In: Szalay, F., (Ed.). Approaches to Primate Paleobiology. Contributions to Primatology 5, 218–242.

Falk, D., 1983. Cerebral cortices of East African early hominids. Science 221, 1072–1074.

Falk, D., Hildebolt, C., Smith, K., Morwood, M.J., Sutkina, T., Brown, P., Jatmiko, Saptomo, E.W., Brunsden, B., Prior, F., 2005. The brain of LB1, *Homo floresiensis*. Science 308, 242–245.

Gabunia, L., Vekua, A., 1995. A Plio-Plieistocene hominid from Dmanisi, East Georgia, Caucasus. Nature 373, 509–512.

Gabunia, L., Vekua, A., Lordkipanidze, D., Swisher, C.C., Ferring, R., Justus, A., Nioradze, M., Tvalchrelidze, M., Antón, S.C., Bosinski, G., Joris, O., de Lumley, M.-A., Majsuradze, G., Mouskhelishvili, A., 2000. Earliest Pleistocene cranial remains from Dmanisi, Republic of Georgia: taxonomy, geological setting, and age. Science 288, 1019–1025.

Gebo, D.L., Schwartz, G.T., 2006. Foot bones from Omo: implications for hominin evolution. American Journal of Physical Anthropology 129, 499–511.

Godfrey, L., Samonds, K.E., Jungers, W.L., Sutherland, M.R., 2003. Dental development and primate life histories. In: Kappeler, P.M., Pereira, M. (Eds.), Primate Life Histories and Socioecology. University of Chicago Press, Chicago, IL, pp. 177–203.

Goodman, M., Porter, C.A., Czelusniak, J., Page, S.L., Schneider, H., Shoshani, J., Gunnell, G., Groves, C.P., 1998. Towards a phylogenetic classification of primates based on DNA evidence complemented by fossil evidence. Molecular Phylogenetics and Evolution 9, 585–598.

Goodman, M., Czelusniak, J., Page, S.L., Meireles, C.M., 2001. Where DNA sequences place *Homo sapiens* in a phylogenetic classification of primates. In: Tobias, P.V., Raath, M.A., Moggi-Cecchi, J., Doyle, G.A. (Eds.), Humanity from African Naissance to Coming Millenia. Firenze University Press, Firenze, pp. 274–289.

Gordon, A, D., Nevell, L., Wood, B., 2008. The *Homo floresiensis* cranium (LB1): size, scaling and early *Homo* affinities. Proceedings of the National Academy of Sciences USA 105, 4650–4655.

Grine, F.E., Strait, D.S., 1994. New hominid fossils from Member 1 "Hanging Remnant", Swartkrans Formation, South Africa. Journal of Human Evolution 26, 57–75.

Grine, F.E., Demes, B., Jungers, W.L., Cole, T.M., 1993. Taxonomic affinity of the early *Homo* cranium from Swartkrans, South Africa. American Journal of Physical Anthropology 92, 411–426.

Grine, F.E., Jungers, W.L., Schultz, J., 1996. Phenetic affinities among early *Homo* crania from East and South Africa. Journal of Human Evolution 30, 189–225.

Groves, C.P., 2001. Primate Taxonomy. Smithsonian Institution Press, Washington, DC.

Guatelli-Steinberg, D., Reid, D.J., Bishop, T.A., Larson, C.S., 2005. Anterior tooth growth periods in Neandertals were comparable

to those of modern humans. Proceedings of the National Academy of Sciences U.S.A. 102(40), 14197–14202.
Haeusler, M., McHenry, H.M., 2004. Body proportions of *Homo habilis* reviewed. Journal of Human Evolution 46, 433–465.
Haeusler, M., McHenry, H.M., 2007. Evolutionary reversals of limb proportions in early hominids? Evidence from KNM-ER 3735 (*Homo habilis*). Journal of Human Evolution 53, 383–405.
Hartwig-Scherer, S., Martin, R.D., 1991. Was 'Lucy' more human than the 'child'? Observations on early hominid postcranial skeletons. Journal of Human Evolution 21, 439–449.
Hennig, W., 1966. Phylogenetic Systematics. University of Illinois Press, Chicago, IL.
Hill, A., Ward, S., Deino, A., Curtis, G., Drake, R., 1992. Earliest *Homo*. Nature 355, 719–722.
Hillis, D.M., Bull, J.J., 1993. An empirical test of bootstrapping as a method of assessing confidence in phylogenetic analysis. Systematic Biology 42, 182–192.
Hofman, M.A., 1984. On the presumed coevolution of brain size and longevity in hominids. Journal of Human Evolution 13, 371–376.
Howell, F.C., Coppens, Y., Isaac, G.L., Leakey, R.E.F., 1976. An overview of Hominidae from the Omo succession, Ethiopia. In: Howell, F.C., Coppens, Y., Isaac, G.L., Leakey, R.E.F. (Eds.), Earliest Man and Environments in the Lake Rudolf Basin. University of Chicago Press, Chicago, IL, pp. 522–532.
Hughes, A.R., Tobias, P.V., 1977. A fossil skull probably of the genus *Homo* from Sterkfontein, Transvaal. Nature 265, 310–312.
Johanson, D.C., Masao, F.T., Eck, G.G., White, T.D., Walter, R.C., Kimbel, W.H., Asfaw, B., Manega, P., 1987. New partial skeleton of *Homo habilis* from Olduvai gorge, Tanzania. Nature 327, 205–209.
Jungers, W.L., 2009. Interlimb proportions in humans and fossil hominins: variability and scaling. In: Grine, F.E., Fleagle, J.G., Leakey, R.E. (Eds.), The First Humans: Origin and Early Evolution of the Genus *Homo*. Springer, Dordrecht, pp. 93–98.
Kimbel, W.H., 2009. The origin of *Homo*. In: Grine, F.E., Fleagle, J.G., Leakey, R.E. (Eds.), The First Humans: Origin and Early Evolution of the Genus *Homo*. Springer, Dordrecht, pp. 31–37.
Kimbel, W.H., Rak, Y., 1993. The importance of species taxa in paleoanthropology and an argument for the phylogenetic concept of the species. In: Kimbel, W.H., Martin, L. (Eds.), Species, Species Concepts, and Primate Evolution. Plenum, New York, pp. 461–484.
Kimbel, W.H., Walter, R.C., Johanson, D.C., Reed, K.E., Aaronson, J.L., Assefa, Z., Marean, C.W., Eck, G.G., Bobe, R., Hovers, E., Rak, Y., Vondra, C., Yemane, T., York, D., Chen, Y., Evensen, N.M., Smith, P.E., 1996. Late Pliocene *Homo* and Oldowan tools from the Hadar Formation (Kadar Hadar Member), Ethiopia. Journal of Human Evolution 31, 549–561.
King, W., 1864. The reputed fossil man of the Neanderthal. Quarterly Journal of Science 1, 88–97.
Le Gros Clark, W.E., 1955. The Fossil Evidence for Human Evolution: An Introduction to the Study of Paleoanthropology. The Scientist's Library: Biology and Medicine. University of Chicago Press, Chicago, IL.
Le Gros Clark, W.E., 1964. The Fossil Evidence for Human Evolution: An Introduction to the Study of Paleoanthropology, Second edition. The Scientist's Library: Biology and Medicine. University of Chicago Press, Chicago, IL.
Leakey, L.S.B., Tobias, P.V., Napier, J.R., 1964. A new species of the genus *Homo* from Olduvai Gorge. Nature 202, 7–9.
Leakey, M.G., Spoor, F., Brown, F.H., Gathogo, P.N., Kiarie, C., Leakey, L.N., McDougall, I., 2001. New hominin genus from eastern Africa shows diverse middle Pliocene lineages. Nature 410, 433–440.
Leakey, M.G., Spoor, F., Brown, F.H., Gathogo, P.N., Leakey, L.N., 2003. A new hominin cranium from Ileret (Kenya). American Journal of Physical Anthropology Supplement 36, 136.
Leakey, R.E.F., 1973. Evidence for an advanced Plio-Pleistocene hominid from East Rudolf, Kenya. Nature 242, 447–450.
Lieberman, D.E., Wood, B.A., Pilbeam, D.R., 1996. Homoplasy and early *Homo*: an analysis of the evolutionary relationships of *H. habilis sensu stricto* and *H. rudolfensis*. Journal of Human Evolution 30, 97–120.
Linnaeus, C., 1758. Systema Naturae. Laurentii Salvii, Stockholm.
Lordkipanidze, D., Vekua, A., Ferring, R., Rightmire, G.P., Agusti, J., Kiladze, G., Mouskhelishvili, A., Nioradze, M., Ponce de León, M.S., Tappen, M., Zollikofer, C.P.E., 2005. The earliest toothless hominin skull. Nature 434, 717–718.
Lordkipanidze, D., Jashashvili, T., Vekua, A., Ponce de León, M.S., Zollikofer, C.P.E., Rightmire, G.P., Pontzer, H., Ferring, R., Oms, O., Tappen, M., Bukhsianidze, M., Agusti, J., Kahlke, R., Kiladze, G., Martinez-Navarro, B. Mouskhelishvili, A., Nioradze, M., and Rook., L. 2007. Postcranial evidence from early *Homo* from Dmanisi, Georgia. Nature 449, 305–310.
Macchiarelli, R., Bondioli, L., Debenath, A., Mazurier, A., 2006. How Neanderthal molar teeth grew. Nature 444, 748–751.
Martin, R.D., 1981. Relative brain size and basal metabolic rate in terrestrial vertebrates. Nature 293, 57–60.
Martin, R.D., 1983. Human brain evolution in an ecological context. American Museum of Natural History, 1–58.
Mayr, E., 1944. On the concepts and terminology of vertical subspecies and species. National Research Council Committee on Common Problems of Genetics, Paleontology, and Systematics Bulletin 2, 11–16.
Mayr, E., 1950. Taxonomic categories in fossil hominids. Cold Spring Harbor Symposium on Quantitative Biology 15, 109–118.
Mayr, E., Linsley, E.G., Usinger, R.L., 1953. Methods and Principles of Systematic Zoology. McGraw-Hill, New York.
McHenry, H.M., 1988. New estimates of body weight in early hominids and their significance to encephalization and megadontia in robust australopithecines. In: Grine, F.E. (Ed.), Evolutionary History of the "Robust" Australopithecines. Aldine de Gruyter, New York, pp. 133–148.
McHenry, H.M., 1991. Femoral length and stature in Plio-Pleistocene hominids. American Journal of Physical Anthropology 85, 149–158.
McHenry, H.M., Coffing, K., 2000. *Australopithecus* to *Homo*: transformations in body and mind. Annual Review of Anthropology 29, 125–46.
Morwood, M.J., Soejono, R.P., Roberts, R.G., Sutikna, T., Turney, C.S.M., Westaway, K.E., Rink, W.J., Zhao, J.-x., van der Bergh, G.D., Due, R.A., Hobbs, D.R., Moore, M.W., Bird, M.I., Fifield, L.K., 2004. Archaeology and age of a new hominin from Flores in eastern Indonesia. Nature 431, 1097–1091.
Morwood, M.J., Brown, P., Jatmiko, Sutikna, T., Saptomo, E.W., Westaway, K.E., Due, R.A., Roberts, R.G., Maeda, T., Wasisto, S., Djubiantono, T., 2005. Further evidence for small-bodied hominins from the Late Pleistocene of Flores, Indonesia. Nature 437, 1012–1017.
Oppenoorth, W.F.F., 1932. *Homo* (*Javanthropus*) *soloensis* een Plistoceene mensch van Java. Wetenschappelijike Mededeelingen 20, 49–74.
Oppenoorth, W.F.F., 1937. The place of *Homo soloensis* among fossil men. In: MacCurdy, G.G. (Ed.), Early Man. J.B. Lippincott, New York, pp. 349–360.
Potts, R. Behrensmeyer, A.K., Deino, A., Ditchfield, P., Clark, J., 2004. Small mid-Pleistocene hominin associated with East African Acheulean technology. Science 305, 75–78.
Prat, S., Brugal, J.-P., Tiercelin, J.-J., Barrat, J.-A., Bohn, M., Delagnes, A., Harmand, S., Kimeu, K., Kibunjia, M., Texier, P.-J., Roche, H., 2005. First occurrence of early *Homo* in the Nachukui Formation (West Turkana, Kenya) at 2.3–2.4 Myr. Journal of Human Evolution 49, 230–240.

Quam, R., Bailey, S., Wood, B.A., in press. Evolution of M^1 crown size and cusp proportions in the genus *Homo*. Journal of Anatomy
Rak, Y., 1987. The unusual face of KNM-ER 1470. In: Giacobini, G. (Ed.), Abstracts 2nd International Congress Human Paleontology. Jaca, Milan, pp. 90–91.
Ramirez Rozzi, F., Bermudez de Castro, J.-M., 2004. Surprisingly rapid growth in Neanderthals. Nature 428, 936–939.
Reno, P.L., DeGusta, D., Serrat, M.A., Meindl, R.S., White, T.D., Eckhardt, R.B., Kuperavage, A.J., Galik, K., Lovejoy, C.O., 2005. Plio-Pleistocene hominid limb proportions: evolutionary reversals or estimation errors? Current Anthropology 46, 575–588.
Richmond, B.G., Aiello, L.C., Wood, B.A., 2002. Early hominin limb proportions. Journal of Human Evolution 43, 529–548.
Rightmire, G.P., Lordkipanidze, D., Vekua, A., 2006. Anatomical descriptions, comparative studies and evolutionary significance of the hominin skulls from Dmanisi, Republic of Georgia. Journal of Human Evolution 50, 115–141.
Robinson, J.T., 1961. The australopithecines and their bearing on the origin of man and of stone tool-making. South African Journal of Science 57, 3–13.
Robson, S.L., Wood, B.A., 2008. Hominin life history: reconstruction and evolution. Journal of Anatomy 219, 394–425.
Sacher, G.A., 1975. Maturation and longevity in relation to cranial capacity in hominid evolution. In: Tuttle, R.H. (Ed.), Primate Functional Morphology and Evolution. Mouton, The Hague, pp. 418–441.
Schoetensack, O., 1908. Der Unterkiefer des *Homo heidelbergensis* aus den Sanden von Mauer bei Heidelberg. W. Engelmann, Leipzig.
Skinner, M., 1997. Dental wear in immature Late Pleistocene hominines. Journal of Archeological Science 24, 677–700.
Skinner, M., Wood, B.A, 2006. The evolution of modern human life history: a paleontological perspective. In: Hawkes, K., Paine, R.R. (Eds.), The Evolution of Modern Human Life History. School of American Research, Sante Fe, pp. 331–400.
Smith, B.H., 1989. Dental development as a measure of life history in primates. Evolution 43, 683–688.
Smith, B.H., 1992. Life history and the evolution of human maturation. Evolutionary Anthropology 1, 134–142.
Smith, B.H., Tompkins, R.L., 1995. Toward a life history of the Hominidae. Annual Review of Anthropology 24, 257–279.
Smith, T. M., Tafforeau, P., Reid, D. J., Grün, R., Eggins, S., Boutakiout, M., Hublin, J.-J., 2007. Earliest evidence of modern human life history in North African early *Homo sapiens*. Proceedings of the National Academy of Sciences USA 104, 6128–6133.
Smith, R.J., Gannon, P.J., Smith, B.H., 1995. Ontogeny of australopithecines and early *Homo*: evidence from cranial capacity and dental eruption. Journal of Human Evolution 29, 155–168.
Spoor, C.F., Wood, B., Zonneveld, F.W., 1994. Implications of early hominid labyrinthine morphology for evolution of human bipedal locomotion. Nature 369, 645–648.
Spoor, F., 1994. The bony labyrinth in *Homo erectus*: a preliminary report. Courier Forschungs Institut Senckenberg 171, 251–256.
Spoor, F., Leakey, M.G., Gathogo, P.N., Brown, F.H., Antón, S.C., McDougall, I., Kiarie, C., Manthi, F.K., Leakey, L.N., 2007. Implications of new early *Homo* fossils from Ileret, east of Lake Turkana, Kenya. Nature 448, 688–691.
Strait, D.S., Grine, F.E., 2004. Inferring hominoid and early hominid phylogeny using craniodental characters: the role of fossil taxa. Journal of Human Evolution 47, 399–452.
Strait, D.S., Grine, F.E., Moniz, M.A., 1997. A reappraisal of early hominid phylogeny. Journal of Human Evolution 32, 17–82.
Tocheri, M.W., 2007. Three-dimensional riddles of the radial wrist: derived carpal and carpometacarpal joint morphology in the genus *Homo* and the implications for understanding the evolution of stone tool-related behaviors in hominins. Ph.D. Thesis, Arizona State University, Tempe, Arizona.
Tocheri, M.W., Orr, C., Larson, S.G., Sutikna, T., Jatmiko, Saptomo, E.W., Due, R.A., Djubiantono, Morwood, M.J. and Jungers, W.L., 2007. The primitive wrist of *Homo floresiensis* and its implications for hominin evolution. Science 317, 1743–1745.
Ungar, P.S., 2004. Dental topography and diets of *Australopithecus afarensis* and early *Homo*. Journal of Human Evolution 46, 605–622.
Ungar, P.S., Scott, R.S., 2009. Dental evidence for diets of early *Homo*. In: Grine, F.E., Fleagle, J.G., Leakey, R.E.F. (Eds.), The First Humans: Origin and Early Evolution of the Genus *Homo*. Springer, Dordrecht, pp. 121–134.
Ungar, P.S., Grine, F.E., Teaford, M.F., Zaatari, S.E., 2006. Dental microwear and diets of African early *Homo*. Journal of Human Evolution 50, 78–95.
Vekua, A., Lordkipanidze, D., Rightmire, G.P., Agusti, J., Ferring, R., Maisuradze, G., Mouskhelishvili, A., Nioradze, M., Ponce de Leon, M., Tappen, M., Tvalchrelidze, M., Zollikofer, C., 2002. A new skull of early *Homo* from Dmanisi, Georgia. Science 297, 85–89.
Walker, A., 1976. Remains attributable to *Australopithecus* in the East Rudolf succession. In: Howell, F.C., Coppens, Y., Isaac, G.L., Leakey, R.E.F. (Eds.), Earliest Man and Environments in the Lake Rudolf Basin. University of Chicago Press, Chicago, IL, pp. 484–489.
Watson, E.E., Easteal, S., Penny, D., 2001. *Homo* genus: a review of the classification of humans and the great apes. In: Tobias, P.V., Raath, M.A., Moggi-Cecchi, J., Doyle, G.A. (Eds.), Humanity from African Naissance to coming Millenia. Firenze University Press, Firenze, pp. 307–318.
Weidenreich, F., 1940. Some problems dealing with ancient man. American Anthropologist 42, 375–383.
Wildman, D., Uddin, M., Liu, G., Grossman, L.I., Goodman, M., 2003. Implications of natural selection in shaping 99.4% nonsynonymous DNA identity between humans and chimpanzees: enlarging genus *Homo*. Proceedings of the National Academy of Sciences U.S.A. 100, 7181–7188.
Wood, B.A., 1974. Olduvai Bed I post-cranial fossils: a reassessment. Journal of Human Evolution 3, 373–378.
Wood, B.A., 1991. Koobi Fora Research Project. Volume 4. Hominid Cranial Remains. Koobi Fora Research Project. Clarendon Press, Oxford.
Wood, B.A., 1992. Origin and evolution of the genus *Homo*. Nature 355, 783–790.
Wood, B.A., Aiello, L.C., 1998. Taxonomic and functional implications of mandibular scaling in early hominins. American Journal of Physical Anthropology 105, 523–538.
Wood, B.A., Collard, M.C., 1999. The human genus. Science 284, 65–71.
Wood, B., Constantino, P., 2007. *Paranthropus boisei*: fifty years of fossil evidence and analysis. Yearbook of Physical Anthropology 50, 106–132.
Woodward, A.S., 1921. A new cave man from Rhodesia, South Africa. Nature 108, 371–372.

Part II
Craniodental Perspectives on Taxonomy and Systematics

Chapter 4
The Origin of *Homo*

William H. Kimbel

Keywords Taxonomy • *Homo habilis* • *Homo rudolfensis* • *Homo erectus*

Introduction

Despite a steady increase in the number and diversity of African Middle Pliocene hominin fossils, paleoanthropologists are not now substantially closer to understanding the temporal, geographical or ecological contexts of the origin of the *Homo* clade than was the case in 1964, when Louis Leakey, Phillip Tobias and John Napier introduced *Homo habilis* as the earliest species of the human genus. The reasons for this circumstance are several. The first is a still data-poor record in the relevant time periods. Although the half-million year interval between 2.5 and 3 Ma witnessed heightened morphological, taxonomic and diet-driven adaptive diversity in extinct hominins, all of the known taxa from this interval are of australopith grade (i.e., *Australopithecus africanus, A. garhi, A. aethiopicus,* late *A. afarensis*) and none shares derived morphological characteristics exclusively with geologically younger, securely attributed representatives of the *Homo* lineage. A few specimens that at one time or another have been attributed to *Homo* fall in the time period 2.5–2.0 Ma but most of these are fragmentary and/or of questionable (or at least debated) diagnostic value (see discussion below). The earliest well-sampled record of fossil hominins that bear characters found only in the genus *Homo* date to ca. 1.7–1.8 Ma. The fact that as many as three species with unique ties to *Homo* (*H. habilis, H. rudolfensis, H. erectus*) have their apparent FADs at about this time (Kimbel, 1991; Wood, 1991; Spoor et al., 2007) speaks to an earlier differentiation of the lineage, but other than the Hadar maxilla A.L. 666-1, with an age of ca. 2.3 Ma (Kimbel et al., 1997), the record of specimens bearing on the earliest phases of the lineage's evolution is mostly uninformative.

The second reason for our poor comprehension of *Homo* origins is conceptual. Historically, the search for the origins of *Homo* has been conflated with defining the morphological and behavioral boundaries between ape and human. In 1871, Darwin postulated an evolutionary scenario linking terrestrial bipedality, canine reduction, tool manufacture, and brain enlargement at the base of the human lineage. When growing knowledge of *Australopithecus* (including *Paranthropus*) made it clear by the 1940s that bipedality and canine reduction evolved well before hominins became significantly encephalized and began to leave a record of their lithic technology, large brains and stone tools were elevated to a foundational role in the divergence of the *Homo* lineage from some generalized australopith form. Thus, Oakley's (1959) "Man the Tool-maker" joined Leakey et al.'s (1964) recognition of the "handy man" *Homo habilis* in framing thinking about the emergence of *Homo* for the latter half of the twentieth century (Tattersall, 1998).

The earliest known stone tools are about 2.5 million years old (Semaw, 2000), but the identity of their manufacturers is unknown. The idea that the co-occurrence of Oldowan lithics, stone-tool-cut-marked mammalian bone (implying access to meat protein), and fossil hominin remains on surfaces of eroded African outcrops connotes a causal relationship has enormous intuitive appeal. Certainly, by ca. 1.5 Ma, there is circumstantial evidence for a link between these phenomena and evidence for hominin brain enlargement beyond that of the australopiths, which has figured influentially in hypotheses about the emergence and early evolution of the *Homo* lineage (e.g., Tobias, 1987; Falk, 1992; Stanley, 1992; Aiello and Wheeler, 1995; Bramble and Lieberman, 2004).

If we grant that the co-occurrence of encephalization, stone tools and paleontological evidence for meat consumption is functionally meaningful, is it reasonable to take these presumptively derived differences between *Homo* and the australopiths as explanatory of the origin of the lineage leading to modern humans? To do so is to take a logical path notably similar to the one Darwin took in proposing an explanation for the divergence of humans from an African

W.H. Kimbel (✉)
Institute of Human Origins, School of Human Evolution and Social Change, Arizona State University, P.O. Box 874101, Tempe, AZ 85287-4101, USA
e-mail: wkimbel.iho@asu.edu

F.E. Grine et al. (eds.), *The First Humans: Origin and Early Evolution of the Genus* Homo,
Vertebrate Paleobiology and Paleoanthropology,

great ape ancestor. In both cases, the absence of the chronicle of events – the lack of a fossil record – reduces explanation to projecting the linkage of particular anatomical and behavioral characters in descendants backwards in time to hypothetical ancestors. The problem is that today we do not have the evidence to test hypotheses about the relative timing of the origin of these characters because we have not adequately sampled taxa at or near the base of the lineage. It is quite possible, given our current understanding, that none of the anatomical or behavioral characteristics implied by the paleontological and archaeological evidence in Beds I and II at Olduvai Gorge, for example, was important in the processes operating on mid-Pliocene hominin populations that were uniquely ancestral to *H. habilis* and other early *Homo* species around the Plio-Pleistocene boundary.

DOES THIS MAKE SENSE

Taxonomic Issues

In one sense, the scope of the problem of the origin of *Homo* will correspond to the delineation of the taxonomic boundary of the group of species included within it. In contrast to the species category, the evolutionary process does not narrow the range of acceptable definitions of the genus, and for this reason it has not been the subject of much discussion. Moreover, genera are "discovered" using phylogenetic grouping criteria (synapomorphies) that may be irrelevant to the identification of their contained species. Wood and Collard (1999; Collard and Wood, 2007; Wood, 2009) promoted a definition of the genus category as a monophyletic group of species occupying a single adaptive zone. In their construal of the evidence, neither *Homo habilis* nor *H. rudolfensis* meet the criteria for membership in the same genus as *H. sapiens* (they place both species in *Australopithecus*, adding to the evidence that this genus that is almost certainly non-monophyletic [Strait et al., 1997, 2007; Strait and Grine, 2004]). For Wood and Collard, the origins of the genus *Homo* lie somewhere in the early Pleistocene with the appearance of *H. ergaster* (= *H. erectus*). Of course, one might question whether even *H. erectus* (including samples sometimes attributed to *H. ergaster* and *H. georgicus*) and *H. sapiens* ever shared a single adaptive zone, given the differences in absolute and relative brain size, patterns of growth and development, cultural complexity, etc.

NEVER EVOLVING ADAPTIVE ZONE

Monophyly establishes the relative timing of the appearance of the lineage bearing characters shared uniquely with living humans, whereas adaptive unity addresses the cause(s) and/or adaptive correlates of diversification among groups of monophyletic species (the evidence for each need not be mutually exclusive, of course). One implication of Wood and Collard's definition is that the "lower" boundary of a genus is not the same thing as the origin of the lineage (or clade) to which it belongs, as a monophyletic group of species may comprise two or more genera with different adaptive strategies. Thus, the issues of monophyly and adaptive unity are separable, but the monophyly criterion must be primary in evaluating the evidence for the origin of the *lineage* to which *H. sapiens* belongs.

Monophyly of Early *Homo* Species

Wood and Collard (1999; Collard and Wood, 2007; Wood, 2009) noted that while most recent phylogenetic analyses of *Homo* find a monophyletic clade that includes the species *H. habilis* and *H. rudolfensis* at or near its base, the statistical support for this arrangement is weak. This, however, does not *falsify* the hypothesis of monophyly, and the fact that multiple analyses using different data sets come to pretty much the same or very similar conclusions about the relationships of these two species must count in favor of the monophyly hypothesis (Strait et al., 2007). Indeed, it is not difficult to compose a list of derived cranial characters that *H. habilis* and *H. rudolfensis* share with later, undoubted species in the *Homo* lineage (although it must be remembered that the hypodigm for *H. rudolfensis* is pitifully small for most taxonomically informative areas of the skull and dentition).

Face

Both *H. habilis* and *H. rudolfensis* share with undoubted species of the genus *Homo* a lesser degree of subnasal prognathism than in *Australopithecus* (including the "robust" species sometimes attributed to *Paranthropus*). The inferiorly flexed subnasal plane is sharply demarcated from the floor of the nasal cavity, which is often in the form of a raised platform situated between the anterior nasal spine and the insertion of the vomer into the nasal septum. The facial plates of the maxillary frontal processes are everted, supporting an anteriorly prominent nasal bridge (it is clearly evident even in the KNM-ER 1470 specimen of *H. rudolfensis*, with its forwardly positioned zygomatic bones, and irrespective of how the facial fragment is positioned in relation to the braincase, and is also present in *H. habilis* specimen KNM-ER 1805). In *H. habilis* (but not *H. rudolfensis*) the frontal process of the zygomatic bone faces anterolaterally (as opposed to anteriorly, the primitive condition in the australopiths). The palate is broad in relation to its length.

Calvaria

There is a "true" supraorbital torus in *H. habilis*, which means that the torus is independent of temporal line incursion and protrudes superiorly above the floor of a supratoral sulcus (though not in *H. rudolfensis*, which is australopith-like in the lack of strong topographic demarcation between the superior surface of the supraorbital elements and the external surface of the frontal squama). In relation to facial breadths, the frontal bone is wide across the postorbital constriction, and the temporal lines are usually well-separated here as well as further posteriorly on the braincase. The occipital sagittal arc occupies a large percentage of the total calvarial arc. The mandibular fossa tends to be compressed anteroposteriorly.

Mandible

Mandibular synapomorphies that link *H. habilis* and *H. rudolfensis* exclusively to later species of the genus *Homo* are much more difficult to locate than in the cranium, which may be due in part to a very small sample of morphologically informative specimens (*H. habilis* is represented only by OH 13 and KNM-ER 3734). Overall gracility of the mandibular corpus is usually cited as a derived feature of *Homo*, but this characterization would not capture the two more complete mandibles attributed to *H. rudolfensis* (KNM-ER 1802 and UR 501).

Dentition

Dental synapomorphies include a skewed upper M2 occlusal outline, buccolingual compression of upper and lower M1, buccolingual narrowing and occlusal simplification of lower premolars (not in *H. rudolfensis*), increased symmetry of canine crown. (Buccolingually narrow cheek teeth may serve to link *H. habilis* and *H. rudolfensis* to one another, but it is unclear whether this constitutes strong evidence of an exclusive relationship to later *Homo*.)

The foregoing list does not pretend to be exhaustive. But it does illustrate the extent to which homoplasy needs to be invoked in different parts of the cranium and dentition if these species are not considered to represent the lineage that includes African *H. erectus*, for example. What is interesting about these putative synapomorphies is that it is difficult to perceive in them a strong pattern of relationships to the adaptive shifts we typically associate with early *Homo* after around 1.8 Ma (encephalization, reduced megadonty, etc.). Indeed, while the postcanine teeth of early *Homo* are smaller than those of contemporaneous australopiths – and endocranial volumes typically larger – the tiny sample of informative postcranial remains associated with diagnostic skull parts leaves room for widely divergent interpretations of their functional anatomy (e.g., Tobias, 1991; Wood and Collard, 1999). The significance of the OH 62 female partial skeleton of *H. habilis* has less to do with limb proportions, which, due to poor preservation of critical parts, may be "unknowable" (Reno et al., 2005), than with its small apparent body size, for, given a roughly Lucy-sized body, and *assuming* within-species proportionality, megadonty may have been pronounced in this species (Johanson et al., 1987), and at least as great as that observed in *A. afarensis* and *A. africanus* (McHenry and Coffing, 2000; see also Wood, 2009). In contrast, if the endocranial volume of OH 62, which cannot be estimated from the preserved fragments, was typical of Olduvai *H. habilis* (ca. 600–680 cc; Holloway et al., 2004), then this species may have been more encephalized relative to *Australopithecus* than absolute volumes alone would indicate. However, because we have no certain evidence for male body size and thus for the extent of body size dimorphism in *H. habilis*, we can place no great confidence in statements about megadonty and encephalization in this species. For *H. rudolfensis* we have no idea at all about body size because there are no postcranial bones associated with any of the few cranial or mandibular specimens usually attributed to this species. (It has been common to assign to *H. rudolfensis* large isolated limb bones from sub-KBS Tuff horizons at Koobi Fora approximately contemporary with KNM-ER 1470, such as femora KNM-1481 and KNM-ER 1472, but this is sheer speculation.)

In sum, the synapomorphies that link *H. habilis* and (less securely) *H. rudolfensis* to later *Homo* may not overlap the set of features constituting the adaptive complex commonly attributed to *H. erectus* and subsequent species of the genus. Thus, it would be difficult to argue, on the basis of currently available evidence, that the origin of the *Homo* clade was synonymous with the first appearance of the adaptive complex that epitomizes the more recent species of the lineage. It is not an unexpected outcome of descent with modification for basal species of a clade to lack the synapomorphies (adaptive or otherwise) of the clade's later representatives. However, if the Wood and Collard (1999; Collard and Wood, 2007) definition of the genus category is adopted, then I would argue that the appropriate solution would not be to assign the species *habilis* and *rudolfensis* to *Australopithecus*, a grade-level taxon that almost certainly includes species with no unique ties to *Homo*, but rather to recognize the monophyletic clade incorporating all those extinct species more closely related to living humans than to any species of *Australopithecus* or *Paranthropus* as a taxon (ranked at the level of tribe) comprising at least two adaptively distinct genera, one *Homo*, and the other, as yet unnamed, for the currently recognized basal species of the clade. A significant downside to this solution would stem from the paraphyly introduced by a determination that one or the other of the species, *habilis* or *rudolfensis*, in the new genus was the actual ancestor of the species *erectus* in the genus *Homo* (though not likely as a descendant via anagenesis from populations represented by the known east African samples in ca. 1.7–1.8 Ma-old time horizons – a Middle Pliocene cladogenetic event would need to be postulated; see below). For this reason, it would be preferable to keep the basal species of the clade that includes *H. sapiens* within the genus *Homo*.

Candidates for Earliest *Homo*

A small group of fossils from the time period between ca. 2.5 and 2.3 Ma includes specimens that have at one time or another been proposed as earliest known *Homo*, or as an ancestor of late Pliocene *Homo* species. The review presented below illustrates the impoverished nature of the fossil evidence bearing on the Pliocene origins of the *Homo* lineage.

Sts 19 (Sterkfontein, Member 4, South Africa, ca. 2.7–2.5 Ma?)

This specimen is a well preserved cranial base found in a rubble dump associated with early twentieth century lime-mining activities at Sterkfontein (Member 4). Broom et al. (1950) were impressed with its humanlike morphology, and, later, Clarke (1977) considered whether it should be attributed to *Homo* sp. rather than *A. africanus* based on a suite of temporal and sphenoid bone characters. This possibility was reviewed but rejected by Dean and Wood (1982) in favor of a wide range of variation for *A. africanus*, but Kimbel and Rak (1993) formally proposed that it be assigned to *Homo* (but see Ahern et al., 1998). The temporal bone characters of Sts 19, which mostly involve a humanlike position and form of the tympanic and petrous elements and their relationships to adjacent basicranial anatomy, are not observed in combination in any other cranium from Sterkfontein, but they are qualitatively similar to configurations in some "robust" australopiths (upright tympanic with well differentiated petrous crest; moderate vaginal process of the styloid; close approximation of the tympanic and the mastoid process; relatively coronal orientation of the petrous, etc.). The cranial base anatomy of Sts 19 is less derived than that commonly observed in post-2.0 Ma-old *Homo habilis* (e.g., OH 13, OH 16, OH 24; KNM-ER 3891, Omo L. 894-1, Stw 53), in which the mandibular fossa is compressed anteroposteriorly, and the preglenoid plane is very steep.

KNM-BC 1 (Chemeron Formation, Kenya, ca. 2.4 Ma)

Hill et al. (1992; Sherwood et al., 2002) promoted this fragmentary temporal bone, found in 1966, as the earliest known example of the *Homo* lineage based mainly on the supposed "extreme" medial position of the mandibular fossa relative to the lateral wall of the braincase, which they thought reflected brain expansion, as well as several qualitative features of the glenoid region and petrous element (tegmen tympani exposed in ceiling of mandibular fossa, anteromedial recess present, steep and restricted preglenoid plane, sagittally convex tympanic laterally). The phylogenetic valence of these characters has been questioned – their presence and expression varies widely across fossil hominin taxa and/or can be interpreted as symplesiomorphic – and morphometrically the specimen does not exhibit unique affinities with *Homo* temporal bones (Lockwood et al., 2002; Lockwood and Kimbel, in preparation). As Martyn and Tobias (1967) appreciated, the affinities of the Chemeron temporal are mixed and there is little prospect of improving on their assignment of it to Hominidae gen. et sp. indet., although Asfaw et al. (1999) raised the possibility that it could represent *Australopithecus garhi* (see below).

UR 501 (Chiwondo Beds, Uraha, Malawi, ca. 2.5–1.9 Ma?)

The geological age of this relatively complete mandibular body with teeth is uncertain. Although Bromage et al. (1995) settled on an age of 2.3–2.5 Ma for the specimen, the surface from which it comes contained a temporally mixed fauna with elements potentially ranging in age from approximately 3 Ma to less than 2 Ma (K. Reed, pers. comm., 2007, considers the fauna to indicate an age of anywhere between 2.5 and 1.9 Ma). Notwithstanding doubts about its "earliest" status, the mandible has been affiliated with mandibles of *H. rudolfensis* (i.e., KNM-ER 1802; Bromage et al., 1995) based on the absolutely large size of the anterior and postcanine teeth, relatively thick molar enamel, and double, plate-like lower P4 roots.

A.L. 666-1 (Hadar Formation, Ethiopia, ca. 2.3 Ma)

This specimen, from ca. 2.3 Ma-old sediments at Hadar, Ethiopia, is a maxilla, most likely of a male individual, with most of the dentition. In its low subnasal prognathism, everted frontal processes, relatively wide palate, symmetric upper canine crowns, high upper P3/P4 crown area ratio, mesiodistally elongated upper M1, and rhomboidal upper M2 occlusal outline, this specimen follows the morphological pattern of post-2.0 Ma-old *H. habilis* specimens, such as KNM-ER 1813, L. 894-1, and material from Bed I and lower Bed II, Olduvai Gorge. It lacks the derived zygomatico-alveolar, infraorbital and subnasal morphology of *H. rudolfensis* specimen KNM-ER 1470 (Kimbel et al., 1997). However, the A.L. 666-1 upper P4 is buccolingually broad compared to these more recent specimens, similar to "non-robust" hominin premolars from pre-2.0 Ma levels in the Shungura Formation, Ethiopia (Suwa, 1990).

Teeth (Members E-G, Shungura Formation, Ethiopia, ca. 2.4–2.0 Ma)

A small number of postcanine teeth from Members E-G of the Shungura Formation, ca. 2.4–2.0 Ma, show derived morphological features observed otherwise only in post-2.0 Ma-old *H. habilis* and *H. rudolfensis* (Suwa et al., 1996). Lower M1s are buccolingually narrow, while lower P3 crowns are likewise narrow and also display a less asymmetric occlusal outline and a better developed mesial marginal ridge than in most early australopith homologues (*A. afarensis* and *A. africanus*). Suwa et al. (1996) noted that the lower P3-M1 crowns from this time horizon in the Shungura Formation are relatively large, with expanded premolar talonids, resembling the condition in *H. rudolfensis* specimens KNM-ER 1802 (Wood, 1991) and UR 501 (Bromage et al., 1995). They suggested that this dental pattern may have been primitive for the *Homo* lineage. On the other hand, a lower first molar crown (KNM-WT 42718) from the Nachukui Formation at West Turkana, dated to ca. 2.3 Ma, appears to be similar, especially in its size and high degree of buccolingual compression, to teeth from Olduvai Gorge and Koobi Fora attributed to *H. habilis* (Prat et al., 2005). Combined with the *H. habilis*-like morphology of the Hadar maxilla A.L. 666-1, this evidence hints at a divergence of early *Homo* morphotypes prior to 2.3 Ma.

Remains Assigned to Australopithecus garhi *(Hata Member, Bouri Formation, Ethiopia, ca. 2.5 Ma)*

The ca. 2.5 Ma-old partial cranium and dentition (BOU-VP 12/130) of *A. garhi* (Asfaw et al., 1999) reveal a primitive maxillofacial configuration similar to that of *A. afarensis* combined with pronounced premolar–molar row expansion reminiscent of *A. africanus* and later Pliocene "robust" australopith species, though apparently without the occlusal or enamel thickness specializations of the latter group's postcanine dental battery. Little if anything in the published morphological evidence links the species uniquely to the *Homo* lineage (Strait and Grine, 1999, 2001), but the presence in the same Middle Awash stratigraphic unit of stone-tool-modified bovid skeletal remains clearly influenced Asfaw et al.'s (1999) suggestion that, among other options, *A. garhi* may have been exclusively ancestral to *Homo* (as did, perhaps, a set of associated but taxonomically unattributed hominin limb bones with a low – humanlike – estimated humerofemoral length index).

Conclusions

By ca. 1.7–1.8 Ma the East African fossil record reveals as many as three distinct species representing the *Homo* lineage (*H. habilis, H. rudolfensis, H. erectus*). However, while the fragmentary evidence from Hadar, the Omo River basin, and Uraha (assuming an older rather than a younger age for the hominin-bearing deposits) might, in combination, suggest that the morphological configurations diagnostic of *H. habilis* and *H. rudolfensis* have earliest known appearances in the 2.4–2.3 Ma interval, the fossil record between 2.0 and 3.0 Ma remains so poorly sampled across anatomy, time and space that hypotheses concerning the taxonomic distinctiveness or phylogenetic relationships of early *Homo* species in this time period are not currently amenable to meaningful tests. This uncertainty is compounded by the poorly documented morphological variability of *H. rudolfensis* and the under-sampled anatomy of some craniodental regions (i.e., the mandible) of *H. habilis*.

Though the ecological context in which *Homo* originated is not known at present, one question that might be addressed with the evidence in hand concerns the ecological circumstances of the known early *Homo* species. How should we explain the potential sympatry and synchrony of two or three *Homo* species around the Plio-Pleistocene boundary (at Koobi Fora, for example; Spoor et al., 2007) – assuming that they inhabited similar ecological niches featuring the exploitation of meat protein? I am not here raising the specter of the single species hypothesis, but if these taxa indeed overlapped on ecological (as opposed to geological) scales of time and space, then their morphology and inferred behavior should lead to testable hypotheses about the causes of divergence and subsequent adaptive evolution and should spur research aimed at clarifying the ecologically-related contextual information associated with fossils of each of these taxa.

The Sterkfontein Member 4 specimen Sts 19 may predate the earliest known East African *Homo* fossils, but given the ambiguous chronological position of this South African faunal assemblage, it is impossible to determine this with any certainty. The temporal bone anatomy of Sts 19 is certainly more derived in the direction of later *Homo* than that of any other cranium from *A. africanus*-bearing sediments at Makapansgat or Sterkfontein. If Sts 19 does not itself represent an early species of *Homo*, and instead belongs to *A. africanus*, as many workers continue to hold, then the considerable variation thereby encapsulated by this species would almost certainly include unique conditions expected in an early representative of the *Homo* clade. In this respect, the Sterkfontein *A. africanus* sample would differ from samples of other australopith-grade middle Pliocene hominins so far known.

Acquisition of knowledge about the adaptations of early *Homo* species has been severely hampered by the poor or undocumented record of body size for *H. habilis* and *H. rudolfensis*, respectively. While it is possible to identify cranial and dental synapomorphies that tie these species exclusively to later, securely attributed representatives of the *Homo* lineage, debate over whether these species should be attributed to the same *genus* as *H. sapiens* – given an adaptation-based definition of this taxonomic category – is not currently resolvable. And it may be less critical than securing better evidence bearing on the monophyly of the lineage through new additions to the fossil record. On balance, it is preferable to keep the basal species of the *Homo* clade within the genus *Homo*.

We are confronted with a "black box" with respect to the cause(s) of the origin of the *Homo* lineage. The (often unstated) assumption has usually been that the root of the *Homo* lineage would be found in the morphological-behavioral complex whose elements prominently include stone-tool manufacture, meat-protein consumption, and encephalization. Early Pleistocene *H. erectus* bears the stamp of this classic Darwinian formulation, but just how far back into the Pliocene this complex can be projected is a matter of conjecture. The suggestion that mid-Pliocene *A. garhi* was exclusively ancestral to the early *Homo* species of the late Pliocene illustrates just how important this adaptive scenario has been to marking the "lower" boundary of the human genus.

None of the recognized australopith species from the mid-Pliocene (*A. garhi, A. aethiopicus, A. africanus* – but see above) appears to be linked cladistically to these Plio-Pleistocene *Homo* species. The tenuous suggestion that the ca. 3.5 Ma *Kenyanthropus platyops* cranium KNM-WT 40000 is linked by derived features with the much younger *H. rudolfensis* (i.e., flat midface with advanced zygomatics and low subnasal projection; Leakey et al., 2001) would, if bolstered by further discoveries in the intervening 1.5 Myr time gap, establish a minimum age for the origin of the *Homo* lineage by linking these two taxa as sister-species. However, the same derived features shared by *K. platyops* and *H. rudolfensis* have been implicated in a case of convergent evolution between *H. rudolfensis* and "robust" australopiths, and otherwise the cranial morphology of *K. platyops* is prevailingly primitive with respect to the earliest known and subsequent species in the *Homo* lineage. It is not beyond question that the *Homo* lineage was rooted in the same Pliocene radiation of post-3 Ma australopith species – the diverse craniodental morphologies of which may represent alternative approaches to solving the problem of changing Pliocene dietary resources via "heavy mastication" – that gave rise to the robust australopiths.

New fossil evidence from the 3.0–2.0 Ma time period is urgently needed to address these and other critical issues surrounding the origin of the *Homo* lineage.

Acknowledgments Thanks to John Fleagle, Fred Grine and Richard Leakey for inviting me to participate in the Stony Brook symposium, out of which this essay was developed. Thanks also to John Fleagle and two anonymous referees for constructive comments on the manuscript and to Luke Delezene for helping me mull over some of the issues and critically reading the penultimate draft.

References

Aiello, L.C, Wheeler, P. E., 1995. The expensive-tissue hypothesis: the brain and the digestive system in human and primate evolution. Current Anthropology 36, 199–221.

Asfaw, B., White, T., Lovejoy, C.O., Latimer, B., Simpson, S., Suwa, G., 1999. *Australopithecus garhi:* a new species of early hominid from Ethiopia. Science 284, 629–635.

Bramble, D.M., Lieberman, D.E., 2004. Endurance running and the evolution of *Homo*. Nature 432, 345–352.

Bromage, T.G., Schrenk, F., Zonnevel, F.W., 1995. Paleoanthropology of the Malawi Rift: an early hominid mandible from the Chiwondo Beds, northern Malawi. Journal of Human Evolution 28, 71–108.

Broom, R., Robinson, J.T., Schepers, G., 1950. The Sterkfontein Ape-Man *Plesianthropus transvaalensis*. Transvaal Museum Memoir 4, Pretoria.

Clarke, R.J., 1977. A juvenile cranium and some adult teeth of early *Homo* from Swartkrans, Transvaal. South African Journal of Science 73, 46–49.

Collard, M., Wood, B.A., 2007. Defining the genus *Homo*. In: Henke, W., Tattersall, I. (Eds), Handbook of Paleoanthropology, Volume 3: Phylogeny of Hominids. Springer, Berlin, pp. 1575–1610.

Dean, M.C., Wood, B.A., 1982. Basicranial anatomy of Plio-Pleistocene hominids from East and South Africa. American Journal of Physical Anthropology 59, 157–174.

Falk, D., 1992. Evolution of the brain and cognition in hominids. James Arthur Lectures on the Evolution of the Human Brain, Number 62. American Museum of Natural History, New York.

Hill, A., Ward, S.C., Deino, A.L., Curtis, G.H., Drake, R.E.,1992. Earliest *Homo*. Nature 355, 719–722.

Holloway, R.L., Broadfield, D., Yuan, M., 2004. Brain Endocasts: The Paleoneurological Record. The Human Fossil Record, Volume 3. Wiley-Liss, Hoboken, NJ.

Johanson, D.C., Masao, F. T., Eck, G.G., White, T.D., Walter, R.C., Kimbel, W.H., 1987. New partial skeleton of *Homo habilis* from Olduvai Gorge, Tanzania. Nature 327, 205–209.

Kimbel, W.H., 1991. Species, species concepts and hominid evolution. Journal of Human Evolution 20, 355–371.

Kimbel, W.H., Rak, Y.,1993. The importance of species taxa in paleoanthropology and an argument for the phylogenetic concept of the species category. In Kimbel, W.H., Martin, L.B. (Eds), Species, Species Concepts, and Primate Evolution. Plenum, New York, pp. 461–484.

Kimbel, W.H., Johanson, D.C., Rak, Y., 1997. Systematic assessment of a maxilla of *Homo* from Hadar, Ethiopia. American Journal of Physical Anthropology 103, 235–262.

Leakey, L.S.B., Tobias, P.V., Napier, J.,1964. A new species of the genus *Homo* from Olduvai Gorge. Nature 202, 7–9.

Leakey, M., Spoor, F., Brown, F.H., Gathogo, P.N., Kiarie, C., Leakey, L.N., MacDougall, I., 2001. New hominin genus from eastern Africa shows diverse middle Pliocene lineages. Nature 410, 433–440.

Lockwood, C.A., Kimbel, W.H., n.d. The temporal bone of early *Homo*. In preparation.

Lockwood, C.A., Kimbel, W.H., Lynch, J., 2002. Temporal bone morphology and earliest *Homo*. American Journal of Physical Anthropology S34, 102.

Martyn, J., Tobias, P.V., 1967. Pleistocene deposits and new fossil localities in Kenya. Nature 215, 476–480.

McHenry, H.M., Coffing, K., 2000. *Australopithecus* to *Homo:* transformations in body and mind. Annual Review of Anthropology 29, 125–146.

Oakley, K.P., 1959. Man the Tool-Maker. University of Chicago Press, Chicago, IL.

Prat, S., Brugal, J-P., Tierceling, J-J., Barrat, J-A., Bohn, M., Delagnes, A., Harmand, S., Kimeu, K., Kibunjia, M., Texier, P-J., Roche, H., 2005. First occurrence of early *Homo* in the Nachukui Formation (West Turkana, Kenya) at 2.3–2.4 Myr. Journal of Human Evolution 49, 230–240.

Reno, P., DeGusta, D., Serrat, M., Meindl, R., White, T., Eckhardt, R., Kuperavage, A., Galik, K., Lovejoy, C.O., 2005. Plio-Pleistocene hominid limb proportions. Current Anthropology 46, 575–588.

Semaw, S., 2000. The world's oldest stone artifacts from Gona, Ethiopia: their implications for understanding stone technology and patterns of human evolution between 2.6–1.5 million years ago. Journal of Archaeological Science 27, 1197–1214.

Sherwood, R., Ward, S.C., Hill, A., 2002. The taxonomic status of the Chemeron temporal (KNM-BC 1). Journal of Human Evolution 42, 153–184.

Spoor, F., Leakey, M.G., Gathogo, P.N., Brown, F.H., Anton, S.C., McDougall, I., Kiarie, C., Manthi, F.K., Leakey, L.N., 2007. Implications of new early *Homo* fossils from Ileret, east of Lake Turkana, Kenya. Nature 448, 688–691.

Stanley, S.M., 1992. An ecological theory for the origin of *Homo*. Paleobiology 18, 237–257.

Strait, D.S., Grine, F.E., Moniz, M.A., 1997. A reappraisal of early hominid phylogeny. Journal of Human Evolution 32, 17–82.

Strait, D.S., Grine, F.E., 1999. Cladistics and early hominid phylogeny. Science 285, 1210.

Strait, D.S., Grine, F.E., 2001. The systematics of *Australopithecus garhi*. Ludus Vitalis, 9, 109–135.

Strait, D.S, Grine, F.E., 2004. Inferring hominoid and early hominid phylogeny using craniodental characters: the role of fossil taxa. Journal of Human Evolution 47, 399–452.

Strait, D.S., Grine, F.E., Fleagle, J.G., 2007. Analyzing hominid phylogeny. In: W. Henke, I. Tattersall (Eds.), Handbook of Paleoanthropology, Volume 3: Phylogeny of Hominids. Springer, Berlin, pp. 1781–1806.

Suwa, G., 1990. A comparative analysis of hominid dental remains from the Shungura and Usno Formations, Omo Valley, Ethiopia. Ph.D. dissertation, University of California, Berkeley.

Suwa, G., White, T.D., Howell, F.C., 1996. Mandibular postcanine dentition from the Shungura Formation, Ethiopia: crown morphology, taxonomic allocations, and Plio-Pleistocene hominid evolution. American Journal of Physical Anthropology 101, 247–282.

Tattersall, I., 1998. Becoming Human. Harcourt Brace, New York.

Tobias, P.V., 1987. The brain of *Homo habilis*: a new level of organization in cerebral evolution. Journal of Human Evolution 16, 741–761.

Tobias, P.V., 1991. The Skulls, Endocasts and Teeth of *Homo habilis*. Olduvai Gorge, Volume 4. Cambridge University Press, Cambridge.

Wood, B.A., 1991. Hominid Cranial Remains. Koobi Fora Research Project, Volume 4. Oxford University Press, Oxford.

Wood, B.A., 2009. Where does the genus *Homo* begin, and how would we know? In: Grine, F.E., Fleagle, J.G., Leakey, R.E. (Eds), The First Humans: Origin and Early Evolution of the Genus *Homo*. Springer, Dordrecht, pp. 17–28.

Wood, B.A., Collard, M., 1999. The human genus. Science 284, 65–71.

Chapter 5
Comparisons of Early Pleistocene Skulls from East Africa and the Georgian Caucasus: Evidence Bearing on the Origin and Systematics of Genus *Homo*

G. Philip Rightmire and David Lordkipanidze

Keywords Species • morphology • variation • *Homo habilis* • *Homo rudolfensis* • *Homo erectus* • Dmanisi • human evolution

Introduction

A half century ago, when there were fewer fossils (and not so many paleoanthropologists), characterizing the genus *Homo* was relatively straightforward. In addition to modern humans, Neanderthals could be included, along with other archaics such as Broken Hill (now Kabwe) from Zambia and the Ngandong assemblage from Java. Also, it was becoming clear that *Atlanthropus* from northern Africa, and *Pithecanthropus* and *Sinanthropus* from eastern Asia, should be lumped into this same taxon. Phillip Tobias (2009) has noted that the situation changed rather dramatically in 1964, with the naming of *Homo habilis*. Accommodating the newly discovered Olduvai remains in *Homo* required expanding the existing definition of the genus. This trend has continued.

Along with the traditional emphasis on morphology of skulls, teeth, and postcranial bones, investigators have added criteria relating to energetics and diet, tool-making, and the ability to communicate using spoken language. Recently, issues of life history and the timing of development have been raised as well. Bernard Wood (2009) has reminded us that some of these characters help in exploring phylogeny (*Homo* as a clade) and others assist in assessing the grade status of individual species.

Discussions at this workshop repeatedly returned to the topic of characterizing *Homo* and recognizing its first appearance in the fossil record. A key question concerns the identification of the oldest fossils that seem to be *Homo*-like, rather than similar to australopiths. On the basis of apparent brain expansion, rounding of the vault, reduction of the masticatory complex, and narrowing of the cheek teeth, a number of specimens from both Olduvai Gorge and the Koobi Fora localities have been attributed to *Homo habilis*. Since the early 1990s, a subset of these remains has been set apart as *Homo rudolfensis*. Craniodental fragments from the Omo, a maxilla from Hadar, a tooth from West Turkana, and a mandible from Uraha have been compared to both of these hypodigms. Also, it is widely agreed that a partial cranium and other fossils from Swartkrans are properly assigned to *Homo*, although there is still uncertainty about the species represented (Grine et al., 2009). Despite this broad consensus, doubts as to the status of these early hominins have been expressed. Wood and Collard (1999) and Wood (2009) consider that in important criteria including body build, encephalization, and size of the teeth and jaws, neither *Homo habilis* nor *Homo rudolfensis* is more similar to *Homo sapiens* than to australopiths. On these grounds, the observers argue that both species should be removed from *Homo*.

An obvious problem here is the scarcity of fossils that are reasonably complete and/or possess useful morphology. Especially in the case of *Homo rudolfensis*, for which there is one good cranium (KNM-ER 1470) but only a handful of other remains, sample size is small. Indeed it has been difficult to devise a solid basis for diagnosing one early species relative to another, and the material is so inadequate as to preclude answering all of the questions that are relevant to the separation of "early *Homo*" from the australopiths. Without more evidence, many workers will be reluctant to remove the Olduvai and Koobi Fora fossils from the hypodigm of *Homo*. At the same time, it is recognized that there is so much variation in these assemblages that it is probably inappropriate to lump the specimens together in one species.

A third hominin is known from eastern and northwestern Africa and probably from South Africa as well. Since the 1960s and 1970s, much information bearing on the origin and evolution of *Homo erectus* has been recovered from Olduvai Gorge and the Koobi Fora sites. Also, the nearly complete subadult skeleton (KNM-WT 15000) from Nariokotome has provided insight into growth and body form of one early

G.P. Rightmire (✉)
Department of Anthropology, Peabody Museum, Harvard University, Cambridge, MA 02138, USA
e-mail: gprightm@fas.harvard.edu

D. Lordkipanidze
Georgian National Museum, 3 Purtseladze Street, 0105 Tbilisi, Republic of Georgia
e-mail: dlordkipanidze@museum.ge

member of this group (Dean and Smith, 2009). More recently, there have been intriguing new finds from Ileret and Olorgesailie in Kenya, and Bouri (Daka) in Ethiopia. The Ileret and Olorgesailie crania (along with OH 12) confirm that some *Homo erectus* individuals were quite small (Leakey et al., 2003; Potts et al., 2004; Spoor et al., 2007), and it is apparent that size variation within the African populations was substantial. The larger and more robust Daka calvaria shares numerous features with *Homo erectus* in Asia, supporting the view that the African and Far Eastern hominins belonged to a single, widely dispersed paleospecies (Asfaw et al., 2002).

Additional discoveries bearing on the evolution of *Homo* come from the Georgian Caucasus. Since 1991, excavations at Dmanisi have produced crania, lower jaws with teeth, postcranial bones, and numerous crude stone artifacts. Studies of the sedimentary context in which the fossils occur indicate that the material was sealed in the deposits over a brief interval (<10,000 years), ca. 1.77 million years ago (Ma) (Lordkipanidze et al., 2006, 2007). Thus, in paleontological terms (Howell, 1999), the Dmanisi assemblage documents a single paleodeme. It is important to point out that this situation is very rare. At Plio-Pleistocene sites such as Koobi Fora or Sangiran, the fossils are scattered through sediments accumulated over hundreds of thousands of years, and variation due to sampling from different time periods cannot be ignored. At Dmanisi, however, morphological differences among the individuals can be attributed more confidently to intragroup variation.

The goal of this paper is not to review all that has been said about *Homo habilis* over more than 4 decades. Instead, it is helpful to take a step back from the existing systematic framework, so that the several hypodigms proposed for early *Homo* in Africa can be viewed in the light of recent finds from western Asia. This exercise does not immediately demonstrate that taxonomic changes are required, but it does bring out the fact that we know very little about sex dimorphism and other sources of variation in these ancient populations. It is hardly surprising that experienced workers have not been able to agree on just where species boundaries lie. Also, there is no consensus as to which of the early species might have evolved toward *Homo erectus*. The fossils from Dmanisi offer a fresh perspective in this regard.

What Do We Know on the Basis of the Early African Record?

From discoveries in the Olduvai and Turkana Basins, along with a few other finds from Ethiopia, Malawi, and South Africa, it can be inferred that *Homo* evolved in Africa. The well-preserved Hadar maxilla with teeth differs from the australopith condition (Kimbel et al., 1997), as does an isolated lower molar from West Turkana (Prat et al., 2005). These hominins are dated to ca. 2.3 Ma. The Uraha mandible also exhibits some *Homo*-like morphology, but here the dating is far less precise. Faunal correlations to radiometrically dated horizons in East Africa point to an age of 2.5–2.3 Ma, but the fossil can be younger by 0.5 Ma or more (Bromage et al., 1995). It is unlikely that these specimens mark the actual first appearance of the *Homo* clade. Kimbel (2009) notes that the A.L. 666-1 maxilla shares a number of derived traits with later hominins from Olduvai and Koobi Fora, so it is probable that the Hadar individual postdates the origin of the genus, perhaps by a substantial interval. Whether any of the currently available candidate species of *Australopithecus* might have been ancestral to *Homo* is still not clear. In the past, detailed phylogenetic arguments have been presented for *Australopithecus africanus* (Robinson, 1967; Tobias, 1967) and for *Australopithecus afarensis* (Johanson and White, 1979), while more recently, it has been suggested that *Australopithecus garhi* is a taxon from which *Homo* could have been derived (Asfaw et al., 1999). Another possibility is that the *Homo* lineage arose instead from a generalized mid-Pliocene antecedent that has not yet been discovered. In any event, the Hadar and West Turkana fossils (also the materials from the upper Burgi Member at Koobi Fora) are older than any remains found outside of Africa. Even after many decades of prospecting, no morphologically convincing, comparably ancient hominins have been unearthed in southern Asia or the Far East.

Hypodigms of Early *Homo*

Although a case can be made for lumping all of the early *Homo* fossils together as members of a single highly dimorphic species (Tobias, 1991), there is agreement, reflected among workshop participants, that the resulting hypodigm is so variable that partitioning is warranted. Sorting the fossils to two or more groups has been done differently by a number of workers (Groves, 1989; Wood, 1991, 1992; Rightmire, 1993; Blumenschine et al., 2003), partly as a consequence of the varying emphasis placed on characters such as brain size, facial prognathism, and tooth morphology. Interestingly, it is OH 7 (the type specimen of *Homo habilis* as defined by Leakey, Tobias and Napier in 1964) that has several times been shifted between a hypodigm composed of relatively small crania, and a larger-brained group best documented by KNM-ER 1470. As a result of this uncertainty regarding OH 7, the nomen *Homo habilis* has in the past been associated with rather different sets of fossils. Sorting OH 7 with the other Olduvai hominins (including OH 65) and lumping these with gracile individuals such as KNM-ER 1813 from Koobi Fora, following Wood (1991, 1992), is the course favored by most at the workshop.

Morphology of *Homo habilis*

On the basis of several reasonably complete crania, and jaws with teeth, *Homo habilis* (*sensu stricto*) can be characterized as having a mean endocranial capacity of ca. 610 cm³, thin vault bones, a forwardly placed foramen magnum, nasal bones that widen inferiorly, a relatively narrow midface with a near-vertical infraorbital (malar) region, and overall reduction of tooth size in comparison to australopiths (Wood, 1991). Postcranial bones thought to belong with this hypodigm show adaptations for bipedal locomotion. The OH 8 foot has a human-like metatarsal robusticity pattern, and there is convincing evidence that the first metatarsal was adducted, even if the foot retained a degree of grasping ability (Aiello and Dean, 1990). However, the presence of a longitudinal arch cannot be confirmed, and the OH 8 talus differs in key features from the human pattern (Lewis, 1989). The argument for bipedalism is strengthened if the OH 8 fossils are paired with the OH 35 tibia, as suggested by Susman and Stern (1982). This association now seems secure on the basis of matching taphonomic traces (Roche et al., 2009). Comparison with the tibia shows the foot to be relatively short as in modern humans, rather than elongated as in apes. Unfortunately, the more complete but very fragmentary skeleton of OH 62 is of only limited utility. A preliminary estimate for the humerofemoral index suggested primitive proportions and a robust arm (Johanson et al., 1987), but later studies of OH 62 and KNM-ER 3735 conclude that *Homo habilis* probably possessed a hindlimb that was long relative to that of australopiths (Haeusler and McHenry, 2004).

Crania, Mandibles and Teeth Attributed to *Homo rudolfensis*

Homo rudolfensis is significantly less well documented. Because it is the most complete cranium in the hypodigm, KNM-ER 1470 has routinely been emphasized in published descriptions of the species. This practice has perhaps exaggerated the impression that *Homo rudolfensis* is big-brained, relative to *Homo habilis*. In fact, KNM-ER 1470 is the sole individual for which brain volume (750 cm³) can be measured reliably. Part of the KNM-ER 3732 cranium is also preserved, but the frontal and parietal bones alone provide only a general indication of increased capacity. Both of the Koobi Fora fossils show that the facial skeleton is markedly orthognathic, with a massive, anteriorly inclined malar surface (Wood, 1991). KNM-ER 1470 also exhibits flattening of the maxillary clivus below the nose. Here the facial morphology contrasts with that in *Homo habilis*, where there is greater subnasal projection. This difference is unlikely to reflect sex dimorphism, as it is just the reverse of the condition observed in living hominoid primates, where (smaller) females tend to have flatter lower faces than the (larger) more prognathic males.

No tooth crowns are preserved for KNM-ER 1470. The anterior and posterior permanent teeth of another Koobi Fora specimen (KNM-ER 1590), presumed to belong to the same species, are very large, but the basis for attributing this fragmentary subadult to *Homo rudolfensis* is far from firm. The same holds for the OH 65 palate with dentition (but see Blumenschine et al., 2003). If this Olduvai fossil is linked with *Homo habilis* rather than with the hypodigm of KNM-ER 1470, then information relating to the upper teeth of *Homo rudolfensis* is very scarce. Mandibles are somewhat more plentiful, as there are five specimens from Koobi Fora. Several are fragmentary, but KNM-ER 1802 is reasonably complete, showing marked relief on the external corpus and eversion of the base. The robust UR 501 mandible does not display either of these traits, and it carries premolars and molars with large crown areas. Thus it is mainly on the evidence of premolar crown shape (relatively broad) and premolar root morphology (plate-like) that the Malawi mandible has been assigned to *Homo rudolfensis* (Bromage et al., 1995). Several postcranial bones that are often discussed in relation to this species, such as the KNM-ER 1472 and KNM-ER 1481 femora, were not found with skulls or diagnostic dental remains. Given this lack of associated limb bones, there are no solid estimates for body size or encephalization, and few clues concerning the locomotor capabilities, of *Homo rudolfensis*.

Characters Diagnostic for Genus *Homo*

When all of the early African crania, mandibles and teeth are considered, it is possible to list a series of features that characterize emerging *Homo*. As already noted, such lists are not new, and definitions of the genus have been formulated by Le Gros Clark (1964), Leakey et al. (1964), Howell (1978), and subsequent workers including Wood and Collard (1999). In Table 5.1, only observations that can be made directly from the fossils are presented, without inferences as to behavior or life history. The list is relatively brief, as it omits features (e.g., a bony chin, canine size) included in some earlier definitions that have proved to be uninformative. Given the obvious problems encountered in interpreting the significance of the few postcranial remains associated (more or less securely) with early *Homo* at Olduvai and Koobi Fora, the limbs, hand, and foot are set aside. Table 5.1 summarizes aspects of brain size, the face, and the masticatory apparatus that distinguish *Homo* from australopiths and thus have diagnostic value.

Table 5.1 Characters of the skull and teeth diagnostic for genus *Homo*

- Brain size (absolute and relative) increased in comparison to other groups
- Parietal sagittal curvature reduced, relative to australopiths (Leakey et al., 1964)
- Occipital angle large, reflecting open curvature of posterior vault (Leakey et al., 1964)
- Masticatory complex (maxilla and mandible) smaller than in *Australopithecus* (Leakey et al., 1964)
- In anterior view, the maxillary walls appear vertical and do not taper superomedially. This gives the maxilla a "squared off" appearance (different from the triangular outline seen in *A. afarensis* or *A. africanus*) (Kimbel et al., 1997)
- Nasal sill platform-like, and there may be transverse (lateral or spinal) crests. The sill is set at nearly an acute angle to the subnasal part of the maxilla
- Nasoalveolar clivus flattened from side to side, and there is less subnasal prognathism than in other groups
- Palate relatively broad
- M^1 crown shape square (length/width index ca. 1.0), rather than buccolingually broad as in *Australopithecus* (length/width index reduced)
- M^2 mesial cusps dominate the distal cusps, and the paracone bulges buccally relative to the metacone, giving the tooth an asymmetric (rhomboidal) outline (Brown and Walker, 1993; Kimbel et al., 1996)
- M_1 buccolingually narrow in relation to its length (Tobias, 1991)

Early African *Homo erectus*

For *Homo erectus*, there is considerably more material. Eastern and southern African assemblages include several well preserved crania, mandibles, and parts of an adult skeleton (affected by pathology) from Koobi Fora, the boy from Nariokotome, cranial and postcranial remains from Olduvai, and additional specimens from Olorgesailie, Konso, Bouri (Daka), Buia, and probably Swartkrans. The KNM-ER 3733 cranium has been assigned a date of 1.78 Ma (Feibel et al., 1989), but recent fieldwork at Koobi Fora has produced support for a revised stratigraphy, placing many of the *Homo erectus* fossils at 1.65–1.45 Ma (Gathogo and Brown, 2006). Part of a thickened occipital bone (KNM-ER 2598), bearing a prominent and rounded transverse torus, is often cited as evidence for a first appearance of *Homo erectus* at ca. 1.9 Ma, but if this specimen was collected from a lag surface, it may well have come originally from a higher level in the deposits, as noted by Suwa et al. (2007). Even if the earlier dates cannot be confirmed, it may be inferred that *Homo erectus* was present in the Turkana Basin as a contemporary to *Homo habilis* (Spoor et al., 2007). It is also possible that one (or both) species overlapped in time with *Homo rudolfensis*. This is important information bearing on ecology and behavior. It may also rule out scenarios of anagenetic transformation within an early *Homo* lineage, but it does not offer a basis for choosing among competing cladistic (branching) hypotheses.

Compared to that of early *Homo*, the *Homo erectus* braincase is enlarged. Cranial capacities range from 727 cm^3 (OH 12) to 1,067 cm^3 (OH 9). The skull is relatively long and low in contour, with projecting supraorbital tori, often a midline keel, well-developed crests in the mastoid region, and an angled occipital. The cranial base and mandibular fossa exhibit a number of (primitive?) features shared with *Homo habilis*, along with some aspects of tympanic plate and petrous temporal morphology that appear to be specialized. The midface is relatively broad. The nasal aperture is wider than in *Homo habilis*, and the nasal saddle tends to be more prominent (Rightmire, 1990; Antón, 2003). Mandibles on record from Konso, the Turkana sites, Olduvai, and Swartkrans are similar in proportions to those of some *Homo habilis* (e.g., OH 13), but the corpus is less thickened than in KNM-ER 1802 or UR 501. The symphyseal profile is receding. Whether measured in absolute terms (crown surface area) or in relation to estimated body mass (megadontia), postcanine tooth size is reduced in *Homo erectus* (McHenry and Coffing, 2000).

Estimates for stature, weight, and various limb indices are available for KNM-WT 15000, and this individual was quite tall and linear, with proportions like those of modern humans adapted to tropical environments (Ruff and Walker, 1993). Given its subadult status, the Nariokotome skeleton was surprisingly large in comparison to most other conspecific individuals from Koobi Fora and Olduvai (Dean and Smith, 2009). Nevertheless, it is apparent that average body mass for *Homo erectus* was greater than that of *Homo habilis* or the australopiths. It has been argued that this increase reflects a change in diet (Aiello and Wheeler, 1995). Also, body proportions and other musculoskeletal specializations can be read to show that *Homo erectus* was capable of endurance running over long distances (Bramble and Lieberman, 2004).

What New Information Is Provided by the Discoveries at Dmanisi?

Given the radiometric, paleomagnetic, stratigraphic, and biochronological evidence compiled from Dmanisi, there is no doubt that hominins were present in the Caucasus 1.8–1.7 Ma ago. This site preserves an extraordinary record of what is presently the earliest known occupation outside of Africa. However, it is unlikely that Dmanisi documents the very first human excursions from Africa into Eurasia, and most probably there were earlier dispersal events, perhaps largely unsuccessful, for which archaeological traces have not been found. In any case, it can be established that the West Asian populations were broadly coeval with both *Homo habilis* and *Homo erectus* in East Africa.

The very complete crania, entire mandibles, and teeth from Dmanisi display the suite of diagnostic characters listed in Table 5.1, and the fossils are clearly representative of *Homo*. Five individuals can be identified. One (D2700/D2735) is subadult, while another (D3444/D3900), displaying severe resorption of the alveolar processes and retaining only a single (lower) tooth at death, is probably an older adult. There is considerable variation in morphology within the Dmanisi assemblage. Indeed, differences among the specimens have led some workers to claim that multiple species may be documented (Schwartz and Tattersall, 2002), or that size variation exceeds the level expected for populations belonging to genus *Homo* (Skinner et al., 2006). To a degree, these conclusions are driven by the appearance of the D2600 mandible. This large individual has been described as possessing a novel combination of features, not observed in *Homo rudolfensis*, *Homo habilis*, or *Homo erectus* (Gabunia et al., 2002). However, there are difficulties with such an argument. D2600 presents obvious signs of periodontal disease, and this has affected the original morphology. Also, it has not been demonstrated that corpus size precludes placing D2600 with the other lower jaws (Rightmire et al., 2008).

Variation in the Dmanisi Paleodeme

Capacities obtained for the four Dmanisi crania range from 600 to 775 cm^3. In order to draw comparisons with other samples, it is appropriate to employ a measure of relative variation. A size-independent statistic that has been demonstrated to be useful in paleontological situations is the coefficient of variation (CV). Where the number of individuals is small, the CV may be modified as $(1 + 1/4N) \times (100s/\bar{x})$, following Sokal and Braumann (1980). The resulting unbiased statistic is V*. For endocranial volume, V* is 12.3 at Dmanisi, 13.0 for *Homo habilis* (N = 6, including KNM-ER 1470) (Tobias, 1991), and 14.1 for *Homo erectus* (N = 32) (Rightmire, 2004), while the CV is ca. 12–15 for modern humans. Here, there is no indication that variation within the Dmanisi paleodeme is excessive, relative to that in other hominins. The same conclusion has been reached in a resampling study, showing sex dimorphism to be a sufficient explanation for brain size differences in the Caucasus population (Lee, 2005).

Linear dimensions of the braincase that appear to be especially variable include cranial length (V* = 7.3), maximum frontal breadth (V* = 10.4), and the parietal sagittal arc (V* = 9.5). These values for Dmanisi approach the V*s calculated for a larger sample of African and Asian *Homo erectus*, and the coefficients are greater than in recent *Homo sapiens*. Within modern hominoid reference groups, measurements from regions of the skull (such as the neurocranium) that are not subject to masticatory strain tend to have low CVs (Wood and Lieberman, 2001). So, the variation expressed at Dmanisi is unexpected, under the assumption that this sample is composed of individuals belonging to a single species. Craniofacial traits exhibiting substantial variation are supraorbital torus thickness (V* = 14.1) and cheek height (V* = 9.8). Both of these dimensions also show high intragroup variability in other ancient *Homo* taxa. This is predictable, given the finding that facial structures subject to high magnitudes of mechanical strain tend to be more variable than the neurocranium and skull base (Wood and Lieberman, 2001). Indeed, cheek height is particularly variable in recent humans. Conversely, biorbital breadth, nasal breadth, and several angles measuring subnasal protrusion in the Dmanisi faces seem to be less variable than anticipated, probably because the number of specimens is quite small.

Evidence for One Taxon at Dmanisi

If the Dmanisi hominins exhibit a pattern of intragroup variation that does not conform to what has been observed in extant reference samples, and particularly if some measurements predicted to have low variability are associated instead with relatively high values of V*, there are several reasonable explanations. Dimensions of the vault and skull base are usually held to have high taxonomic valence and to be useful in testing hypotheses about heterogeneity in samples. Therefore, greater than expected variation in neurocranial size or shape may constitute evidence that more than one species is documented in the Dmanisi assemblage. Alternatively, ancient *Homo* in the Caucasus may display a pattern of craniofacial variation different from that characteristic of modern humans and apes. A third possibility is simply that the Dmanisi paleodeme is so small that CVs cannot be assessed reliably.

At Dmanisi, as in other situations where only a few fossils have been recovered, statistical problems associated with sampling can never be ruled out. Also, our understanding of variation in the skulls and postcranial skeletons of fossil hominins is still quite incomplete, and it may be the case that some early groups exhibited levels of sex dimorphism or inter-individual variation not recorded in recent primates. However, the totality of evidence considered here does not support an argument for two species at Dmanisi. Neither cranial volume nor dimensions of the face exceed the amount of intraspecies variation expected on the basis of appropriate comparisons. To date, our detailed anatomical studies have suggested that the Dmanisi skulls share a common bauplan and can be accommodated in one taxon (Lordkipanidze et al., 2006; Rightmire et al., 2006; see also Baab, 2008a). Bootstrap analyses of craniofacial and mandibular measurements indicate that size differences at Dmanisi may be large in comparison to modern human and chimpanzee standards but not significantly greater than in other ape reference groups (Macaluso, 2006; Van Arsdale, 2006; Rightmire et al., 2008).

Dmanisi – *Homo habilis*?

In some respects, the Dmanisi skulls resemble *Homo habilis* (Rightmire et al., 2006). Similarities to *Homo rudolfensis* have also been noted by de Lumley et al. (2006), although these authors refer the Dmanisi assemblage to a new species (*Homo georgicus*). The average endocranial volume (663 cm^3) is slightly greater than the mean (610 cm^3) reported for *Homo habilis* by Wood (1992), but individual values for D2280, D2282, D2700, and D3444 fall near the lower limit of the range observed for *Homo erectus* ($\overline{x}$ = 975 cm^3) (Rightmire, 2004). In an evolutionary context, absolute size should be considered in relation to body mass. Encephalization quotients (EQ) are sensitive to assumptions about scaling, and there is always variation within and between populations. Nevertheless, EQ remains a useful way of quantifying differences among hominin taxa. For the four Dmanisi specimens, EQ calculated from average brain mass and body weight following Ruff et al. (1997) and Martin (1981) is 2.8–2.9 (2.9–3.0 if the D2700 subadult is omitted). The corresponding figure for KNM-WT 15000 is 3.28 (closer to 3.0 if there is no correction for brain growth). The EQ given for one possible hypodigm of *Homo habilis* by McHenry and Coffing (2000) is 3.6. However, estimates of body size for this species are probably too low, and this bias will increase measures of relative brain size. In any case, it is evident that the Dmanisi population is marginally less encephalized than early African *Homo erectus*.

As with *Homo habilis*, the supraorbital tori are well defined but not greatly thickened, and there is very marked postorbital constriction. The mastoid region is inflated and laterally projecting. There may be little or no expression of a transverse torus on the occiput. Also, D2700 displays orbital proportions, nasal bone shape, a midfacial profile (the "nasocanine contour" of Kimbel et al., 1984), and a forward sloping maxillary clivus that are similar to the condition in KNM-ER 1813. The zygomaticoalveolar crest is curved to form an incisure as in *Homo habilis, Homo erectus* and most (but not all) later hominins. The hard palate is shallow rather than deeply arched. The well-preserved palate of D2700 is again like that of KNM-ER 1813. The Dmanisi mandibles show a thickened, vertical symphysis with little expression of a mental eminence, and internal transverse tori.

This comparative anatomical evidence must be weighed carefully. Many of the resemblances to *Homo habilis* appear to be either size related, or primitive in the sense that they are common not only to earlier *Homo* but also species of *Australopithecus* and/or extant apes (Table 5.2). By themselves, these similarities do not support (or rule out) a close link to the hypodigm containing KNM-ER 1813 and OH 13. Such traits may serve to situate the Dmanisi population near the stem of the *Homo* clade, and they are a useful part of any species definition. But plesiomorphies will not help in answering key questions about phylogenetic affinities. Characters that have changed during evolution provide the information that is needed to determine how the Dmanisi paleodeme is related to other groups of hominins.

Table 5.2 States for characters identified in the Dmanisi crania and mandibles

Plesiomorphies for *Homo*
– Cranial capacity low (600–775 cm^3)
– Frontal narrowing pronounced (postorbital constriction index 68.7–74.4)
– Mastoid region inflated and laterally projecting
– Occipital transverse torus poorly expressed or absent
– Occipital scale index low (85.0–102.1)
– Midfacial contour projecting
– Canine jugum prominent and bounded laterally by maxillary sulcus
– Zygomaticoalveolar incisure present
– Hard palate shallow
– Palatal opening to incisive canal situated posteriorly
– Mental eminence slight or absent
– Symphysis buttressed internally (superior transverse torus)
Synapomorphies with *Homo erectus* (*sensu lato*)
– Supraorbital torus bar-like and projecting
– Bregmatic eminence with parietal sagittal keel
– Angular torus present
– Temporal squama low with straight upper border
– Mastoid tip inturned and flattened posteriorly
– Petrotympanic angle reduced (140–150^0)
– Petrous pyramid smooth ("dense") in appearance
– Foramen lacerum restricted
– Nasal saddle prominent
Possible synapomorphies with Asian *Homo erectus*
– Parasagittal flattening (depression) of parietal surfaces
– Paramastoid and occipitomastoid crests present
Possible autapomorphies in the Dmanisi paleodeme
– Sagittal keel double
– Tympanic plate relatively delicate
– Supratubarius process absent
– M3 reduced in size

Dmanisi – *Homo erectus*?

There are numerous resemblances to *Homo erectus*. These include the low cranial profile, flattened frontal, sagittal keeling, reduced width of the parietal vault in relation to the cranial base, cresting at the parietal angle and mastoid region, shape of the temporal squama, angled occiput (D2280,

D3444), depth and architecture of the mandibular fossa, and orientation of the petrous axis. In the elevation of the nasal saddle, lack of surface relief on the nasal sill, and posterior location of the palatal incisive canal, the facial skeleton is also like that of *Homo erectus*. Additional similarities include the shape of the mandibular corpus and the occurrence of multiple mental foramina (in D211 and D2600). Most measurements of tooth size place the Dmanisi hominins within the range observed for *Homo erectus*.

Once again, not all of these characters are equally useful. Some are distributed widely in other taxa and thus have descriptive value without being diagnostic. Aspects of facial morphology show variation within groups of *Homo erectus*, most probably as a consequence of sex dimorphism, and traits that are highly variable pose problems for alpha taxonomy. Also, polarities are not easily determined. Despite the patchy nature of the fossil record, it can be ascertained that several of the features linking the Dmanisi population with *Homo erectus* are probably primitive retentions. This caution applies to the mandibular fossa, for example. The proportions of this cavity, the hollowed articular "tubercle" and the flattened postglenoid process are alike in *Homo erectus* and earlier *Homo*. The same can be said of the canine jugum and the associated maxillary sulcus.

Other characters described for the Dmanisi hominins are more clearly diagnostic for African or Asian *Homo erectus* (Table 5.2). The supraorbital tori of D2280 and D3444 differ from those of *Homo habilis* and are best matched in populations of *H. erectus*. An eminence at bregma accompanied by sagittal keeling on the parietals seems to be a derived condition. A low temporal squama with a straight upper border passing downward toward asterion is a consistent marker for the species. A prominent supramastoid crest coupled with a mastoid process that is inturned at its tip and flattened posteriorly so as to make this face coplanar with the adjacent nuchal area of the occipital may also be diagnostic. Although there is variation among the Dmanisi specimens, the D2280 occiput is flexed to about the same extent seen in African and Asian populations. Prominence of the nasal saddle is characteristic of *Homo erectus* and later hominins.

Variation, Systematics, Biogeography and the Evolution of Ancient *Homo*

Geology and taphonomy, along with anatomical and metric studies of the Dmanisi fossils, support the hypothesis that just one species is present at this Caucasus locality. There are differences among the individuals, and indeed the amount of cranial and mandibular size variation is substantial. Certain dimensions of the braincase show high CVs, even though it has been demonstrated on the basis of modern reference populations that the neurocranium and cranial base are associated with low relative variation (Wood and Lieberman, 2001). At the same time, neither gross cranial capacity nor measures of the facial skeleton appear to be more variable than expected, when the Dmanisi sample is compared to other groups. From these findings, it may be inferred that the Caucasus population is characterized by a level of (sex?) dimorphism at least as great as that seen in humans and chimpanzees, and probably also a pattern of craniofacial variation differing from that in modern reference taxa.

Dmanisi and Hominin Taxonomy

Given the nature/extent of morphological diversity documented at Dmanisi, it is to be anticipated that hominin assemblages containing both robust and also lightly constructed individuals, differing in supraorbital prominence and vault dimensions as well as facial proportions, may be encountered in the ancient record. Here, it is important to sound a note of caution. Particularly where samples are limited and/or composed of fragmentary specimens, the sorting process will not be straightforward. There is a real risk of assigning individuals to separate populations, when in fact the level and pattern of variation are consistent with what may be expected within a single species. An obvious example is afforded by the finds from Olduvai and Koobi Fora that have been referred to *Homo habilis* and *Homo rudolfensis*. Here, only a few fossils are relatively complete, and there is a long history of disagreement as to how they should be identified. The evidence has been read to support different species definitions based on varying hypodigms, but a fair assessment is that the information available is insufficient for this purpose. For the moment, it is difficult to offer any final resolution of the taxonomy for early *Homo*.

The Dmanisi skulls display some traits that are shared with early *Homo* sp., along with others that are more clearly diagnostic for *Homo erectus*. Given the morphology of the cranial base and face, it is appropriate to refer the Dmanisi assemblage to *Homo erectus* (contra de Lumley et al., 2006). However, it should be recognized that this decision has the effect of expanding the species hypodigm, by adding adult crania (D2282, D3444) that have low capacities, and a subadult (D2700) that resembles KNM-ER 1813. Such a change will alter the definition of *Homo erectus* and erode the distinction between *erectus* and other early taxa. Thus, as new discoveries are made, there will be a question as to whether small skulls should be identified as *Homo habilis*, or placed instead within the range of variation accepted for *Homo erectus*. Precisely this issue has come up, in respect to a diminutive cranium from Ileret. Recently, Spoor et al. (2007) have elected to group KNM-ER 42700 with *Homo erectus*, although Baab (2008b) has suggested otherwise.

African or Asian Origin for *Homo erectus*?

The Caucasus hominins differ from both African and Far Eastern conspecifics. Although there are overall resemblances to KNM-ER 3733 and other specimens from Koobi Fora, aspects of vault shape and facial morphology set these groups apart. Similarities to fossils from Sangiran (Java) are less numerous. Also, the Caucasus population presents a few characters that appear to be unique (Rightmire et al., 2006). These findings are in accord with the view that *Homo erectus* was a geographically dispersed, polytypic species. One view that has been widely favored holds this species to have originated in eastern Africa. According to this hypothesis, a population of *Homo habilis*, or *Homo rudolfensis*, gave rise to *Homo erectus*, as suggested by the time-stratigraphic sequences at Olduvai Gorge and in the Turkana Basin. Groups of *Homo erectus* then ventured out of Africa, leaving abundant traces of their passing in the Jordan Valley, and to the north, in the Georgian Caucasus. From sites such as Dmanisi, the hominins could presumably have spread westward into Europe and also across southern Asia to the Far East. This scenario implies that differences between African *Homo erectus* and the Dmanisi fossils reflect geographic distance, adaptation to new environments in western Asia, or drift in small isolates.

Problems with such an "African origins" hypothesis have been noted, and it is increasingly clear that alternatives must be explored (see Dennell and Roebroeks, 2005 for a review). As has been emphasized, the Georgian crania are small and lack strong crests or tori. There are numerous resemblances to *Homo habilis* from East Africa. Skulls presently included within this latter hypodigm constitute plausible structural antecedents to *Homo erectus* (Lieberman et al., 1996; Strait et al., 1997; Kimbel et al., 2004). Therefore, it can be argued that a population composed of such small-brained and lightly built individuals was ancestral to the Dmanisi hominins. In this view, early (pre-*erectus*) *Homo* dispersed from Africa into western Asia, sometime prior to 1.8 Ma. The travelers were able to make simple chopping tools and utilize flakes and cobbles to obtain meat and marrow from animal carcasses. These skills would have been important to survival at higher latitudes, especially during the cold winters, when plant foods were probably scarce (Lordkipanidze et al., 2006). There is presently little hard evidence to support such a claim, as fossils and stone artifacts of the requisite age have not yet been documented unequivocally in the Levant or in Arabia. Nevertheless, a *H. habilis*-like founding population could later have evolved the anatomical bauplan seen at Dmanisi.

This "Asian origins" hypothesis fits comfortably within the constraints imposed by geochronology. Dates for Koobi Fora (Feibel et al., 1989; Gathogo and Brown, 2006) allow the possibility that *Homo erectus* evolved in western Eurasia and only later reached Africa, where the species is sampled in the Turkana Basin, Olduvai Gorge, Konso in southern Ethiopia, Bouri (Daka) in the Middle Awash region, and Buia in the Eritrean Danakil Depression. Here again, it is reasonable to suppose that the Dmanisi paleodeme is related to the populations of the Far East. Dates emerging from fieldwork at Sangiran and in the Nihewan Basin of China show that these areas were inhabited 1.7–1.6 Ma ago by hominins who must have moved through the southern parts of Asia (Swisher et al., 1998; Larick et al., 2001; Zhu et al., 2004). Here caution is appropriate, however. While there is clear evidence for a ca. 1.8–1.7 Ma human presence at Dmanisi, there is no certainty that people managed to colonize this region on a long-term basis. Indeed, it seems likely that many of the earliest dispersals eastward into Asia resulted in occupations that were ephemeral, and the Early Pleistocene record does not document any continuity of populations through southern Asia to the Far East (Dennell, 2003).

Summary

Fossils differing from australopiths appear in the Turkana Basin and in the Middle Awash region ca. 2.3 Ma ago, and it is very likely that *Homo* evolved in Africa. The earliest representatives of our genus are still poorly documented. Following the announcement of *Homo habilis* in 1964, a number of specimens from Olduvai and other localities have been referred to this species. Also, it has been argued that a second, larger-brained taxon is present in the record. Unfortunately, most of the remains are incomplete or damaged. It has proved especially difficult to define *Homo rudolfensis*, as the proposed hypodigm includes only one intact cranium, and nothing is known of the postcranial skeleton. Whether one or two species should be recognized, and whether they are sufficiently like later humans to merit placement within *Homo*, are questions on which there is no clear consensus.

Discoveries from Dmanisi add important new information. The more complete skulls making up this western Asian assemblage demonstrate that substantial morphological variation is present within a single hominin paleodeme. Thus the level and pattern of variation in ancient populations may not always conform to what is expected from studies of modern reference samples. This finding further clouds the issue of sorting the fossils from Olduvai and Koobi Fora. If anything, it is now less clear than before how intragroup variation is to be partitioned from differences that mark species boundaries. At present, the evidence available is simply not good enough to provide a final resolution of the taxonomy of early *Homo*.

The Dmanisi population can be referred to *Homo erectus*. Nevertheless, the Caucasus hominins differ from their African and Far Eastern conspecifics, and it is apparent that they share primitive features with small-brained, gracile skulls such as KNM-ER 1813 from Koobi Fora and OH 13 from Olduvai. This morphology suggests that the Dmanisi individuals had a *habilis*-like ancestor, and it is possible to argue that an early *Homo* population was the first to disperse from eastern Africa. *Homo erectus* might then have originated in western Asia, and only later evolved the larger brain and greater body size that characterize African and East Asian representatives of the species. This "Asian origins" hypothesis differs from the generally accepted view that *Homo erectus* evolved in Africa, before dispersing into other regions of the Old World. Apart from the Dmanisi discoveries, there is currently little hard evidence on which to base such an alternative, but further fieldwork and analysis of the fossils should be designed to refine and test these biogeographic hypotheses.

Acknowledgments We thank all of the students and professionals who have joined in the excavations and worked in the laboratory at Dmanisi. For assistance with preparation of the hominin fossils at the Georgian National Museum and guidance in our analyses, we are indebted to M. Bukhsianidze, T. Jashashvili, G. Kiladze and A. Vekua. We much appreciate the opportunity to attend the Stony Brook workshop, organized by Richard Leakey, John Fleagle and Frederick Grine. G.P.R. thanks the American Philosophical Society, the Eckler Fund of Binghamton University, the Leakey Foundation and the National Science Foundation. D.L. acknowledges support from the Leakey Foundation and the National Geographic Society.

References

Aiello, L., Dean, M.C., 1990. An Introduction to Human Evolutionary Anatomy. Academic, London.

Aiello, L., Wheeler, P., 1995. The expensive tissue hypothesis: the brain and digestive system in human and primate evolution. Current Anthropology 36, 199–221.

Antón, S., 2003. Natural history of *Homo erectus*. Yearbook of Physical Anthropology 46, 126–170.

Asfaw, B., White, T., Lovejoy, O., Latimer, B., Simpson, S., Suwa, G., 1999. *Australopithecus garhi*: a new species of early hominid from Ethiopia. Science 284, 629–635.

Asfaw, B., Gilbert, W.H., Beyene, Y., Hart, W.K., Renne, P.R., WoldeGabriel, G., Vrba, E., White, T.D., 2002. Remains of *Homo erectus* from Bouri, Middle Awash, Ethiopia. Nature 416, 317–320.

Baab, K.L., 2008a. The taxonomic implications of cranial shape variation in *Homo erectus*. Journal of Human Evolution 54, 827–847.

Baab, K.L., 2008b. A re-evaluation of the taxonomic affinities of the early *Homo* cranium KNM-ER 42700. Journal of Human Evolution 55, 741–746.

Blumenschine, R.J., Peters, C.R., Masao, F.T., Clarke, R.J., Deino, A.L., Hay, R.L., Swisher, C.C., Stanistreet, I.G., Ashley, G.M., McHenry, L.J., Sikes, N.E., van der Merwe, N.J., Tactikos, J.C., Cushing, A.E., Deocampo, D.M., Njau, J.K., Ebert, J. I., 2003. Late Pliocene *Homo* and hominid land use from western Olduvai Gorge, Tanzania. Science 299, 1217–1221.

Bramble, D.M., Lieberman, D.E., 2004. Endurance running and the evolution of *Homo*. Nature 432, 345–352.

Bromage, T.G., Schrenk, F., Zonnefeld, F.W., 1995. Paleoanthropology of the Malawi Rift: an early hominid mandible from the Chiwondo Beds, northern Malawi. Journal of Human Evolution 28, 71–108.

Brown, B., Walker A., 1993. The dentition. In: Walker, A., Leakey, R. (Eds), The Nariokotome *Homo erectus* Skeleton. Harvard University Press, Cambridge, pp. 161–192.

Dean, M.C., Smith, B.H., 2009. Growth and development of the Nariokotome youth, KNM-WT 15000. In: Grine, F.E., Fleagle, J.G., Leakey, R.E. (Eds), The First Humans: Origin and Early Evolution of the Genus *Homo*. Springer, Dordrecht, pp. 101–120.

de Lumley, M.A., Gabunia, L., Vekua, A., Lordkipanidze, D., 2006. Les restes humains du Pliocène final et du début du Pléistocène inférieur de Dmanissi, Géorgie (1991–2000) I – Les cranes, D2280, D2282, D2700. L'Anthropologie 110, 1–110.

Dennell, R., 2003. Dispersal and colonization, long and short chronologies: how continuous is the Early Pleistocene record for hominids outside East Africa? Journal of Human Evolution 45, 421–440.

Dennell, R., Roebroeks, W., 2005. An Asian perspective on early human dispersal from Africa. Nature 438, 1099–1104.

Feibel, C.S., Brown, F.H., McDougall, I., 1989. Stratigraphic context of fossil hominids from the Omo Group deposits, northern Turkana Basin, Kenya and Ethiopia. American Journal of Physical Anthropology 78, 595–622.

Gabunia, L., de Lumley, M.A., Vekua, A., Lordkipanidze, D., de Lumley, H., 2002. Découverte d'un nouvel hominidé a Dmanissi (Transcaucasie, Géorgie). Comptes Rendus Palevol 1, 242–253.

Gathogo, P.N., Brown, F.H., 2006. Revised stratigraphy of Area 123, Koobi Fora, Kenya, and new age estimates of its fossil mammals, including hominins. Journal of Human Evolution 51, 472–479.

Grine, F.E., Smith, H.F., Heesy, C.P., Smith, E.J., 2009. Phenetic affinities of Plio-Pleistocene *Homo* fossils from South Africa: molar cusp proportions. In: Grine, F.E., Fleagle, J.G., Leakey, R.E. (Eds), The First Humans: Origin and Early Evolution of the Genus *Homo*. Springer, Dordrecht, pp. 49–62.

Groves, C.P., 1989. A Theory of Primate and Human Evolution. Oxford University Press, Oxford.

Haeusler, M., McHenry, H.M., 2004. Body proportions of *Homo habilis* reviewed. Journal of Human Evolution 46, 433–465.

Howell, F.C., 1978. Hominidae. In: Maglio, V.J., Cooke, H.B.S. (Eds), Evolution of African Mammals. Harvard University Press, Cambridge, MA, pp. 154–248.

Howell, F.C., 1999. Paleo-demes, species clades, and extinctions in the Pleistocene hominin record. Journal of Anthropological Research 55, 191–243.

Johanson, D.C., White, T.D., 1979. A systematic assessment of early African hominids. Science 202, 321–330.

Johanson, D.C., Masao, F.T., Eck, G.G., White, T.D., Walter, R.C., Kimbel, W.H., Asfaw, B., Manega, P., Ndessokia, P., Suwa, G., 1987. New partial skeleton of *Homo habilis* from Olduvai Gorge, Tanzania. Nature 327, 205–209.

Kimbel, W.H., 2009. The origin of *Homo*. In: Grine, F.E., Fleagle, J.G., Leakey, R.E. (Eds), The First Humans: Origin and Early Evolution of the Genus *Homo*. Springer, Dordrecht, pp. 31–37.

Kimbel, W.H., White, T.D., Johanson, D.C., 1984. Cranial morphology of *Australopithecus afarensis*: a comparative study based on a composite reconstruction of the adult skull. American Journal of Physical Anthropology 64, 337–388.

Kimbel, W.H., Walter, R.C., Johanson, D.C., Reed, K.E., Aronson, J.L., Assefa, Z., Marean, C.W., Eck, G.G., Bobe, R., Hovers, E., Rak, Y., Vondra, C., Yemane, T., York, D., Chen, Y., Evensen, N.M., Smith, P.E., 1996. Late Pliocene *Homo* and Oldowan tools from the

Hadar Formation (Kada Hadar Member), Ethiopia. Journal of Human Evolution 31, 549–561.

Kimbel, W.H., Johanson, D.C., Rak, Y., 1997. Systematic assessment of a maxilla of *Homo* from Hadar, Ethiopia. American Journal of Physical Anthropology 103, 235–262.

Kimbel, W.H., Rak, Y., Johanson, D.C., 2004. The Skull of *Australopithecus afarensis*. Oxford University Press, Oxford.

Larick, R., Ciochon, R.L., Zaim, Y., Sudijono, Suminto, Rizal, Y., Aziz, F., Reagan, M., Heizler, M., 2001. Early Pleistocene $^{40}Ar/^{39}Ar$ ages for Bapang Formation hominins, Central Jawa, Indonesia. Proceedings of the National Academy of Sciences 98, 4866–4871.

Leakey, L.S.B., Tobias, P.V., Napier, J.R., 1964. A new species of the genus *Homo* from Olduvai Gorge. Nature 202, 7–9.

Leakey, M.G., Spoor, F., Brown, F.H., Gathogo, P.N., Leakey, L.N., 2003. A new hominin calvaria from Ileret (Kenya). American Journal of Physical Anthropology S36, 136.

Lee, S.H., 2005. Is variation in the cranial capacity of the Dmanisi sample too high to be from a single species? American Journal of Physical Anthropology 127, 263–266.

Le Gros Clark, W.E., 1964. The Fossil Evidence for Human Evolution. University of Chicago Press, Chicago, IL.

Lewis, O.J., 1989. Functional Morphology of the Evolving Hand and Foot. Oxford University Press, Oxford.

Lieberman, D.E., Wood, B.A., Pilbeam, D.R., 1996. Homoplasy and early *Homo*: an analysis of the evolutionary relationships of *Homo habilis sensu stricto* and *H. rudolfensis*. Journal of Human Evolution 30, 97–120.

Lordkipanidze, D., Vekua, A., Ferring, R., Rightmire, G.P., Agusti, J., Kiladze, G., Mouskhelishvili, A., Nioradze, M., Ponce de Leon, M., Tappen, M., Zollikofer, C., 2006. A fourth hominin skull from Dmanisi, Georgia. Anatomical Record 288A, 1146–1157.

Lordkipanidze, D., Jashashvili, T., Vekua, A., Ponce de Leon, M.S., Zollikofer, C.P.E., Rightmire, G.P., Pontzer, H., Ferring, R., Oms, O., Tappen, M., Bukhsianidze, M., Agusti, J., Kahlke, R., Kiladze, G., Martinez-Navarro, B., Mouskhelishvili, A., Nioradze, M., Rook, L., 2007. Postcranial evidence from early *Homo* from Dmanisi, Georgia. Nature 449, 305–310.

Macaluso, P.J., 2006. Descriptions and comparative studies of the hominin dental remains from Dmanisi, Georgia: 1991–2002 collections. Ph.D. dissertation, Binghamton University (SUNY).

Martin, R.D., 1981. Relative brain size and basal metabolic rate in terrestrial vertebrates. Nature 293, 57–60.

McHenry, H.M., Coffing, K., 2000. *Australopithecus* to *Homo*: transformations in body and mind. Annual Review of Anthropology 29, 125–146.

Potts, R., Behrensmeyer, A.K., Deino, A., Ditchfield, P., Clark, J., 2004. Small mid-Pleistocene hominin associated with East African Acheulean technology. Science 305, 75–78.

Prat, S., Brugal, J.P., Tiercelin, J.J., Barrat, J.A., Bohn, M., Delagnes, A., Harmand, S., Kimeu, K., Kibunjia, M., Texier, P.J., Roche, H., 2005. First occurrence of early *Homo* in the Nachukui Formation (West Turkana, Kenya) at 2.3–2.4 Myr. Journal of Human Evolution 49, 230–240.

Rightmire, G.P., 1990. The Evolution of *Homo erectus*. Comparative Anatomical Studies of an Extinct Human Species. Cambridge University Press, Cambridge.

Rightmire, G.P., 1993. Variation among early *Homo* crania from Olduvai Gorge and the Koobi Fora region. American Journal of Physical Anthropology 90, 1–33.

Rightmire, G.P., 2004. Brain size and encephalization in early to mid-Pleistocene *Homo*. American Journal of Physical Anthropology 106, 61–85.

Rightmire, G.P. Lordkipanidze, D., Vekua, A., 2006. Anatomical descriptions, comparative studies and evolutionary significance of the hominin skulls from Dmanisi, Republic of Georgia. Journal of Human Evolution 50, 115–141.

Rightmire, G.P., Van Arsdale, A.P., Lordkipanidze, D., 2008. Variation in the mandibles from Dmanisi, Georgia. Journal of Human Evolution 54, 904–908.

Robinson, J.T., 1967. Variation and the taxonomy of the early hominids. In: Dobzhansky, T., Hecht, M.K., Steere, W. (Eds), Evolutionary Biology. Appleton-Century-Crofts, New York, pp. 69–99.

Roche, H., Blumenschine, R.J, Shea, J.J., 2009. Origins and adaptations of early genus *Homo*: what archeology tells us. In: Grine, F.E., Fleagle, J.G., Leakey, R.E. (Eds), The First Humans: Origin and Early Evolution of the Genus *Homo*. Springer, Dordrecht, pp. 135–147.

Ruff, C.B., Walker, A., 1993. Body size and body shape. In: Walker, A., Leakey, R. (Eds), The Nariokotome *Homo erectus* Skeleton. Harvard University Press, Cambridge, MA, pp. 234–265.

Ruff, C.B., Trinkaus, E., Holliday, T.W., 1997. Body mass and encephalization in Pleistocene *Homo*. Nature 387, 173–176.

Schwartz, J.H., Tattersall, I., 2002. The Human Fossil Record. Volume 1. Terminology and Craniodental Morphology of Genus *Homo* (Europe). Wiley-Liss, New York.

Skinner, M.M., Gordon, A.D., Collard, N.J., 2006. Mandibular size and shape variation in the hominins at Dmanisi, Republic of Georgia. Journal of Human Evolution 51, 36–49.

Sokal, R.R., Braumann, C.A., 1980. Significance tests for coefficients of variation and variability profiles. Systematic Zoology 29, 50–66.

Spoor, F., Leakey, M.G., Gathogo, P.N., Brown, F.H., Antón, S.C., McDougall, I., Kairie, C., Manthi, F.K., Leakey, L.N., 2007. Implications of new early *Homo* fossils from Ileret, east of Lake Turkana, Kenya. Nature 448, 688–891.

Strait, D.S., Grine, F.E., Moniz, M.A., 1997. A reappraisal of early hominid phylogeny. Journal of Human Evolution 32, 17–82.

Susman, R.L., Stern, J.T., 1982. Functional morphology of *Homo habilis*. Science 217, 931–934.

Suwa, G., Asfaw, B., Haile-Selassie, Y., White, T., Katoh, S., WoldeGabriel, G., Hart, W.K., Nakaya, H., Beyene, Y., 2007. Early Pleistocene *Homo erectus* fossils from Konso, southern Ethiopia. Anthropological Science 115, 133–151.

Swisher, C.C., Scott, G.R., Curtis, G.H., Butterworth, J., Antón, S.C., Jacob, T., Suprijo, A., Widiasmoro, Sukandar, Koeshardjono, 1998. The Antiquity of *Homo erectus* in Java: $^{40}Ar/^{39}Ar$ Dating and Paleomagnetic Study of the Sangiran Area. Dual Congress 1998 (abstract), Sun City.

Tobias, P.V., 1967. Olduvai Gorge. Volume 2. The Cranium of *Australopithecus (Zinjanthropus) boisei*. Cambridge University Press, Cambridge.

Tobias, P.V., 1991. Olduvai Gorge. Volume 4. The Skulls, Endocasts and Teeth of *Homo habilis*. Cambridge University Press, Cambridge.

Tobias, P.V., 2009. *Homo habilis*: a premature discovery remembered by one of its founding fathers, 42 years later. In: Grine, F.E., Fleagle, J.G., Leakey, R.E. (Eds), The First Humans: Origin and Early Evolution of the Genus *Homo*. Springer, Dordrecht, pp. 7–15.

Van Arsdale, A., 2006. Mandibular variation in early *Homo* from Dmanisi, Georgia. Ph.D. dissertation, University of Michigan.

Wood, B., 1991. Koobi Fora Research Project. Volume 4. Hominid Cranial Remains. Clarendon Press, Oxford.

Wood, B., 1992. Origin and evolution of the genus *Homo*. Nature 355, 783–790.

Wood, B, 2009. Where does the genus *Homo* begin, and how would we know? In: Grine, F.E., Fleagle, J.G., Leakey, R.E. (Eds), The First Humans: Origin and Early Evolution of the Genus *Homo*. Springer, Dordrecht, pp. 17–28.

Wood, B., Collard, M., 1999. The human genus. Science 284, 65–71.

Wood, B., Lieberman, D.E., 2001. Craniodental variation in *Paranthropus boisei*: a developmental and functional perspective. American Journal of Physical Anthropology 116, 13–25.

Zhu, R., Potts, R., Xie, F., Hoffman, K.A., Deng, C.L., Shi, C.D., Pan, Y.X., Wang, H.Q., Shi, R.P., Wang, Y.C., Shi, G.H., Wu, N.Q., 2004. New evidence on the earliest human presence at high northern latitudes in northeast China. Nature 431, 559–562.

Chapter 6
Phenetic Affinities of Plio-Pleistocene *Homo* Fossils from South Africa: Molar Cusp Proportions

Frederick E. Grine, Heather F. Smith, Christopher P. Heesy and Emma J. Smith

Keywords Taxonomy • molar morphology • *Homo habilis* • *Homo erectus* • South Africa • Sterkfontein • Swartkrans • Drimolen

Introduction

As aptly observed by Aiello et al. (2000), any attempt to reconstruct human evolutionary history depends upon reliable hypotheses pertaining to the species groups that are represented in the hominin fossil record. There has been a notable lack of consensus regarding the species of *Homo* that are represented in the Plio-Pleistocene karst cave deposits of South Africa (Howell, 1978; Clarke, 1977a, b, 1985a, b, 1994; Chamberlain, 1987; Rightmire, 1990; Wood, 1991; Tobias, 1991; Kimbel and Rak, 1993; Grine et al., 1993, 1996; Ahern, 1998; Kuman and Clarke, 2000; Grine, 2001, 2005; Curnoe, 2001, 2002, 2008; Prat, 2002; Dunsworth and Walker, 2002; Schwartz and Tattersall, 2003; Curnoe and Tobias, 2006; Smith and Grine, 2008). To some degree, this is because these discussions have been restricted to a few incomplete mandibles and crania from the sites of Sterkfontein and Swartkrans. However, there are larger collections of teeth from these and other sites that have been attributed to this genus and which might shed light on this issue. We here attempt to assess the phenetic affinities of some of these dental remains by comparison with penecontemporaneous East African fossils that have been referred to *H. rudolfensis*, *H. habilis* and *H. erectus* (=*H. ergaster*).

Overall crown size and shape may sometimes serve to distinguish the molar teeth *Homo* from those of *Australopithecus* and *Paranthropus*, but it is not clear that these attributes are necessarily useful in differentiating among species of *Homo*. At the same time, however, the utility of molar cusp sizes and proportions in species-level distinctions has been explored by a number of workers (Corruccini, 1977; Lavelle, 1978; Hills et al., 1983; Hartman, 1989; Uchida, 1991, 1992, 1998a, b; Wood and Xu, 1991; Macho and Moggi-Cecchi, 1992; Matsumura et al., 1992; Smith, 1999; Bailey, 2004). Such data have been shown to be of some use in taxonomic evaluations of the hominin fossils from East Africa (Wood et al., 1983; Wood and Uytterschaut, 1987; Wood and Engleman, 1988; Suwa, 1988, 1990; Suwa et al., 1994, 1996).

The attribution of some of the South African fossils to *Homo*, *Australopithecus* or *Paranthropus* has been a matter of disagreement. The notion that the Swartkrans specimens SK 47, SK 843, SK 846 and SKX 4446 represent *Homo* rather than *Paranthropus* (Olson, 1978; Schwartz and Tattersall, 2003) has been effectively dismissed (Dean and Wood, 1982; Grine, 2005). Similarly, Braga and Thackeray (2003) have posited that the Kromdraai specimen KB 5223 represents *Homo* rather than *Paranthropus*, but their analysis simply confirms that the teeth of some *P. robustus* specimens are smaller than those of others. Indeed, the morphological attributes of KB 5223 cited in support of its attribution to *Homo* are exhibited by other Kromdraai *P. robustus* specimens (e.g., TM 1536) (Grine, 1982), and while some of the features relating to its molar enamel development may be unique, its evenly spaced incisor perikymata are characteristic of *Paranthropus* rather than *Homo* (Lacruz, 2007). The attribution of other fossils, such as the Sts 19 cranial base from Sterkfontein, continues to be debated without satisfactory resolution (Kimbel and Rak, 1993; Ahern, 1998). This specimen was attributed to *Australopithecus africanus* by

F.E. Grine (✉)
Departments of Anthropology and Anatomical Sciences,
Stony Brook University, Stony Brook, NY 11794-4364, USA
e-mail: Frederick.Grine@stonybrook.edu

H.F. Smith
School of Human Evolution and Social Change,
Arizona State University, P.O. Box 872402,
Tempe, AZ 85287-2402, USA
e-mail: Heather.F.Smith@asu.edu

C.P. Heesy
Department of Anatomy, Arizona College of Osteopathic Medicine,
Midwestern University, 19555 North 59th Avenue, Glendale,
AZ 85308, USA
e-mail: metacephalon@gmail.com

E.J. Smith
Pharmaceutical Product Development, Inc., 929 North Front Street,
Wilmington, NC 28401, USA
e-mail: Emma.Smith@wilm.ppdi.com

Broom and Robinson (1950), who noted that it attested to a "considerable degree of variation" in that taxon. Following Kimbel and Rak (1993), Strait et al. (1997; Strait and Grine, 2001) included Sts 19 in *H. habilis*, but it alone accounts for over a third of the variable characters in their hypodigm of that species, and in three characters it differs from all East African *H. habilis* fossils (Grine, 2001).

Perhaps the most compelling evidence for the presence of *Homo* in the Sterkfontein deposits is the Stw 53 partial cranium. In their initial description of it, Hughes and Tobias (1977) cited five features that they believed provided "convincing evidence that [Stw 53] belonged to the genus *Homo* and either to *H. habilis* or to a form closely related (*H.* aff. *habilis*)." Howell (1978) also assigned Stw 53 to *H. habilis,* and both Clarke (1985a) and Tobias (1991) noted its resemblance to OH 24, a specimen that they attribute to *H. habilis*. Subsequently, Kuman and Clarke (2000) have argued that Stw 53 represents *Australopithecus* because of its small cranial capacity, narrow frontal, flattened nasal skeleton, and its large teeth, with M^3 exceeding M^1 in size. Curnoe and Tobias (2006) have reiterated Tobias's original position that this specimen represents *H. habilis*.

Just over a dozen specimens from Sterkfontein have been attributed to *Homo*, and most of these consist of isolated teeth and small jaw fragments (Moggi-Cecchi et al., 2006; Ungar et al., 2006) (Table 6.1). Tobias (1965) suggested that five isolated teeth (SE 255, SE 1508, SE 1579, SE 1937, SE 2396) excavated from the "Middle Breccia" of the "Extension Site" in the late 1950's by Robinson (1961, 1962) indicated the presence of a hominin that he likened to *H. habilis*. Curnoe and Tobias (2006) have reiterated that assignment for SE 255, but have opined that SE 1508 is attributable to *Homo* aff. *H. sapiens* (*H. sapiens sensu lato* is Curnoe's sometime appellation for *H. erectus*), on the basis of overall crown size and shape. Clarke (1985b: 295) listed several additional hominin fossils from the Sterkfontein Member 5 deposits, noting that among them "Stw 80 and Stw 84 can be assigned to *Homo habilis*, whilst there is no reason to exclude the other remains from that species." Kuman and Clarke (2000) have stated that Stw 80 has "strong similarities" to the SK 15 mandible from Swartkrans Member 2, which Clarke (1994) assigned to *H. ergaster*. Spoor (1993) and Moggi-Cecchi et al. (1998) have suggested that Stw 151 may represent early *Homo*, or at least a species "more derived towards an early *Homo* condition" than *A. africanus*. Finally, Moggi-Cecchi et al. (2006) have described dental remains excavated from Sterkfontein between 1966 and 1996, attributing several to *Homo*, although none was ascribed to a particular species. Curnoe and Tobias (2006) have referred several of these to *H. habilis* and others to *Homo* aff. *H. sapiens* on the basis of overall crown dimensions.

Sixteen craniodental specimens from Swartkrans have been attributed to *Homo* (Table 6.1). Grine (2005) has provided a recent review of the fossil evidence for *Homo* at this

Table 6.1 Craniodental specimens from the South African sites of Sterkfontein, Swartkrans and Drimolen attributable to the genus *Homo*

Specimen		Part	Provenance	
SE	255	dm^1-M^1	Member 5C[a]	**Sterkfontein**
SE	1937	C	Member 5C	
SE	1508	M^2	Member 5C	
SE	1579	M^2	Member 5C	
SE	2396	P^3	Member 5C	
?Sts	19	Cranium	Uncertain	
Stw	19	P^3-P^4, M^2-M^3	Uncertain	
Stw	42	I^2	Member 5C	
Stw	53	Cranium	Member 5A[b]	
Stw	75	I^1–P^3	Member 5A	
Stw	80	Mandible	Member 5C	
Stw	84	Mandible	Member 5C	
?Stw	87	RP_4	Uncertain	
?Stw	151	Skull	Uncertain	
SK	15	Mandible	Member 2	**Swartkrans**
SK	18a	P_3	Member 2	
SK	27	Cranium	Member 1 Hanging Remnant	
SK	45	Mandible	Member 1 Hanging Remnant	
SK	68	I^1	Member 1 Hanging Remnant	
SK	74b	I_1	Member 1 Hanging Remnant	
SK	847	Cranium	Member 1 Hanging Remnant	
SK	2635	C–M^3	Member 1 Hanging Remnant	
SKW	3114	Maxilla	Member 1 Hanging Remnant	
SKX	257	M_1	Member 2	
SKX	267	dm^2, C, M^1	Member 2	
SKX	339	I^1	Member 2	
SKX	610	I^2	Member 2	
SKX	2354	dm^1, I_1–I_2, C	Member 1–2 interface; probably 2	
SKX	21204	I_2,P_4	Member 1 Lower Bank	
DNH	24	di^2	Unstated	**Drimolen**
DNH	35	dm_1–M_1	Unstated	
DNH	45	I^2	Unstated	
DNH	49	di^2	Unstated	
DNH	70/71	I^1, M^1	Unstated	

Individual dental associations at Sterkfontein follow Moggi Cecchi et al. (2006); those at Swartkrans follow Grine (1989); those at Drimolen follow Keyser et al. (2000).

[a] Member 5C (Partridge, 2000) = Member 5 West (Kuman and Clarke, 2000) = Extension Site (Robinson, 1962).

[b] Member 5A (Partridge, 2000) = "Stw 53 Infill" (Kuman and Clarke, 2000).

site, and the taxonomic discussions about these specimens. Apart from two mandibles (SK 15 and SK 45) and two fragmentary crania (SK 27 and SK 847), these specimens comprise isolated teeth or tiny fragments of jaws with teeth. Curnoe and Tobias (2006) have recently attributed SK 27, SK 268 and SKX 339 to *H. habilis* on the basis of overall crown metrics.

Some half dozen specimens (all consisting of isolated teeth) from the site of Drimolen have been tentatively referred to *Homo* by Keyser et al. (2000) (Table 6.1), although evidence for this attribution was provided only for DNH 35. Curnoe and Tobias (2006) have stated that DNH 45 and DNH 70 have affinities with *H. habilis* on the basis of crown size.

Because dental remains constitute the vast majority of South African fossils that have been attributed to *Homo* (Table 6.1), we have chosen to examine molar cusp proportions in an attempt to evaluate their specific affinities. In so doing, we first re-examine the reliability of this type of information to discriminate among samples of extant, large-bodied apes at different levels of taxonomic distinction. We then compare the cusp proportions of the South African early *Homo* fossils with East African homologues attributed to *H. habilis*, *H. rudolfensis*, and *H. erectus* (=*H. ergaster*). This choice of comparators is reasonable not only because various of the South African fossils have been referred to one or another of these species by different workers, but also because the South African specimens are roughly penecontemporaneous with various of these East African *Homo* fossils.

Material and Methods

This study involves the analysis of cusp size data for permanent maxillary and mandibular molars for large samples of four living ape species (*Pongo pygmaeus*, *Gorilla gorilla*, *Pan troglodytes* and *Pan paniscus*), and small samples of fossils attributed to *Homo* from the Plio-Pleistocene deposits of East and South Africa. In all instances, only a single tooth of any one molar type was included for each individual (i.e., antimeres were not averaged).

Extant Ape Samples

The molars of the living ape specimens were measured by one of us (EJS), and the sample and method of data recording are the same as that reported by her elsewhere (Smith, 1999) (Table 6.2). Measurements were recorded from video images of the occlusal surfaces of the molar crowns according to the methodology described by Wood et al. (1983). The majority of the chimpanzee sample is represented by *P. troglodytes troglodytes*, with a smaller number of *P. troglodytes schweinfurthi*; because only the M1 samples for the latter were of adequate size for statistical analysis (n = 10 M^1s and 11 M_1s), we did not analyze this subspecies sample separately. Because our sample contained only a few individuals of *P. troglodytes verus*, it was not treated as a separate entity despite its apparent distinctiveness (Morin et al., 1994). Similarly, the orangutan sample included specimens of *Pongo pygmaeus pygmaeus* and *P. pygmaeus abelii*, which have been argued to warrant specific status (Ruvolo et al., 1994; Steiper, 2006). However, because only the M3 samples for the Sumatran orang were even marginally adequate for separate statistical analysis, we combined the two.

Table 6.2 Numbers of molars of extant ape taxa employed in the present study

Taxon	M^1	M^2	M^3	M_1	M_2	M_3
Gorilla gorilla gorilla	60	65	49	53	56	45
Gorilla gorilla graueri	39	27	28	35	35	33
Gorilla gorilla beringei	35	22	18	28	26	22
Pan paniscus	84	53	20	81	51	23
Pan troglodytes	72	62	51	70	63	43
Pongo pygmaeus	64	55	32	61	55	31

Early Homo *Samples*

The proposal that two taxa, namely *H. habilis* and *H. rudolfensis*, can be distinguished among the fossils traditionally regarded as representing *Homo habilis sensu lato* is not universally accepted (e.g., Suwa et al., 1996; Miller, 1991, 2000; Dunsworth and Walker, 2002; Lee and Wolpoff, 2005). However, we believe that good evidence has been put forward favoring this distinction, regardless of how the hypodigms of these species are constructed (Lieberman et al., 1988; Wood, 1991, 1992, 1993; Rightmire, 1993; Leakey et al., 2001; Blumenschine et al., 2003). By comparison, the proposal that *H. ergaster* should be recognized as a separate entity from *H. erectus* has received notably less support (Rightmire, 1990, 1993, 1998; Bräuer and Mbua, 1992; Kramer, 1993; Walker, 1993, 1994; Bräuer, 1994; Antón, 2002, 2003, 2004; Asfaw et al., 2002; Dunsworth and Walker, 2002; Gilbert et al., 2003; Antón and Swisher, 2004; Kidder and Durband, 2004; Finlayson, 2005).

Thus, for the purposes of the present study, we recognize three species of *Homo* in the Plio-Pleistocene of East Africa, namely *H. habilis*, *H. rudolfensis* and *H. erectus*. Specimens with measurable permanent molars are recorded in Table 6.3. Cusp size data for most have been recorded by Wood (1991) and Bromage et al. (1995); these values were supplemented by measurements recorded by two of us (HFS, CPH) from occlusal photographs.

Table 6.3 East African specimens attributable to *Homo habilis, H. rudolfensis* and *H. erectus* with permanent molars employed in the present study

Specimen		M^1	M^2	M^3	M^1	M^2	M^3	Derivation
Homo habilis								
A.L.	666-1	X	X					Kadar Hadar Mb, Hadar Fm.
ER	1502				X			upper Burgi Mb., Koobi Fora Fm.
ER	1805						X	KBS Mb., Koobi Fora Fm.
ER	1813	X	X					upper Burgi Mb., Koobi Fora Fm.
OH	4						X	Bed I, Olduvai Gorge
OH	6	X						Bed I, Olduvai Gorge
OH	7				X	X		Bed I, Olduvai Gorge
OH	13	X	X	X	X	X	X	Bed II, Olduvai Gorge
OH	15			X				Bed II, Olduvai Gorge
OH	16	X	X		X	X	X	Bed II, Olduvai Gorge
OH	21	X						Bed ?, Olduvai Gorge
OH	24	X		X				Bed I, Olduvai Gorge
OH	27						X	Bed I, Olduvai Gorge
OH	39	X	X					Bed I, Olduvai Gorge
OH	41	X						Bed I, Olduvai Gorge
OH	44	X						Bed I, Olduvai Gorge
Omo L.	894-1	X						Mb. G-12, Shungura Fm.
Homo rudolfensis								
ER	1590	X	X					KBS Mb., Koobi Fora Fm.
ER	1802				X	X		upper Burgi Mb., Koobi Fora Fm.
UR	501				X	X		Chiwondo Beds, Malawi
Homo erectus								
BK	67					X		Kapthurin Fm.
BK	8518						X	Kapthurin Fm.
ER	806				X	X	X	Okote Mb., Koobi Fora Fm.
ER	820				X			Okote Mb., Koobi Fora Fm.
ER	808	X						Okote Mb., Koobi Fora Fm.
ER	992				X	X	X	Okote Mb., Koobi Fora Fm.
ER	1480						X	KBS Mb., Koobi Fora Fm.
ER	1507				X			KBS Mb., Koobi Fora Fm.
ER	1808					X		KBS Mb., Koobi Fora Fm.
ER	1812						X	upper Burgi Mb., Koobi Fora Fm.
ER	3733		X					KBS Mb., Koobi Fora Fm.
WT	15000	X	X		X	X		Natoo Mb., Nachukui Fm.
OH	22				X	X		Bed IV, Olduvai Gorge
OH	60						X	Bed I, Olduvai Gorge
Omo	K-7-1969-19					X		Mb. L-9, Shungura Fm.

The South African *Homo* sample that is the focus of the present study consists of one specimen from Drimolen, and five each from the sites of Sterkfontein and Swartkrans (Table 6.4). With regard to the Sterkfontein specimens, Stw 80 is a crushed mandibular symphysis and partial left corpus that holds badly damaged molar crowns. It is associated with an isolated RM_3, originally catalogued as Stw 81 (Moggi-Cecchi et al., 2006), which was measured by us. An isolated LM^2, originally catalogued as Stw 34, has been recognized as the antimere of the Stw 19 RM^2 (Moggi-Cecchi et al., 2006), and this was used in the present study. Although Clarke (1990) has asserted that the Stw 73/Sts 22 palate likely belongs to Sts 19, this cannot be substantiated because

Table 6.4 South African *Homo* specimens with permanent molars employed in the present study

Specimen		M^1	M^2	M^3	M_1	M_2	M_3	Derivation
SE	255	X						Sterkfontein Member 5C
SE	1508		X					Sterkfontein Member 5C
Stw	19			X				Sterkfontein, Uncertain
Stw	53			X				Sterkfontein, Member 5A
Stw	80						X	Sterkfontein, Member 5C
SK	27	X	X					Swartkrans, Member 1 HR
SKW	3114	X						Swartkrans, Member 1 HR
SK	15				X	X	X	Swartkrans, Member 2
SKX	257				X			Swartkrans, Member 2
SKX	267	X						Swartkrans, Member 2
DNH	70	X						Drimolen

the only potential point of contact between them is the partial interproximal wear facet on the LM^3 (Kimbel and Rak, 1993). Of the teeth in the Stw 53 maxilla, only the third molars preserve any indication of the occlusal fissures; the RM^3 was used here. With regard to the Swartkrans sample, neither SK 847 nor SK 45 could be included because of occlusal wear. The left molars in the SK 15 mandible were used here, as was the LM^1 of SKX 257.

Methods of Measurement

In all instances, only a single tooth of any one molar type was included for each individual. Thus, unlike the studies by Suwa et al. (1994, 1996), antimeric values were not averaged. As detailed by Smith (1999), cusp area measurements for the living apes were recorded by placing the individual molars in occlusal view below a Canon Hi8 video camera equipped with a 10x macro lens. Each crown was orientated such that the occlusal crown area was maximized. For the fossil specimens, occlusal photographs of either the original specimens or high definition casts were taken following the same method of crown orientation. This method of orientation is likely equivalent to that of Suwa et al. (1994, 1996), in which the area of the occlusal fovea was maximized to define "horizontal." It differs somewhat from the methods employed by Wood et al. (1983), who used "plane of the cervical line," and Bailey (2004), who used the buccal and distal cervices of upper molars for orientation. Nevertheless, such differences in technique result in only slight differences of cusp area measurement that are comparable to intra-observer error rates using only a single method of measurement (Suwa et al., 1994; Bailey et al., 2004). Indeed, Bailey et al. (2004: 329) concluded that "if certain prescribed standards are employed then cusp and crown base areas measured by different workers can be pooled into a single database." Thus, there is no *a priori* reason to preclude the data from these different sources being combined.

Measurements of individual cusps followed the procedure established by Wood et al. (1983). Interstitial wear was conservatively corrected, and accessory cusps (e.g., tuberculum sextum and tuberculum intermedium) were subdivided, with equal parts being added to the areas of the adjacent principal cusps (Wood et al., 1983; Wood, 1991; Suwa et al., 1994, 1996). Crowns on which any one of the occlusal fissures was completely obliterated were excluded from consideration. Other studies of early hominin molars have emphasized relative cusp proportions (Wood et al., 1983; Wood and Engleman, 1988; Suwa et al., 1994, 1996). However, we utilized raw area measurements, which impart information relating to both size and shape (Stumpf et al., 2002), in addition to relative (scale-free) proportional data. Relative cusp areas were calculated simply by dividing the total occlusal area of each crown by the area of each of its constituent cusps.

Statistical Analyses

Discriminant (canonical variates) analyses were used to summarize morphological differences among extant ape taxa and among fossil hominin samples based on cusp proportions. It was employed in preference over principal components analysis (e.g., Suwa et al., 1994, 1996) because group membership was known *a priori* for all extant and for most of the fossil specimens used in this study (Albrecht, 1980; Johnson and Wichen, 1982). This enabled us to assess the canonical loadings of each cusp area for all known groups, and then compare these to the South African fossils that we sought to classify. Wilks' Lambda statistics were employed to determine the number of functions that differentiated known taxa. Except where noted otherwise, Wilks' Lambda values are reported for the first two canonical axes, which explain the majority of variance for each subset of the data. The first two canonical axes were plotted in order to visualize the dispersion of individual fossils relative to each species group centroid, and to examine centroid clustering. The power of these functions/axes to accurately predict group membership for the fossils was assessed first by employing them in analyses of extant ape molars with known taxonomic affiliation. Efficacy for each molar type was assessed by the percentage of correct classifications, and the degree to which either absolute or relative cuspal proportions provided significant differentiation among taxa. As a second method of assessing phenetic similitude among the *Homo* fossils, dendrograms were generated by hierarchical clustering analyses using Mahalanobis D^2 distances.

Results

First, we re-examined the ability of cusp size information to discriminate among samples of extant great ape taxa. Subsequently, we examined their discriminatory power as applied to the much smaller fossil hominin samples.

Extant Apes

Absolute as well as relative cuspal proportions differ significantly among the four species of living great apes (Table 6.5). Absolute cusp dimensions correctly distinguish molars in some 76% of cases, whereas less than 50% of teeth were

Table 6.5 Predicted group membership of extant ape taxa from absolute and relative cuspal proportions

	All taxa – percent correctly classified											
	M^1		M^2		M^3		M_1		M_2		M_3	
Taxon	Abs	Rel	Abs	Rel	Abs	Rel	Abs	Rel	Abs	Rel	Abs	Rel
Gorilla gorilla	96.1	82.4	97.2	92.4	95.2	91.6	92.2	78.9	89.8	82.2	96.3	87.8
Pan troglodytes	94.9	45.8	81.8	40.0	100.0	38.1	90.0	66.7	91.2	10.5	100.0	55.0
Pan paniscus	87.5	25.0	88.1	42.9	0.0	16.7	86.4	43.2	79.5	43.2	13.3	13.3
Pongo pygmaeus	75.5	00.0	82.5	45.0	47.8	13.0	70.8	52.1	79.2	28.3	69.6	4.3
Wilks' Lambda	0.103	0.754	0.089	0.557	0.192	0.742	0.090	0.450	0.134	0.709	0.143	0.682
χ^2	558.6	69.0	578.6	140.0	256.0	46.2	569.5	188.9	535.1	91.6	300.6	59.2
df	12	12	12	12	12	12	15	15	15	15	15	15
p <	0.001	0.001	0.001	0.001	0.001	0.001	0.001	0.001	0.001	0.001	0.001	0.001
Functions	1–3	1–3	1–3	1–3	1–3	1–3	1–3	1–3	1–3	1–3	1–3	1–3
	Pan taxa alone – percent correctly classified											
	M^1		M^2		M^3		M_1		M_2		M_3	
Taxon	Abs	Rel	Abs	Rel	Abs	Rel	Abs	Rel	Abs	Rel	Abs	Rel
Pan troglodytes	89.8	86.4	92.7	82.5	97.6	93.0	90.0	88.3	87.7	86.0	92.5	92.5
Pan paniscus	95.0	52.5	95.2	57.1	75.0	33.3	97.7	84.1	84.1	77.3	93.3	33.3
Wilks' Lambda	0.321	0.812	0.349	0.801	0.543	0.765	0.356	0.478	0.415	0.650	0.438	0.813
χ^2	107.8	19.7	97.9	21.1	30.6	13.6	102.6	73.5	84.9	41.6	41.6	10.4
df	4	4	4	4	4	4	5	5	5	5	5	5
p <	0.001	0.001	0.001	0.001	0.001	0.010	0.001	0.001	0.001	0.001	0.001	ns
Functions	1	1	1	1	1	1	1	1	1	1	1	
	Gorilla taxa alone – percent correctly classified											
	M^1		M^2		M^3		M_1		M_2		M_3	
Taxon	Abs	Rel	Abs	Rel	Abs	Rel	Abs	Rel	Abs	Rel	Abs	Rel
G. g. beringei	42.3	73.1	5.0	10.0	37.5	12.5	63.6	72.7	50.0	53.8	70.0	70.0
G. g. gorilla	92.3	86.3	84.1	85.7	82.9	93.3	88.9	85.1	89.1	76.4	86.5	78.4
G. g. graueri	36.0	28.0	66.7	45.8	38.5	34.6	56.5	47.8	59.5	45.9	32.0	28.0
Wilks' Lambda	0.480	0.574	0.688	0.742	0.749	0.822	0.354	0.473	0.455	0.608	0.535	0.584
χ^2	72.2	54.1	38.3	30.6	22.7	16.2	88.4	65.2	88.9	56.3	48.2	41.4
df	8	8	8	8	8	8	10	10	10	10	10	10
p <	0.001	0.001	0.001	0.001	0.005	0.05	0.001	0.001	0.001	0.001	0.001	0.001
Functions	1–2	1–2	1–2	1–2	1–2	1–2	1–2	1–2	1–2	1–2	1–2	1–2

correctly classified on the basis of relative cuspal proportions. This disparity is understandable given the differences in overall molar crown size between *Gorilla gorilla*, *Pan troglodytes* and *P. paniscus* (Swindler, 1976; Uchida, 1992). At the same time, there is little difference between maxillary and mandibular molars in terms of discriminatory power; absolute cusp sizes of upper and lower molars correctly distinguish among the four taxa in about 79% and 73% of instances respectively.

Absolute and relative cuspal sizes differ significantly between the sibling species of *Pan*, with the exception of M_3 relative cusp proportions, with only one third of *P. paniscus* specimens being correctly categorized. Molar cusp size data differentiate reasonably well among *Gorilla* groups, with significant differences among them in both absolute and relative cuspal proportions. Overall, absolute cusp dimensions correctly distinguish 60% and relative cuspal sizes correctly distinguish 57% of molars of *Gorilla gorilla gorilla*, *G. g. beringei*, and *G. g. graueri*, which is perhaps understandable given the subtle differences in overall molar crown size among them (Uchida, 1992). Maxillary and mandibular molars differ somewhat in terms of their discriminatory power among *Gorilla* samples. On average, absolute and relative dimensions correctly distinguish among upper molars in about 54% and 53% of instances respectively. Among lower molars, absolute and relative dimensions correctly differentiate an average of some 66% and 62% of specimens respectively.

Early Homo *Molars*

The foregoing analyses support previous studies that have suggested molar cusp sizes to be of some utility in the assessment of the taxonomic affinities of hominin fossils (Wood et al., 1983; Wood and Engleman, 1988; Suwa, 1988; Suwa et al., 1994, 1996; Bailey, 2004). Accordingly, it is reasonable to employ such data in the assessment of the species attribution of the early *Homo* specimens from South Africa. However, because of the extremely small comparative fossil samples from East Africa, extreme caution must be exercised in the interpretation of these results.

Data pertaining to the discriminatory power of absolute and relative cusp sizes among East African specimens attributed to *H. habilis*, *H. rudolfensis* and *H. erectus* are presented in Table 6.6. Only in the relative cuspal proportions of the M_1 are the three significantly distinct according to the Wilks' Lambda statistic, but in no instance does the *H. rudolfensis* sample exceed two specimens, and only for the lower molars do the *H. habilis* and *H. erectus* samples exceed two specimens.

When the South African *Homo* fossils are added as a separate sample, the relative cuspal proportions of both the M_1 and M_2 suggest significant inter-group distinctions, although the sample sizes are still exceedingly small (Table 6.7). Nevertheless, it is interesting to note that South African specimens were predicted to have membership in one of the East African samples only with reference to the M^1 and M_3. With regard to the M^1, three of the five South African specimens (SKW 3114, SKX 267, DNH 70) were categorized as belonging with *H. habilis* on the basis of both absolute and relative cusp sizes. With regard to the M_3, absolute cusp sizes predicted one of two South African specimens (Stw 80) to belong with *H. habilis*, while the other (SK 15) was predicted to be a member of the *H. erectus* sample. However, neither the M_1 nor M_2 of SK 15 suggested any such affinity.

Because the sample sizes involved in these analyses are so small as to preclude meaningful statistical inference, it is instructive to examine the distribution of the fossils according to the discriminant functions, and the dendrograms of the Mahalanobis D^2 distances between them.

Among maxillary molars, absolute cusp sizes of the M^1, place all five South African specimens within the *Homo habilis* sample cluster, although three (SK 27, SKW 3114, SE 255) have their closest affinities with one another

Table 6.6 Predicted group membership of East African *Homo* species groups from absolute and relative cuspal proportions

		Species – percent correctly classified											
		M^1		M^2		M^3		M_1		M_2		M_3	
Taxon		Abs	Rel	Abs	Rel	Abs	Rel	Abs	Rel	Abs	Rel	Abs	Rel
H. habilis	n =		10		6		3		4		3		5
H. rudolfensis	n =		1		1		0		2		2		0
H. erectus	n =		2		2		0		5		7		5
H. habilis		90.0	90.0	83.3	100.0	xxx	xxx	100.0	100.0	66.7	100.0	80.0	20.0
H. rudolfensis		100.0	0.0	100.0	100.0	xxx	xxx	50.0	100.0	100.0	100.0	xxx	xxx
H. erectus		50.0	100.0	50.0	100.0	xxx	xxx	100.0	100.0	100.0	85.7	83.3	50.0
Wilks' Lambda		0.379	0.222	0.140	0.402	xxx	xxx	0.189	0.024	0.125	0.151	0.563	0.947
$\chi 2$		8.3	12.8	8.8	4.1	xxx	xxx	10.0	22.4	14.6	13.2	4.1	0.4
df		8	8	8	8	xxx	xxx	10	10	10	10	5	5
p <		ns	ns	ns	ns	xxx	xxx	ns	0.01	ns	ns	ns	ns
Functions					xxx	xxx		1–2					

Table 6.7 Predicted membership of *Homo* groups from absolute and relative cuspal proportions

		Species – percent correctly classified											
		M^1		M^2		M^3		M_1		M_2		M_3	
Taxon		Abs	Rel	Abs	Rel	Abs	Rel	Abs	Rel	Abs	Rel	Abs	Rel
H. habilis	n =		10		6		3		4		3		5
H. rudolfensis	n =		1		1		0		2		2		0
H. erectus	n =		2		2		0		5		7		5
South Africa	n =		5		2		2		2		1		2
H. habilis		90.0	90.0	100.0	83.3	100.0	100.0	100.0	100.0	66.7	100.0	80.0	20.0
H. rudolfensis		100.0	0.0	100.0	100.0	xxx	xxx	50.0	100.0	100.0	100.0	xxx	xxx
H. erectus		50.0	50.0	50.0	0.0	xxx	xxx	100.0	100.0	100.0	85.7	83.3	50.0
South Africa		40.0	40.0	100.0	100.0	100.0	100.0	100.0	100.0	100.0	100.0	0.0	100.0
Wilks' Lambda		0.380	0.353	0.124	0.372	0.099	0.112	0.102	0.029	0.086	0.027	0.470	0.438
$\chi 2$		12.6	13.5	12.5	5.9	3.5	3.3	17.1	26.4	18.4	27.1	6.0	6.6
df		12	12	12	12	3	3	15	15	15	15	10	10
p <		ns	ns	ns	ns	ns	ns	ns	0.05	ns	0.05	ns	ns
Functions									1–3		1–3		

(Fig. 6.1). Relative cusp proportions of the M^1 place three South African fossils within the *H. habilis* sample envelope, while two (SE 255, SKW 3114) are comparatively distant outliers that link with one another. With reference to the M^2 (Fig. 6.2), the two South African specimens (SK 27, SE 1508) fall within or close to the *H. habilis* sample envelopes. In both instances the Sterkfontein and Swartkrans fossils are linked, and separately so for absolute cusp size. Unfortunately, no *H. erectus* or *H. rudolfensis* M^3 is represented in the current sample. Although the only comparison for the two South African fossils (Stw 19, Stw 53) is with *H. habilis* (Fig. 6.3), both Sterkfontein specimens are linked at a distance from the three Olduvai Gorge fossils in their absolute and relative cuspal proportions.

Among mandibular molars, the absolute and relative M_1 cusp proportions (Fig. 6.4) find the two South African specimens (SK 15, SKX 257) falling outside the envelopes for *H. habilis*, *H. erectus* and *H. rudolfensis*. These Swartkrans Member 2 specimens exhibit no special affinity to any one of the East African specimens/samples. With regard to the M_2 (Fig. 6.5), absolute cusp sizes place SK 15 within the *H. erectus* envelope, but although it is nested with that sample, it is an outlier to all of the specimens that comprise it. On the other hand, relative cusp proportions of the M_2 clearly distinguish SK 15 from the *H. habilis*, *H. erectus* and *H. rudolfensis* clusters. Absolute cusp sizes of the M_3 (Fig. 6.6) place the two South African specimens (SK 15, Stw 80) within or just outside the *H. erectus* and *H. habilis* sample envelopes, and they are nested with specimens of both species. Relative cuspal proportions, on the other hand, place the South African fossils outside the *H. habilis* and *H. erectus* sample clusters; neither SK 15 nor Stw 80 has special affinity with either of the East African species samples.

Among the South African specimens, only SK 27 and SK 15 possess more than one molar that is amenable to cusp size analysis. With regard to SK 27, its M^1 suggests close affinity with *H. habilis*, and its M^2 suggests a somewhat more distant relationship with this species. In neither instance do cuspal proportions suggest that SK 27 has special affinity to *H. erectus*. With regard to SK 15, the relative cuspal proportions for all three molars suggest no particular affinity to any species sampled from East Africa; the absolute cuspal sizes of the SK 15 M_2 and M_3 (but not the M_1) suggest affinities with *H. erectus* and/or *H. habilis*.

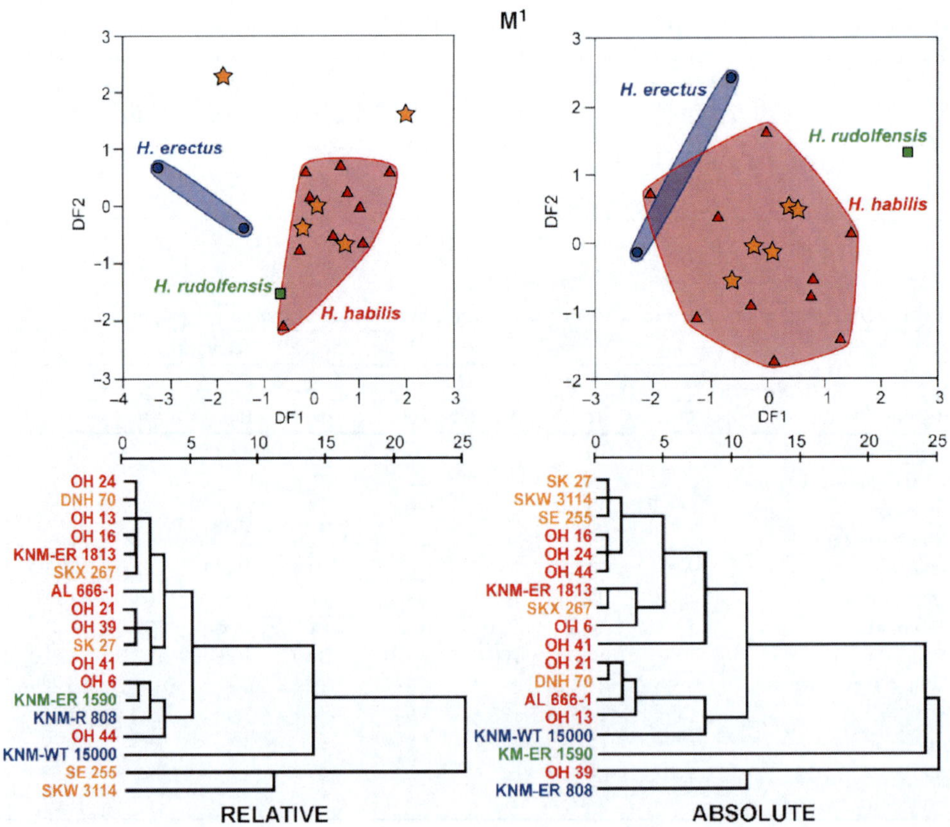

Fig. 6.1 Canonical discriminant functions and dendrograms from relative and absolute cusp areas of early *Homo* M^1's. Triangles = East African *H. habilis* molars; square = East African *H. rudolfensis* molar; circles = East African *H. erectus* molars; stars = South African specimens. Relative area analysis: function 1 (70.4%) is driven by the protocone and hypocone (positive) and paracone (negative); function 2 (22.5%) is driven by the hypocone and paracone (positive), and the metacone and protocone (negative). Absolute area analysis: function 1 (60.3%) is driven by the metacone and hypocone (positive); function 2 (34.2%) is driven by the paracone (positive), and the protocone and hypocone (negative).

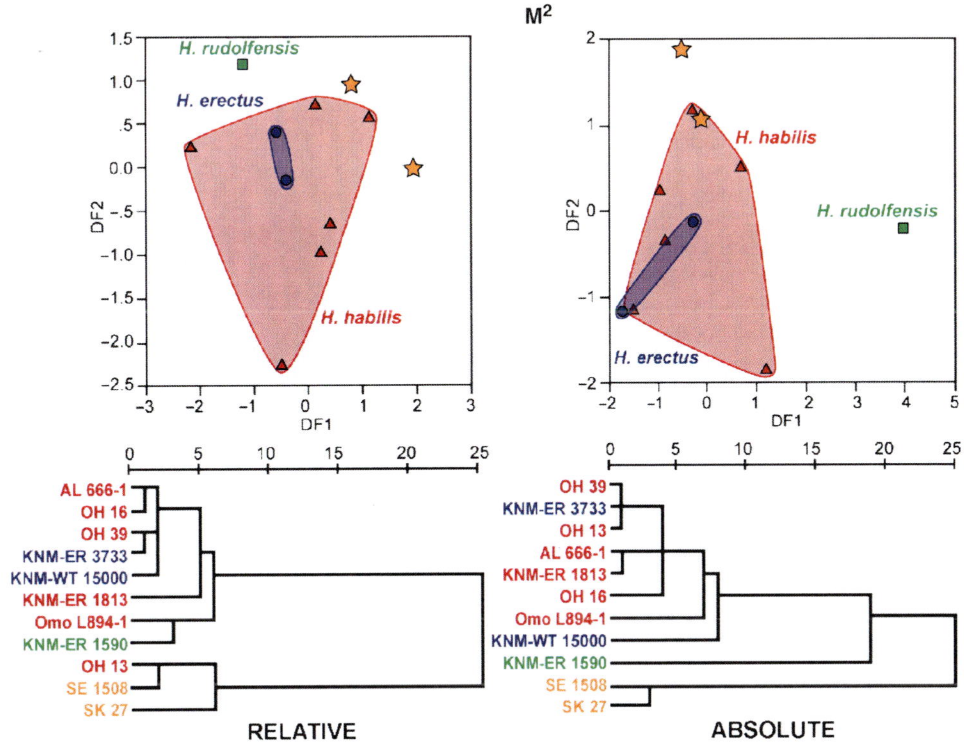

Fig. 6.2 Canonical discriminant functions and dendrograms from relative and absolute cusp areas of early *Homo* M²s. Triangles = East African *H. habilis* molars; square = East African *H. rudolfensis* molar; circles = East African *H. erectus* molars; stars = South African specimens. Relative area analysis: function 1 (47.5%) is driven by the paracone (positive), and the hypocone and metacone (negative); function 2 (33.0%) is driven by the hypocone and metacone (positive), and paracone (negative). Absolute area analysis: function 1 (71.9%) is driven by the metacone and hypocone (positive); function 2 (21.9%) is driven by the paracone (positive).

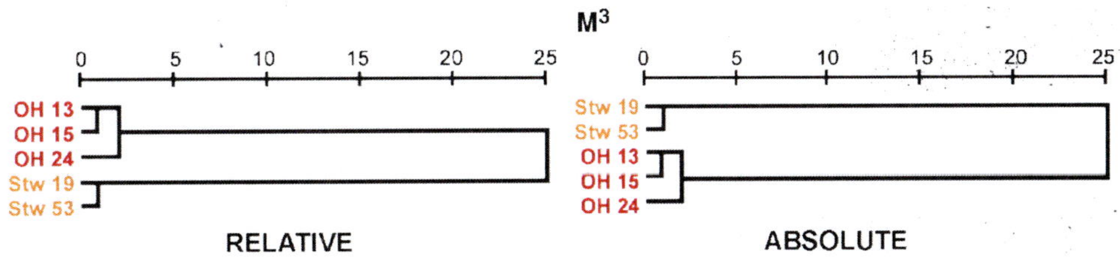

Fig. 6.3 Dendrograms from relative and absolute cusp areas of early *Homo* M³s. OH specimens are all attributed to *H. habilis*; Stw specimens are from Sterkfontein.

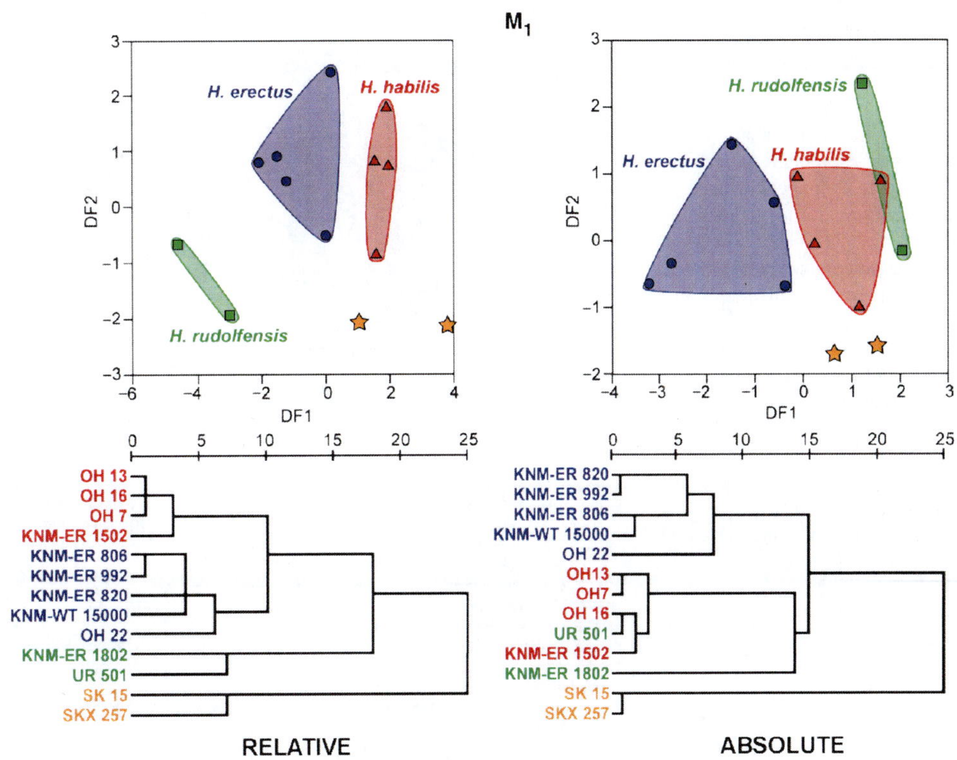

Fig. 6.4 Canonical discriminant functions and dendrograms from relative and absolute cusp areas of early *Homo* M_1s. Triangles = East African *H. habilis* molars; squares = East African *H. rudolfensis* molars; circles = East African *H. erectus* molars; stars = South African specimens. Relative area analysis: function 1 (88.7%) is driven by the entoconid (positive); function 2 (9.3%) is driven by the protoconid (positive), and entoconid (negative). Absolute area analysis: function 1 (66.9%) is driven by the entoconid and hypoconulid (positive); function 2 (22.4%) is driven by the protoconid, metaconid, and hypoconulid (positive).

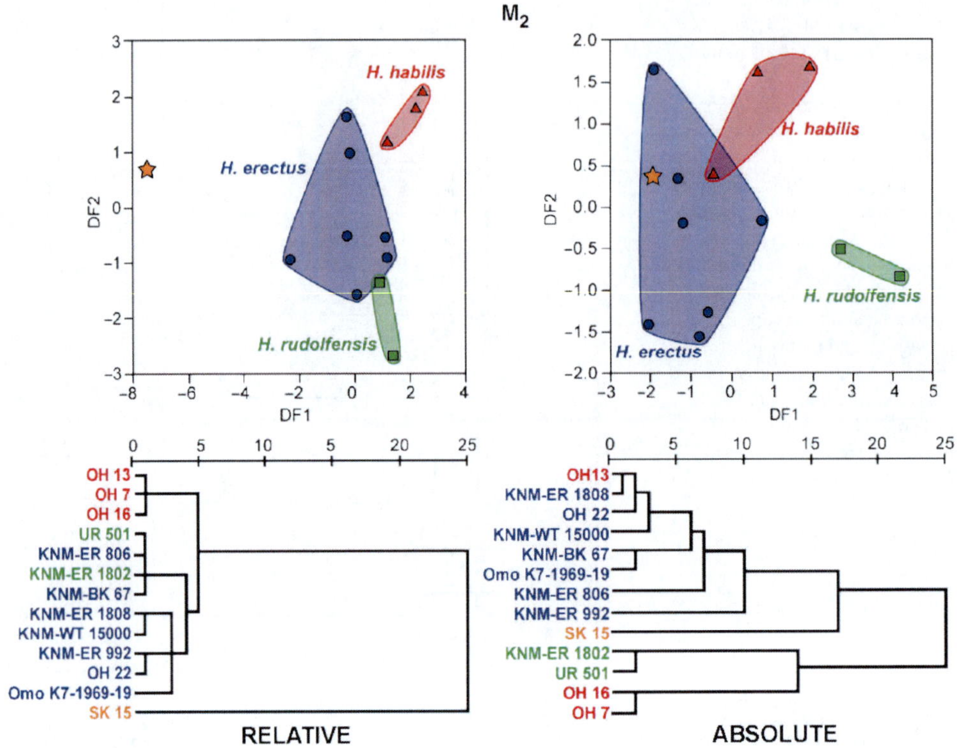

Fig. 6.5 Canonical discriminant functions and dendrograms from relative and absolute cusp areas of early *Homo* M_2s. Triangles = East African *H. habilis* molars; squares = East African *H. rudolfensis* molars; circles = East African *H. erectus* molars; stars = South African specimens. Relative area analysis: function 1 (76.7%) is driven by the metaconid (positive) and the protoconid and hypoconid (negative); function 2 (19.0%) is driven by the protoconid (positive) and hypoconid (negative). Absolute area analysis: function 1 (79.1%) is driven by all five cusps (positive); function 2 (14.5%) is driven by the protoconid and hypoconulid (positive).

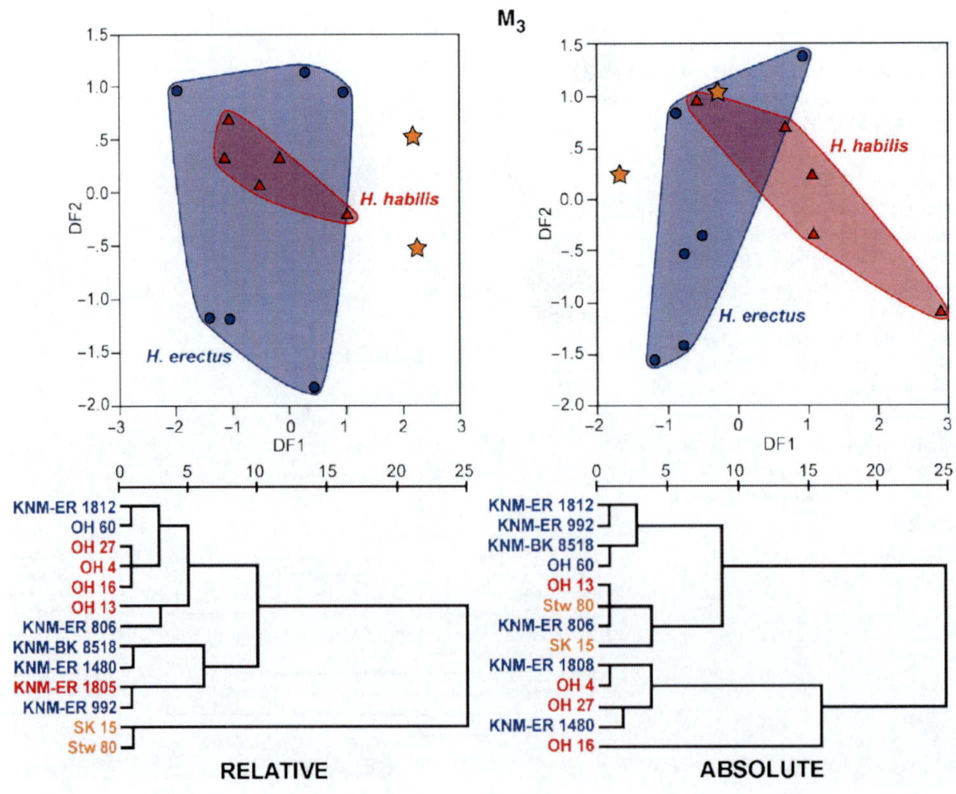

Fig. 6.6 Canonical discriminant functions and dendrograms from absolute and relative cusp areas of early *Homo* M_3s. Triangles = East African *H. habilis* molars; circles = East African *H. erectus* molars; stars = South African specimens. Relative area analysis: function 1 (86.9%) is driven by the entoconid (positive) and the protoconid and hypoconid (negative); function 2 (13.1%) is driven by the metaconid (positive) and the entoconid and hypoconulid (negative). Absolute area analysis: function 1 (95.3%) is driven by the hypoconulid and protoconid (positive); function 2 (4.7%) is driven by the protoconid, hypoconid, and metaconid (positive), and entoconid (negative).

Discussion

Cusp area data for the permanent molars attributed to *Homo* from the sites of Swartkrans, Sterkfontein and Drimolen tend to suggest closer affinities with *H. habilis* than with *H. erectus* homologues from East Africa. At the same time, however, there is generally some degree of linkage among the South African fossils such that a number have closer affinities with one another than with East African specimens. When the South African specimens are considered as a separate (unique) group, relative cuspal proportions of the M_1 and M_2 suggest significant inter-group distinctions. Only five of 14 South African molars were predicted to belong to one of the East African species groups; four (from Sterkfontein Member 5C, Swartkrans Members 1 and 2, and Drimolen) were categorized as belonging with *H. habilis*, and one (from Swartkrans Member 2) was predicted to belong to the *H. erectus* sample.

Specimens from Swartkrans Member 1 (SK 27, SKW 3114) tend to group with some from Sterkfontein Member 5C (SE 255, SE 1508) either as a separate entity, or with East African *H. habilis* teeth. Unfortunately, SK 847 was not among the specimens represented in the present study, but there is no *a priori* reason to suspect that SK 27 and SKW 3114 should not belong to the same taxon as SK 847. At the same time, some specimens from Swartkrans Member 2 and other from Sterkfontein Member 5C (i.e., the SK 15 and Stw 80 M_3s) evince some resemblance to *H. erectus* specimens. In this respect, it is perhaps noteworthy that Kuman and Clarke (2000) have suggested that the Stw 80 and SK 15 mandibles should be attributed to the same species (*H. erectus*) because of their perceived similarities. The possibility that the *Homo* fossils from Swartkrans Members 1 and 2 represent distinct, temporally differentiated species has been suggested by Albarède et al. (2006).

Notwithstanding recent arguments over the level of species diversity that should be expected in the hominin fossil record (e.g., White, 2003), the notion that the South African fossils sample a distinct species lineage should not be too surprising, since South and East Africa share comparatively few Plio-Pleistocene (or recent) mammal species (Grubb et al., 1999). For example, Benefit (1999) records that only some 24% of cercopithecoid species are cosmopolitan to both South and East Africa in the Plio-Pleistocene. The data compiled by Turner et al. (1999) reveal that for sites that date to between about 3.0 and 1.0 Myr, only about 25% of elephants, 29% of horses, 21% of pigs, and between 14% and 17% of bovids are common to both South and East Africa.

With reference to other hominins, South and East Africa share no species that are not attributed to the genus *Homo*. Thus, *Praeanthropus afarensis*, *Kenyanthropus platyops*, *"Australopithecus" anamensis*, *"Australopithecus" garhi*, *Paranthropus boisei* and *Paranthropus aethiopicus* are unknown from South Africa, although some of the South African deposits almost certainly overlap their temporal ranges. Similarly, *Australopithecus africanus* and *Paranthropus robustus* are unknown in East Africa despite the existence of contemporaneous deposits. It is of historical interest that the initial comparisons of the isolated teeth and small jaw fragments found in the late 1960s and early 1970s in the Usno and Shungura Formations of southern Ethiopia suggested the presence of *A. africanus* in the Turkana Basin (e.g., Howell, 1969a, b, 1978; Howell and Coppens, 1976). These analyses, of course, were conducted at a time when contemporaneous hominin fossils from elsewhere in East Africa were only very poorly known, and their conclusions reflect the understandable tendency to initially pigeon-hole fossils of unknown taxonomic affiliation into known categories. Perhaps these same tendencies have affected the interpretation of those fragmentary early *Homo* fossils from South Africa that have been afforded any measure of specific appellation.

Our application of dental morphometrics has expanded the investigation of the morphological affinities of the South African early *Homo* fossils to include a much larger sample including isolated teeth and fragmentary mandibles. As a group, these specimens tend to have closer phenetic resemblances to one another than to East African specimens attributed to *Homo habilis* or *H. erectus*. While more detailed analyses are necessary before the systematic affinities of the fragmentary *Homo* remains from Sterkfontein, Swartkrans and Drimolen can be ascertained with any degree of certainty, multiple lines of evidence seem to suggest that the South African fossils may attest to species (or perhaps a species lineage) that are as yet un-sampled in the Plio-Pleistocene deposits of East Africa.

Conclusions

Permanent molar cusp areas were employed in an effort to evaluate the phenetic affinities of *Homo* specimens from the South African Plio-Pleistocene cave deposits of Sterkfontein, Swartkrans and Drimolen. Quantitative information relating to molar cusp proportions expands the level of investigation of the morphological affinities of these fossils from the two fragmentary crania (SK 847 and Stw 53) that have been the subject of previous investigations to a sample of 11 individuals. The efficacy of such data to successfully discriminate between species groups was demonstrated using samples of extant apes (*Gorilla gorilla*, *Pan troglodytes*, *P. paniscus*, *Pongo pygmaeus*). The commonly small sizes of the East African *Homo habilis* and *H. erectus* samples preclude meaningful statistical analysis in most instances, but when the South African specimens are considered as a separate (unique) group, relative cuspal proportions of the M_1 and M_2 suggest significant inter-group distinctions. Only five of 14 South African molars were predicted to belong to one of the East African species groups; four (from Sterkfontein Member 5C, Swartkrans Members 1 and 2, and Drimolen) were categorized as belonging with *H. habilis*, and one (from Swartkrans Member 2) was predicted to belong to the *H. erectus* sample. As a group, the South African fossils tend to have closer phenetic resemblances among themselves, regardless of derivation, than with East African specimens attributed to either *H. habilis* or *H. erectus*. These data are not inconsistent with suggestions that the early *Homo* fossils from South Africa may represent species (or perhaps a species lineage) un-sampled in the Plio-Pleistocene deposits of East Africa.

Acknowledgments The extant hominoid molars were measured by one of us (E.J.S) at several institutions, including the Powell-Cotton Museum (Birchington), the American Museum of Natural History (New York), the United States Museum of Natural History (Washington, DC), the Museum of Comparative Zoology at Harvard University (Cambridge), and the Koninklijk Museum (Tervuren). This work was supported by grants from the Leakey Foundation and N.S.F. to F.E.G. We thank C.K. Brain and J.F. Thackeray (Transvaal Museum, Pretoria), P.V. Tobias, L. Berger, R.J. Clarke (University of the Witwatersrand, Johannesburg), and A. Keyser for permission to examine the Swartkrans, Sterkfontein and Drimolen dentitions. W.H. Kimbel provided occlusal photographs of the AL 666-1 dentition. We thank B.A. Wood, W.H Kimbel and W.L. Jungers for constructive comments on this manuscript. We are grateful to L. Betti-Nash for the illustrations. We thank the participants in the workshop for lively discussion, and two anonymous referees for constructive comments on the manuscript.

References

Ahern, J.C.M., 1998. Underestimating intraspecific variation: the problem with excluding Sts 19 from *Australopithecus africanus*. American Journal of Physical Anthropology 105, 461–480.

Aiello, L.C., Collard, M., Thackeray, J.F., Wood, B.A., 2000. Assessing exact randomization-based methods for determining the taxonomic significance of variability in the human fossil record. South African Journal of Science 96, 179–183.

Albarède, F., Balter, F. Braga, J. Blichert-Toft, J. Telouk, P., Thackeray, F., 2006. U-Pb dating of enamel from the Swartkrans Cave hominid site (South Africa) by MC-ICP-MS. Geochimica et Cosmochimica Acta 70, A7.

Albrecht, G.H., 1980. Multivariate analysis and the study of form, with special reference to canonical variate analysis. American Zoologist 20, 679–693.

Antón, S.C., 2002. Evolutionary significance of cranial variation in Asian *Homo erectus*. American Journal of Physical Anthropology 118, 301–323.

Antón, S.C., 2003. Natural history of *Homo erectus*. Yearbook of Physical Anthropology 46, 126–169.

Antón, S.C., 2004. The face of Olduvai Hominid 12. Journal of Human Evolution 46, 335–345.

Antón, S.C., Swisher, C.C., 2004. Early dispersals of *Homo* from Africa. Annual Review of Anthropology 33, 271–296.

Asfaw, B., Gilbert, W.H., Beyene, Y., Hart, W.K., Renne, P.R., WoldeGabriel, G., Vrba, E.S., White, T.D., 2002. Remains of *Homo erectus* from Bouri, Middle Awash, Ethiopia. Nature 416, 317–320.

Bailey, S.E., 2004. A morphometric analysis of maxillary molar crowns of Middle-Late Pleistocene hominins. Journal of Human Evolution 47, 183–198.

Bailey, S.E., Pilbrow, V., Wood, B.A., 2004. Interobserver error involved in independent attempts to measure cusp base areas of *Pan* M^1s. Journal of Anatomy 205, 323–331.

Benefit, B.R., 1999. Biogeography, dietary specialization, and the diversification of African Plio-Pleistocene monkeys. In: Bromage, T.G., Schrenk, F. (Eds), African Biogeography, Climate Change, and Human Evolution. Oxford University Press, New York, pp. 172–188.

Blumenschine, R.J., Peters, C.R., Masao, F.T., Clarke, R.J., Deino, A.L., Hay, R.L., Swisher, C.C., Stanistreet, I.G., Ashley, G.M., McHenry, L.J., Sikes, N.E., van der Merwe, N.J., Tactikos, J.C., Cushing, A.E., Deocampo, D.M., Njau, J.K., Ebert, J.I., 2003. Late Pliocene *Homo* and hominid land use from western Olduvai Gorge, Tanzania. Science 299, 1217–1221.

Braga, J., Thackeray, J.F., 2003. Early *Homo* at Kromdraai B: probabilistic and morphological analysis of the lower dentition. Comptes Rendus Palevol 2, 269–279.

Bräuer, G., 1994. How different are Asian and African *Homo erectus*? Courier Forschungsinstitut Senckenberg 171, 301–318.

Bräuer, G., Mbua, E., 1992. *Homo erectus* features used in cladistics and their variability in Asian and African hominids. Journal of Human Evolution 22, 79–108.

Bromage, T.G., Schrenk, F., Zonneveld, F.W., 1995. Paleoanthropology of the Malawi Rift: an early hominid mandible from the Chiwondo Beds, northern Malawi. Journal of Human Evolution 28, 71–108.

Broom, R., Robinson, J.T., 1950. Further evidence on the structure of the Sterkfontein ape-man *Plesianthropus*. Transvaal Museum Memoirs 4, 11–84.

Chamberlain, A.T., 1987. A taxonomic review and phylogenetic analysis of *Homo habilis*. Ph.D. thesis, University of Liverpool.

Clarke, R.J., 1977a. The cranium of the Swartkrans hominid SK 847 and its relevance to human origins. Ph.D. thesis, University of the Witwatersrand, Johannesburg.

Clarke, R.J., 1977b. A juvenile cranium and some adult teeth of early *Homo* from Swartkrans, Transvaal. South African Journal of Science 73, 46–49.

Clarke, R.J., 1985a. *Australopithecus* and early *Homo* in southern Africa. In: Delson, E. (Ed), Ancestors: The Hard Evidence. Alan R. Liss, New York, pp. 171–177.

Clarke, R.J., 1985b. Early Acheulean with *Homo habilis* at Sterkfontein. In: Tobias, P.V. (Ed), Hominid Evolution: Past, Present and Future. Alan R. Liss, New York, pp. 287–298.

Clarke, R.J., 1990. Observations on some restored hominid specimens in the Transvaal Museum, Pretoria. In: Sperber, G.H. (Ed), From Apes to Angels: Essays in Anthropology in Honor of Phillip V. Tobias. Wiley-Liss, New York, pp. 135–151.

Clarke, R.J., 1994. The significance of the Swartkrans *Homo* to the *Homo erectus* problem. Courier Forschungsinstitut Senckenberg 171, 185–193.

Corruccini, R.S., 1977. Crown component variation in hominoid lower third molars. Zeitschrift für Morphologie und Anthropologie 68, 14–25.

Curnoe, D., 2001. Early *Homo* from southern Africa: a cladistic perspective. South African Journal of Science 97, 186–190.

Curnoe, D., 2002. Testing human taxonomic hypotheses with topology-dependent permutation tail tests. South African Journal of Science 98, 310–312.

Curnoe, D., 2008. Affinities of the Swartkrans early *Homo* mandibles. Homo 59, 123–147.

Curnoe, D., Tobias, P.V., 2006. Description, new reconstruction, comparative anatomy, and classification of the Sterkfontein Stw 53 cranium, with discussions about the taxonomy of other southern African early *Homo* remains. Journal of Human Evolution 50, 36–77.

Dean, M.C., Wood, B.A., 1982. Basicranial anatomy of Plio-Pleistocene hominids from East and South Africa. American Journal of Physical Anthropology 59, 157–174.

Dunsworth, H., Walker, A.C., 2002. Early genus *Homo*. In: Hartwig, W. (Ed.), The Primate Fossil Record. Cambridge University Press, New York, pp. 419–435.

Finlayson, C., 2005. Biogeography and evolution of the genus *Homo*. Trends in Ecology and Evolution 20, 457–463.

Gilbert, W.H., White, T.D., Asfaw, B., 2003. *Homo erectus*, *Homo ergaster*, *Homo "cepranensis*," and the Daka cranium. Journal of Human Evolution 45, 255–259.

Grine, F.E., 1982. A new juvenile hominid (Mammalia; Primates) from Member 3, Kromdraai Formation, Transvaal, South Africa. Annals of the Transvaal Museum 33, 165–239.

Grine, F.E., 1989. New hominid fossils from the Swartkrans Formation (1979–1986 excavations): craniodental specimens. American Journal of Physical Anthropology 79, 409–449.

Grine, F.E., 2001. Implications of morphological diversity in early *Homo* crania from eastern and southern Africa. In: Tobias, P.V., Raath, M., Moggi-Cecchi, J., Doyle, G. (Eds), Humanity from the African Naissance to the Coming Millennia. Firenza University Press, Firenza, pp. 107–115.

Grine, F.E., 2005. Early *Homo* at Swartkrans, South Africa: a review of the evidence and an evaluation of recently proposed morphs. South African Journal of Science 101, 43–52.

Grine, F.E., Demes, B., Jungers, W.L., Cole, T.M., 1993. Taxonomic affinity of the early *Homo* cranium from Swartkrans, South Africa. American Journal of Physical Anthropology 92, 411–426.

Grine, F.E., Jungers, W.L., Schultz, J., 1996. Phenetic affinities among early *Homo* crania from East and South Africa. Journal of Human Evolution 30, 189–225.

Grubb, P., Sandrock, O., Kullmer, O., Kaiser, T.M., Schrenk, F., 1999. Relationships between eastern and southern African mammal faunas. In: Bromage, T.G., Schrenk, F. (Eds), African Biogeography, Climate Change, and Human Evolution. Oxford University Press, New York, pp. 107–115.

Hartman, S.E., 1989. Stereophotogrametric analysis of occlusal morphology of extant hominoid molars. American Journal of Physical Anthropology 80, 145–166.

Hills, M., Graham, S.H., Wood, B.A., 1983. The allometry of cusp size in hominoid mandibular molars. American Journal of Physical Anthropology 62, 311–316.

Howell, F.C., 1969a. Remains of Hominidae from Pliocene/Pleistocene Formations in the lower Omo basin, Ethiopia. Nature 223, 1234–1239.

Howell, F.C., 1969b. Hominid teeth from the White Sands and Brown Sands localities, lower Omo basin (Ethiopia). Quaternaria 11, 47–54.

Howell, F.C., 1978. Hominidae. In: Maglio V.J., Cooke, H.B.S. (Eds), Evolution of African Mammals. Harvard University Press, Cambridge, MA., pp. 154–248.

Howell, F.C., Coppens, Y., 1976. An overview of the Hominidae from the Omo succession. In: Coppens, Y., Howell, F.C, Isaac, G.L. (Eds), Earliest Man and Environments in the Lake Rudolf Basin. University of Chicago Press, Chicago, IL, pp. 522–532.

Hughes, A.R., Tobias, P.V., 1977. A fossil skull probably of the genus *Homo* from Sterkfontein, Transvaal. Nature 265, 310–312.

Johnson, R.A., Wichen, D.W., 1982. Applied Multivariate Statistical Analysis. Prentice-Hall, Englewood Cliffs, NJ.

Keyser, A.W., Menter, C.G., Moggi-Cecchi, J., Pickering, T.R., Berger, L.R., 2000. Drimolen: a new hominid-bearing site in Gauteng, South Africa. South African Journal of Science 96, 193–197.

Kidder, J.H., Durband, A.C., 2004. A re-evaluation of the metric diversity within *Homo erectus*. Journal of Human Evolution 46, 299–315.

Kimbel, W.H., Rak, Y., 1993. The importance of species taxa in paleoanthropology and an argument for the phylogenetic concept of the species category. In: Kimbel, W.H, Martin, L.B. (Eds), Species, Species Concepts, and Primate Evolution. Plenum, New York, pp. 461–485.

Kramer, A., 1993. Human taxonomic diversity in the Pleistocene: does *Homo erectus* represent multiple hominid species? American Journal of Physical Anthropology 91, 161–171.

Kuman, K., Clarke, R.J., 2000. Stratigraphy, artefact industries and hominid associations for Sterkfontein, Member 5. Journal of Human Evolution 38, 827–847.

Lacruz, R., 2007. Enamel microstructure of the hominid KB 5223 from Kromdraai, South Africa. American Journal of Physical Anthropology 132, 175–182.

Lavelle, C.L.B., 1978. An analysis of molar tooth form. Acta Anatomica 100, 282–288.

Leakey, M.G., Spoor, F., Brown, F.H., Gathogo, P.N., Kiarie, C., Leakey, L., McDougall, I., 2001. New hominin genus from eastern Africa shows diverse middle Pliocene lineages. Nature 410, 433–440.

Lee, S.H., Wolpoff, M.H., 2005. Habiline variation: a new approach using STET. Theory in Biosciences 124, 25–40.

Lieberman, D.E., Pilbeam, D.R., Wood, B.A., 1988. A probabilistic approach to the problem of sexual dimorphism in *Homo habilis*: a comparison of KNM-ER 1470 and KNM-ER 1813. Journal of Human Evolution 17, 503–512.

Macho, G.A., Moggi-Cecchi, J., 1992. Reduction of maxillary molars in *Homo sapiens*: a different perspective. American Journal of Physical Anthropology 87, 151–159.

Matsumura, H., Nakatsukasa, M., Ishida, H., 1992. Comparative study of crown cusp areas in the upper and lower molars of African apes. Bulletin of the National Science Museum, Series D, 18, 1–15.

Miller, J.M.A., 1991. Does brain size variability provide evidence of multiple species in *Homo habilis*? American Journal of Physical Anthropology 84, 385–398.

Miller, J.M.A., 2000. Craniofacial variation in *Homo habilis*: an analysis of the evidence for multiple species. American Journal of Physical Anthropology 112, 103–128.

Moggi-Cecchi, J., Tobias, P.V., Beynon, A.D., 1998. The mixed dentition and associated skull fragments of a juvenile fossil hominid from Sterkfontein, South Africa. American Journal of Physical Anthropology 106, 425–466.

Moggi-Cecchi, J., Grine, F.E., Tobias, P.V., 2006. Early hominid dental remains from Members 4 and 5 of the Sterkfontein Formation (1966–1996 excavations): catalogue, individual associations, morphological descriptions and initial metrical analysis. Journal of Human Evolution 50, 239–328.

Morin, P.A., Moore, J.J., Chakraborty, R., Jin, L., Goodall, J., Woodruff, D.S., 1994. Kin selection, social structure, gene flow, and the evolution of chimpanzees. Science 265, 1193–1201.

Olson, T.R., 1978. Hominid phylogenetics and the existence of *Homo* in Member I of the Swartkrans Formation, South Africa. Journal of Human Evolution 7, 159–178.

Partridge, T.C., 2000. Hominid-bearing cave and tufa deposits. In: Partridge, T.C., Maud, R.R. (Eds), The Cenozoic of Southern Africa. Oxford University Press, New York, pp.100–125.

Prat, S., 2002. Anatomical study of the skull of the Kenyan specimen KNM-ER 1805: a re-valuation of its taxonomic allocation? Comptes Rendus Palevol 1, 27–33.

Rightmire, G.P., 1990. The Evolution of *Homo erectus*. Cambridge University Press, Cambridge.

Rightmire, G.P., 1993. Variation among early *Homo* crania from Olduvai Gorge and the Koobi Fora region. American Journal of Physical Anthropology 90, 1–33.

Rightmire, G.P., 1998. Evidence from facial morphology for similarity of Asian and African representatives of *Homo erectus*. American Journal of Physical Anthropology 106, 61–85.

Robinson, J.T., 1961. The australopithecines and their bearing on the origin of Man and of stone tool making. South African Journal of Science 57, 3–16.

Robinson, J.T., 1962. Australopithecines and artefacts at Sterkfontein. Part I. Sterkfontein stratigraphy and the significance of the Extension Site. South African Archaeological Bulletin 17, 87–107.

Ruvolo, M., Pan, D., Zehr, S., Goldberg, T., Disotell, T.R., von Dornum, M., 1994. Gene trees and hominoid phylogeny. Proceedings of the National Academy of Sciences of the USA 91, 8900–8904.

Schwartz, J.H., Tattersall, I., 2003. The Human Fossil Record. Volume 2. Craniodental Morphology of Genus *Homo* (Africa and Asia). Wiley-Liss, New York.

Smith, E., 1999. A functional analysis of molar morphometrics in living and fossil hominoids using 2-D digitized images. Ph.D. thesis, University of Toronto.

Smith, H.F., Grine, F.E., 2008. Cladistic analysis of early *Homo* crania from Swartkrans and Sterkfontein, South Africa. Journal of Human Evolution 54, 684–704.

Spoor, C.F., 1993. The comparative morphology and phylogeny of the human bony labyrinth. Doctoral thesis, University of Utrecht.

Steiper, M.E., 2006, Population history, biogeography, and taxonomy of orangutans (Genus: *Pongo*) based on a population genetic meta-analysis of multiple loci. Journal of Human Evolution 50, 509–522.

Strait, D.S., Grine, F.E., 2001. The systematics of *Australopithecus garhi*. Ludus Vitalis 9, 109–35.

Strait, D.S., Grine, F.E., Moniz, M.A., 1997. A reappraisal of early hominid phylogeny. Journal of Human Evolution 32, 17–82.

Stumpf, R.M., Polk, J.D., Oates, J.F., Jungers, W.L., Heesy, C.P., Groves, C.P., Fleagle, J.G., 2002. Patterns of diversity in gorilla cranial morphology. In: Taylor, A., Goldsmith, M. (Eds), Gorilla Biology: A Multidisciplinary Perspective. Cambridge University Press, Cambridge, pp. 35–61.

Suwa, G., 1988. Evolution of the "robust" australopithecines in the Omo succession: evidence from mandibular premolar morphology. In: Grine, F.E. (Ed), Evolutionary History of the "Robust" Australopithecines. Aldine de Gruyter, New York, pp. 199–222.

Suwa, G., 1990. A comparative analysis of hominid dental remains from the Shungura and Usno Formations, Omo Valley, Ethiopia. Ph.D. dissertation, University of California.

Suwa, G., Wood, B.A., White, T.D., 1994. Further analysis of mandibular molar crown and cusp areas in Pliocene and early Pleistocene hominids. American Journal of Physical Anthropology 93, 407–426.

Suwa, G., White, T.D., Howell, F.C., 1996. Mandibular postcanine dentition from the Shungura Formation, Ethiopia: crown morphology, taxonomic allocations, and Plio-Pleistocene hominid evolution. American Journal of Physical Anthropology 101, 247–282.

Swindler, D.R., 1976. Dentition of Living Primates. Academic, New York.

Tobias, P.V., 1965. *Australopithecus*, *Homo habilis*, tool-using, and tool-making. South African Archaeological Bulletin 20, 167–192.

Tobias, P.V., 1991. The Skulls, Endocasts and Teeth of *Homo habilis*. Olduvai Gorge, Vol. 4. Cambridge University Press, Cambridge.

Turner, A., Bishop, L., Denys, C., McKee, J.K., 1999. Appendix: a locality-based listing of African Plio-Pleistocene mammals. In: Bromage, T.G., Schrenk, F. (Eds), African Biogeography, Climate Change, and Human Evolution. Oxford University Press, New York, pp. 369–399.

Uchida, A., 1991. Taxonomy of *Sivapithecus* from the Chinji Formation, Siwalik sequence: morphology of the mandibular third molars. In: Ehara, A., Kimura, T., Takenaka, O., Iwamoto, M. (Eds), Primatology Today. Elsevier, Amsterdam, pp. 521–524.

Uchida, A., 1992. Intra-species variation among the great apes: implications for taxonomy of fossil hominoids. Ph.D. dissertation, Harvard University.

Uchida, A., 1998a. Design of the mandibular molar in the extant great apes and Miocene fossil hominoids. Anthropological Science 106, 119–126.

Uchida, A., 1998b. Variation in tooth morphology of *Gorilla gorilla*. Journal of Human Evolution 34, 55–70.

Ungar, P.S., Grine, F.E., Teaford, M.F., 2006. Diet in early *Homo*: a review of the evidence and a new model of adaptive versatility. Annual Review of Anthropology 35, 209–228.

Walker, A., 1993. Perspectives on the Nariokotome discovery. In: Walker, A., Leakey, R. (Eds), The Nariokotome *Homo erectus* Skeleton. Harvard University Press, Cambridge, MA, pp. 411–430.

Walker, A., 1994. Early *Homo* from 1.8–1.5 million year deposits at lake Turkana, Kenya. Courier Forschungsinstitut Senckenberg 171, 167–173.

White, T.D., 2003. Early hominids - diversity or distortion? Science 299, 1994–1997.

Wood, B.A., 1991. Hominid Cranial Remains. Koobi Fora Research Project, Vol. 4. Clarendon Press, Oxford.

Wood, B.A., 1992. Origin and evolution of the genus *Homo*. Nature 355, 783–790.

Wood, B.A., 1993. Early *Homo*: how many species? In: Kimbel, W.H., Martin, L.B. (Eds), Species, Species Concepts, and Primate Evolution. Plenum, New York, pp. 485–522.

Wood, B.A., Engleman, C.A., 1988. Analysis of the dental morphology of Plio-Pleistocene hominids. V. Maxillary postcanine tooth morphology. Journal of Anatomy 161, 1–35.

Wood, B.A., Uytterschaut, H., 1987. Analysis of the dental morphology of Plio-Pleistocene hominids. III. Mandibular premolar crowns. Journal of Anatomy 154, 121–156.

Wood, B.A., Xu, Q., 1991. Variation in the Lufeng dental remains. Journal of Human Evolution 20, 291–311.

Wood, B.A., Abbott, S.A., Graham, S.H., 1983. Analysis of the dental morphology of Plio- Pleistocene hominids. II. Mandibular molars - study of cusp areas, fissure pattern and cross sectional shape of the crown. Journal of Anatomy 137, 287–314.

Part III
Postcranial Perspectives on Locomotion and Adaptation

Chapter 7
Evolution of the Hominin Shoulder: Early *Homo*

Susan G. Larson

Keywords Shoulder morphology • scapula • clavicle • humerus • *Homo habilis* • *Homo erectus* • *Homo antecessor*

Introduction

The morphology of the human shoulder seems well-suited to serving its role of providing a wide range of mobility for the upper limb. The dorsal position of a highly mobile scapula and the lateral orientation of the small, shallow glenoid fossa articulating with a medially directed, large round humeral head open up nearly a complete sphere of motion allowing the hand to be positioned almost anywhere around the body. For a creature that depends a great deal on its hands to manipulate objects in order to survive, such an ideal pairing of form and function would seem to be a wonderful example of the elegance of design in nature. Of course the human shoulder did not arrive at its present configuration purely through natural selection for its current use. Humans share many features of upper limb morphology with extant apes suggesting that at least some aspects of human shoulder anatomy reflect our phylogenetic heritage. In order to understand which features of the human shoulder have been inherited from a common hominin ancestor, including primitive retentions that continue to serve their original purpose and those that have been co-opted to serve a new function, and which are uniquely derived in humans, it is necessary to determine the course of evolutionary change in the hominin pectoral girdle and shoulder. This paper attempts to trace those changes, beginning with early hominins, but focusing on the characteristics of the shoulder region in early *Homo erectus* (see Larson, 2007, for a more detailed overview of hominin shoulder evolution).

S.G. Larson (✉)
Department of Anatomical Sciences,
Stony Brook University Medical Center, Stony Brook,
NY 11794-8081, USA
e-mail: susan.larson@stonybrook.edu

Early Hominins

Scapula

Most interpretations of early hominin shoulder morphology have been based on two fossil scapular fragments: STS 7 (Broom et al., 1950) attributed to *Australopithecus africanus*, and A.L. 288-1l, (Johanson et al., 1982) attributed to *A. afarensis* (Table 7.1). The feature that has received the most attention in functional analyses of these fossils is the orientation of the glenoid fossa. In hominoid primates the glenoid faces cranially reflecting the importance of overhead limb postures, while in modern humans the fossa faces more laterally reflecting the typical lowered position of the upper limb. Several workers have measured the axillo-glenoid angle of STS 7 (e.g., Campbell, 1966; Oxnard, 1968; Robinson, 1972; Vrba, 1979), and while the reported values vary (ranging between 103° and 125°) all agree that the glenoid faced more cranially than in modern humans suggesting retention of some arboreal adaptations in this early hominin. Stern and Susman (1983) attempted to assess the orientation of the glenoid fossa of A.L. 288-1l relative to the ventral bar, a scapular buttress lying medial to the axillary border, since little of the axillary border is preserved in this specimen. They conclude that the glenoid of this scapular fragment also faces more cranially than in modern humans. Although Inouye and Shea (1997) argue that the cranial orientation of the glenoid of A.L. 288-1 is due to its small size, they base their analysis on ontogenetic scaling of bar-glenoid angle in African apes and modern humans and thus confound age-related changes with possible size effects. Finally, the recently discovered *A. afarensis* juvenile scapula DIK-1-1 from Dikika, Ethiopia (Alemseged et al., 2006), also has a cranially directly glenoid fossa.

The size and shape of the coracoid process is another feature of the STS 7 scapula that has received some attention in the literature (e.g., Campbell, 1966; Oxnard, 1968; Robinson, 1972; Roberts, 1974; Ciochon and Corruccini, 1976; Vrba, 1979). Vrba (1979) notes that the coracoid of STS 7 displays a prominent dorsolateral tubercle placed somewhat more laterally than in modern humans, features which she interprets

Table 7.1 Early hominin pectoral girdle material

	Ardipithecus	*Au. afarensis*	*Au. africanus*	*Australopithecus* sp.	*Homo habilis*
Clavicle	STD-VP-2/893	A.L. 333x-6/9	StW 431	StW 606	OH 48
		A.L. 333-94	StW 582		KNM-ER 3735
		A.L. 288-lbz			
		A.L. 438-lv			
		L.H. 21P			
Scapula		A.L. 288-11	STS 7		
		DIK-1-1	StW 366		
			StW 431		
Proximal Humerus	ARA-VP-7/2	A.L. 288-1r	STS 7	Omo 119-73-2718	
		A.L. 333-87	StW 328	KNM-ER 1473	
		A.L. 333-107	StW 517		
		KNM-BC 1745			

to reflect a scapula positioned high on a funnel-shaped thorax with an oblique clavicle as in modern great apes.

Clavicle

Although there are a number of fossil clavicular specimens known for early hominins (Table 7.1), most are only small segments and attempts to interpret them have been limited. The most complete specimen missing only a portion of its sternal end is A.L. 333x-6/9 (Lovejoy et al., 1982) attributed to *A. afarensis*. On the basis of the orientation of its lateral end on frontal view and the position of the deltoid attachment area, Ohman (1986) concludes that A.L. 333x-6/9 is distinct from extant hominoids, which he interprets as evidence of descent of hominin shoulder by three to four million years ago. However, Voisin's (2006) analysis of clavicular shape in primates indicates that the low scapular position of humans is reflected in curvature of the medial half of the clavicle (in A/P view) rather than that of the lateral end. Since the medial half of A.L. 333x-6/9 is apparently not unlike that of extant hominoids, it is possible that *A. afarensis* still retained a similar high shoulder position. More recently, Partridge et al. (2003) have described a partial clavicle from Jacovec Cavern at Sterkfontein (StW 606), attributed to *Australopithecus* sp., and note that it displays a pronounced conoid tubercle like chimpanzees and unlike other hominin clavicles including modern humans. This suggests that there may be some diversity in clavicular morphology among early hominins. However in general, these clavicular fossils offer very limited information on pectoral girdle/shoulder form and function in early hominins.

Proximal Humerus

To the degree that it can be determined, the head of all known early hominin humeri is elliptical rather than spherical in shape as in apes, and the intertubercular groove is somewhat shallow, rather than deep and tunnel-like as in African apes. Broom et al. (1950) describe the STS 7 proximal humerus as basically humanlike, although the tubercles are somewhat distinct, with a more prominent lesser tubercle than typically seen in modern humans or great apes. The Chemeron proximal humeral fragment KNM-BC 1745 and the A.L. 288-1r proximal humerus are also described as displaying relatively large lesser tubercles (Pickford et al., 1983; Johanson et al., 1982). Robinson (1972) notes that, unlike most modern human humeri, STS 7 displays a prominent ridge on the greater tubercle separating the facets for attachment of supraspinatus and infraspinatus. Although the area of attachment for supraspinatus is not preserved in A.L. 288-1r, there is a ridge that would have separated it from the clearly defined ovoid depression for the attachment of infraspinatus (Johanson et al., 1982). Lovejoy et al. (1982) describe similar separation of the facets for attachment of the dorsal rotator cuff muscles on the greater tubercle of A.L. 333-107.

Humeral torsion refers to the orientation of the humeral head relative to the distal articular surface. Since the degree of humeral torsion is high in both modern humans and African apes (Evans and Krahl, 1945; Krahl and Evans, 1945; Krahl, 1947), this condition is often cited as a shared derived feature of apes and humans (e.g., Le Gros Clark, 1959; Andrews, 1985; Martin, 1986; Harrison, 1987), in which case early hominin humeri should also display a high degree of torsion. Unfortunately, all known early hominin humeri are incomplete, making direct measurement of humeral torsion impossible. However, Larson (1996) has developed methodologies for estimating the degree of humeral torsion on incomplete humeri using regression analysis and a set of alternative references axes. Contrary to expectations, she reports modest to low levels of torsion for A.L. 288-1r, STS 7, Omo 119-73-2718, and KNM-ER 739 (Fig. 7.1). Larson (1996) concludes that the high torsion of modern humans is therefore a more recently acquired characteristic, and its similarity to other hominoids is due to convergence.

In sum, the pectoral girdle/shoulder of early hominins appears to have retained many features of the presumed ancestral condition. The scapula was positioned high on a funnel-shaped thorax (see also Schmid, 1983), the clavicle was

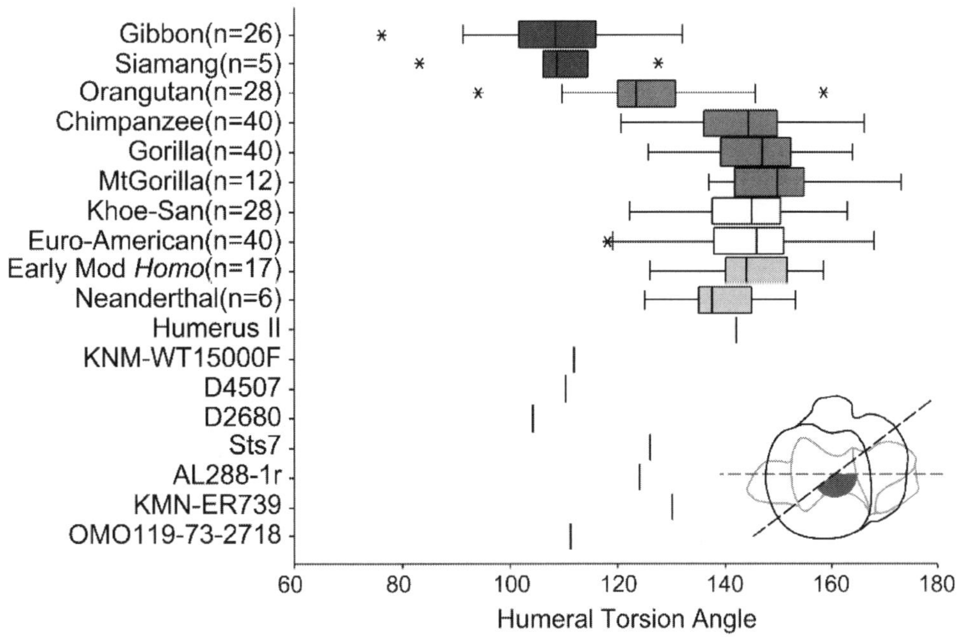

Fig. 7.1 Box and whisker plots of humeral torsion for comparative samples of apes, modern humans, and fossils. Comparative data (except Khoe-San sample) and early hominin fossil torsion estimates are from Larson (1996). Fred Grine and Louise Jacqui Friedling collected the African Khoe-San humeral data. Torsion data for Neandertals (Lezetxiki 1, Régourdou 1, Neanderthal 1, La Chapelle 1, Tabun C1, Kebara 2) and Early Modern *Homo* (Skhul IV, Qafzeh 9, and humeri from 15 Early Upper Paleolithic sites) are from Churchill (1994). Torsion estimates for the juvenile (D2850) and adult (D4507) humeri from Dmanisi are from Lordkipanidze et al. (2007). The humerus II (*H. heidelbergensis*) torsion value is from Carretero et al. (1997). The lesser apes have a fairly low degree of humeral torsion, orangutans display an intermediate amount of torsion, and the African apes have the highest torsion values among hominoids. Modern human samples also display a high degree of torsion. Estimated torsion for early hominins is much lower than modern human values.

oriented obliquely, and the glenoid fossa was cranially directed. However, the intertubercular groove was relatively shallow, the humeral head was elliptical, and the humerus displayed only modest torsion. Putting aside any implications that this morphology may have regarding reconstructing the lifestyle of early hominins, this configuration can be taken as the starting point from which the shoulder of early *Homo* was derived.

Homo habilis

Unfortunately, very little shoulder material is known for *H. habilis* (Table 7.1). The most complete specimen is the OH 48 clavicle, which Napier (1965) describes as basically human-like except for the cross-sectional shape of the medial end. Based on the orientation of the long axis of this cross-section, he concludes that the clavicle would have been rotated slightly around its longitudinal axis and the shoulder positioned higher than in modern humans to sit on a thorax with a steep inlet. Oxnard (1969) reports a significantly higher degree of torsion in the OH 48 clavicle than in modern humans, and concurs that it would have been twisted cranially and the shoulder positioned more superiorly, which he interprets as reflecting some ability for upper limb suspension. Day (1978) however argues that the missing ends of the specimen make any measure of torsion unreliable, and emphasizes the basically human appearance of the fossil, a perspective echoed by Ohman (1986). The only other shoulder remains attributed to *H. habilis* are the lateral portion of a clavicle and a small piece of the scapular spine from the KNM-ER 3735 partial skeleton. Leakey et al. (1989) interpret the thickness of the latter as well as the large size of other forelimb features of KNM-ER 3735 as evidence of substantial climbing ability in *H. habilis*. Based on this very limited sample, therefore, it is possible that early *Homo* continued to possess a somewhat primitive shoulder configuration like that of earlier hominins.

Early *Homo erectus*

The best known early *H. erectus* specimen is the KNM-WT 15000 juvenile partial skeleton from the Nariokotome region of Northern Kenya, which includes both clavicles, one nearly complete and one partial scapula, and a humerus missing only its proximal epiphysis and that of the medial epicondyle (Walker and Leakey, 1993). A second juvenile skeleton of early *H. erectus* as well as the postcranial material from three adult individuals from Dmanisi, Georgia have been recently described by Lordkipanidze et al. (2007). The juvenile partial skeleton includes a nearly complete clavicle (D2724), an incomplete right humerus (D2715) and a nearly complete left humerus missing only its proximal epiphysis (D2680). Most of the adult postcranial material from Dmanisi is attributed to one large individual and includes incomplete right (D4162) and left clavicles (D4161), part of a right scapula (D4166), and a left humerus missing its proximal epiphysis (D4507).

Clavicle

The clavicles from Nariokotome and Dmanisi all display the typical S-shape curvature seen in modern humans and African apes. Walker and Leakey (1993) describe the acromial ends of Nariokotome clavicles as flattened superiorly and somewhat concave inferiorly. Moving medially, the body of the bone twists by about 30° and becomes more rounded in contour. The sternal end is ovoid, and on the inferior surface there is a low, blunt conoid tubercle approximately one quarter of the way from the lateral end. However, Lordkipanidze et al. (2007) report that in mid-shaft and conoid tubercular cross-sectional shape, the Dmanisi clavicles are unlike those of KNM-WT 15000 and more like those of modern humans.

Scapula

The right scapula of the Nariokotome boy (KNM-WT 15000E), though reconstructed from fragments, is the more complete of the pair (Walker and Leakey, 1993). The juvenile age is evident in the missing regions of the superior and inferior angles. According to measurements of a cast for this study, the axillo-glenoid angle is 147°, and glenoid-ventral bar angle is 152°. In contrast, Lordkipanidze et al. (2007) report an axillo-glenoid angle of only 129° the for the Dmanisi D4166 scapular fragment, concluding that the glenoid fossa faced more cranially than in humans. However, much of the axillary border of this specimen is missing making reliable measurement of this angle difficult. In addition, since Lordkipanidze et al. (2007) report an axillo-glenoid angle of only 127° for KNM-WT 15000, considerably below my estimate of 147°; differences in measuring conventions may also be confounding comparisons. Based on a photograph of the D4166 scapular fragment, I estimate the axillo-glenoid angle to be approximately 135°, and therefore within the range of modern humans. While the correct glenoid orientation of D4166 requires further study, at least according to the nearly complete KNM-WT 15000 specimen, the glenoid fossa of early *H. erectus* no longer faced cranially as it did in earlier hominins.

Proximal Humerus

Since the humeri from Nariokotome and Dmanisi are all missing their proximal epiphyses, it is not possible to determine the condition of the head or tubercles. The intertubercular groove of KNM-WT 15000 is wide and shallow, and its shaft, like those of the humeri from Dmanisi, is straight (Walker and Leakey, 1993; Lordkipanidze et al., 2007). Larson (2007) reports that the degree of humeral torsion for KNM-WT15000 was only 111.5°, and Lordkipanidze et al. (2007) similarly report low humeral torsion values for the Dmanisi hominins: 104° for the juvenile humerus (D2850), and 110° for the adult (D4507). These values fall well below average values published for modern humans (see Fig. 7.1).

Strong humeral torsion is most closely associated with the morphological changes in hominoid evolution resulting in a dorsally placed scapula so the glenoid fossa faces more laterally than ventrally, and the humeral head must be directed medially to articulate with it. However, change in the position and orientation of the glenoid only requires alteration of the degree of humeral torsion if it is necessary for the elbow to continue to operate in a parasagittal plane (Inman et al., 1944). Hylobatids, for example, have dorsally placed scapulae and laterally (as well as cranially) facing glenoids, but only limited humeral torsion (Evans and Krahl, 1945; Le Gros Clark and Thomas, 1951; Zapfe, 1960; Larson, 1988, 1996). As a consequence, their elbows have a lateral set; at rest the cubital fossa of their elbow faces more laterally than anteriorly (Larson, 1988). This morphology dramatically increases the range of external rotation at the shoulder and is advantageous during arm-swinging, but is purchased at the price of a reduced range of internal humeral rotation, hence the lateral set to the elbow. Such a non-parasagittally operating elbow joint would seem disadvantageous for a tool-making hominin such a *H. erectus*.

Claviculohumeral Ratio

Because the KNM-WT15000 skeleton includes all of the pectoral girdle/shoulder elements, it is possible to further explore this region to try and determine how a shoulder with such limited humeral torsion could have functioned. The claviculohumeral ratio is a commonly used measure of relative clavicular length (e.g., Schultz, 1930, 1937; McCown and Keith, 1939; Martin and Saller, 1959; Marquer, 1972). The estimated total length of 319 mm for the KNM-WT 15000 humerus reported by Walker and Leakey (1993) yields a claviculohumeral ratio of 40.89 for the Nariokotome boy. Figure 7.2 presents comparative data on claviculohumeral ratios for samples of apes and humans including Euro-Americans and several African populations. KNM-WT15000 falls at the lower fringes of modern human populations and overlaps with apes. However, both clavicular and humeral length change with age, and since the Nariokotome boy is immature, would he have had a more human-like claviculohumeral ratio as an adult? Jungers and Hartman (1988) report that humeral length displays isometric growth allometry in great apes and slight positive growth allometry in humans, while clavicular length displays negative growth allometry in all taxa. Therefore, no matter whether KNM-WT15000 followed a great ape or human growth trajectory, if the Nariokotome boy had reached adulthood these scaling patterns would have resulted in an even shorter relative clavicular length, and the 40.89 claviculohumeral ratio reported in Fig. 7.2 is likely to be an overestimate. Although Lordkipanidze et al. (2007) do not report a claviculohumeral ratio for any of the Dmanisi hominins, they describe the nearly complete subadult clavicle D2724 as being comparatively short, supporting the interpretation of early *H. erectus* displaying a relatively short clavicle.

A relatively short clavicle is surprising for early *H. erectus* since relative clavicular elongation is another trait that is considered to be a shared derived feature of humans and extant hominoids (Le Gros Clark, 1959; Ciochon, 1983; Andrews, 1985; Martin, 1986; Harrison, 1987). It is therefore worth considering whether differences in claviculohumeral ratios between apes and humans could be due to differences in humeral length. Jungers (1994) has shown that humerus length has a conservative scaling relationship to body mass, with relative humeral length being virtually identical in different sized human populations as well as in African apes. To further explore what the primitive condition for relative clavicle length might be, mean clavicular and humeral lengths were collected from the literature and are displayed in Fig. 7.3. Regression analysis of the comparative nonhuman primate data reveals a linear relationship (r = 0.97) passing through the origin, which indicates an isometric scaling relationship between clavicular and humeral length (Mosimann, 1970; Jungers et al., 1995). The close adherence of most nonhuman taxa to this scaling relationship suggests that it may represent the primitive condition for primates. Among the outliers are baboons and *Ateles*, both of which fall well below the line, which could be due to reduced clavicular length in the former and elongation of the humerus in the latter, but this is pure speculation. However, even though the lesser apes also have elongated forelimbs, they appear to follow the isometric scaling relationship quite closely. Among great apes, chimpanzees fall slightly above the line while

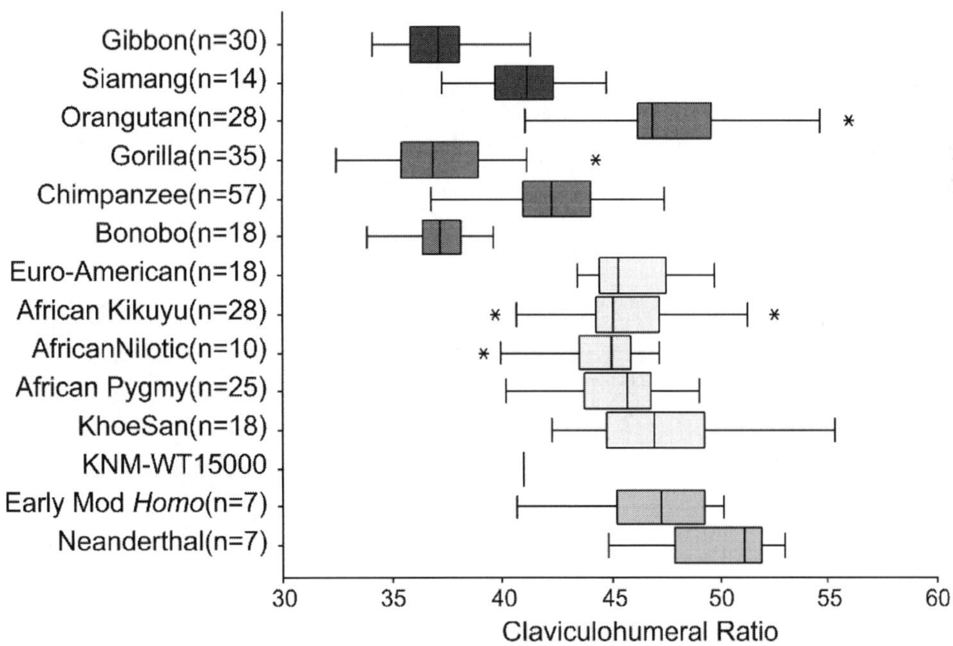

Fig. 7.2 Box and whisker plots of claviculohumeral length ratios for comparative samples and fossils. Comparative ape, African pygmy and Euro-American data were provided by William Jungers. Fred Grine and Louise Jacqui Friedling collected clavicular and humeral data for the African Khoe-San sample. African Kikuyu and Nilotic data were provided by Chris Ruff. Early modern *Homo* sample includes: Abri Pataud 5 (Churchill, 1994), Jebel Sahaba, Wadi Kubbaniya (Angel and Kelley, 1986), Doni Věstonice 13 & 15 (Sládek et al., 2000), and Skuhl IV & V (McCown and Keith, 1939). Neandertal sample includes: Kebara 2 (Churchill, 1994), Shanidar 1 & 3, Régourdou 1, Tabūn C1, La Ferrassie 1 (Trinkaus, 1983), and Neanderthal 1 (McCown and Keith, 1939). The samples of modern humans vary in average in stature yet all have similar claviculohumeral ratios, which are consistently higher than those of apes except for orangutans. Relative clavicular length for KNM-WT 15000 is more similar to apes than modern humans. Neanderthals appear to have the longest clavicles among hominins.

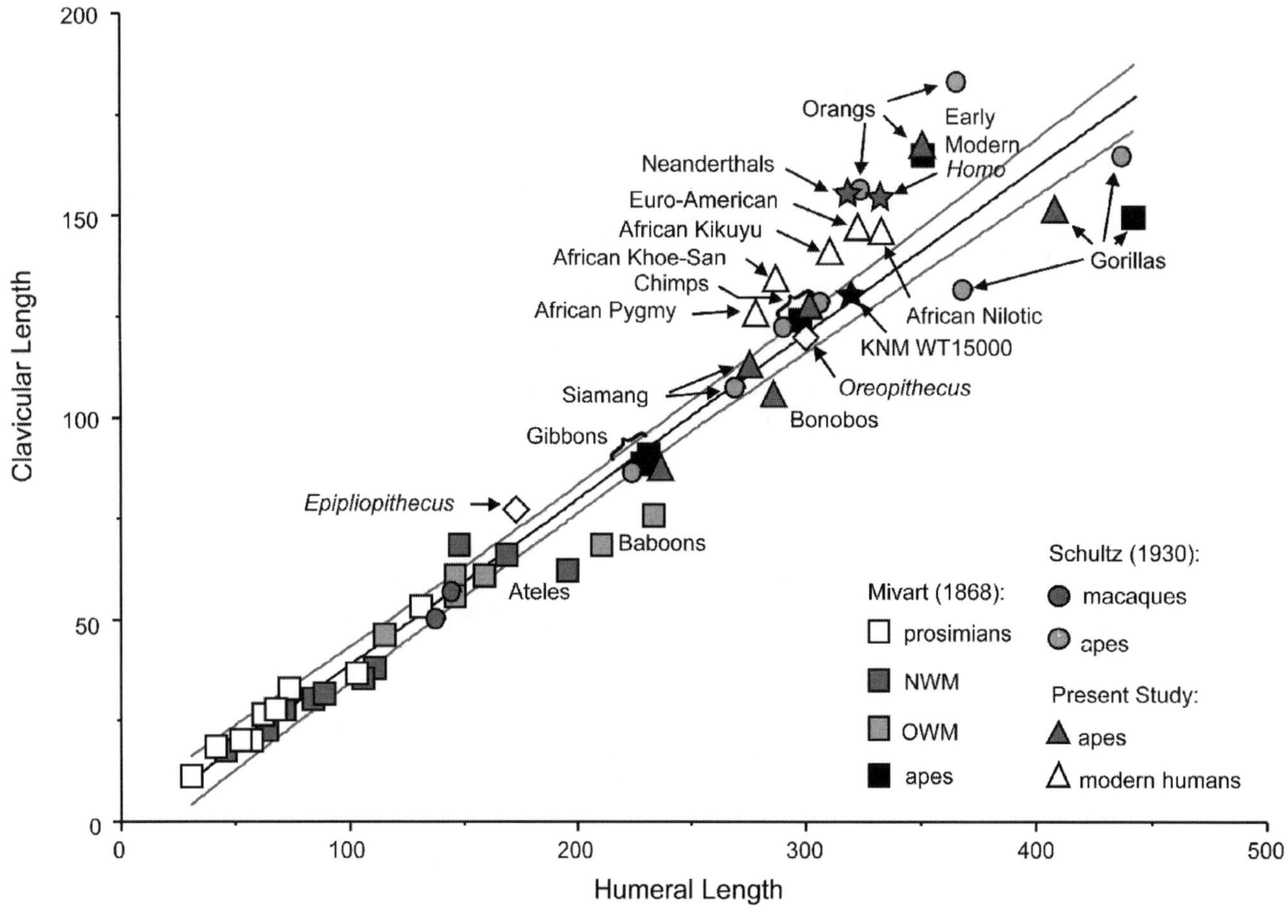

Fig. 7.3 Scatter plot of mean clavicular length against mean humeral length in nonhuman primates, modern human populations, and hominin fossils. Squares indicate data derived from Mivart (1868); circles indicate data from Schultz (1930); triangles represent data provided by William Jungers (apes and African pygmies), Chris Ruff (African Kikuyu and Nilotics), and Fred Grine and Louise Jacqui Friedling (Khoe-San). Early modern *Homo* sample (grey star) includes: Abri Pataud 5 (Churchill, 1994), Jebel Sahaba, Wadi Kubbaniya (Angel and Kelley, 1986), Doni Věstonice 13 & 15 (Sládek et al., 2000), and Skuhl IV & V (McCown and Keith, 1939). Neandertal sample (grey star) includes: Kebara 2 (Churchill, 1994), Shanidar 1 & 3, Régourdou 1, Tabūn C1, La Ferrassie 1 (Trinkaus, 1983), and Neandertal 1 (McCown and Keith, 1939). Black star indicates KNM-WT 15000. Clavicular and humeral lengths for *Epipliopithecus* were measured on casts. An estimate of clavicular length for *Oreopithecus* was provided by Terry Harrison, and humeral length is from Harrison (1986). Ratios for the latter two taxa are included to offer an indication of relative clavicular length in Miocene hominoids. Regression line (with 95% confidence intervals) is for nonhuman primates only and has a correlation coefficient of 0.97. Since it passes through the origin, it indicates an isometric scaling relationship across primate species. Assuming that this linear relationship represents the primitive condition for primates, orangutans and all modern human populations as well as later fossil hominins display relative clavicular elongation. However, KNM-WT 15000 appears to retain the primitive condition.

both bonobos and gorillas lie somewhat below. At the same time, orangutans are highly divergent above the line indicating that they have elongated clavicles relative to their humeri, despite the fact that they too have elongated forelimbs. All of the means for the modern human populations are also above the line, as are the means for samples of early modern *Homo* and Neandertals. If the common isometric scaling relationship seen across nonhuman primates does indeed represent the primitive condition, then modern humans and recent fossil hominins all exhibit the derived condition of relative clavicular elongation. KNM-WT 15000, however, falls very close to the line suggesting that it retains the primitive condition, as do the African and lesser apes.

Pectoral Girdle/Shoulder Configuration in Early *H. erectus*

The shoulder region of the early *H. erectus* presents an unexpected combination of features, including relatively short clavicles and very low humeral torsion, but with a fairly modern looking scapula whose glenoid did not face cranially sitting on a barrel-shaped rib cage (judging from the shape of the ribs of KNM-WT 15000 (Jellema et al., 1993)). Although relative clavicular length is currently unknown for early hominins, if the isometric scaling relationship portrayed in Fig. 7.3 does represent the primitive condition, it seems likely that their clavicles were relatively short as well. In addition, since the estimates for humeral torsion in early hominins offered by Larson (1996) are low (Fig. 7.1), the minimal degree of torsion in the KNM-WT 15000 and Dmanisi humeri may also be a retained primitive condition. However, if these presumed primitive and derived features are combined in a modern human shoulder configuration (i.e., dorsally placed scapula with laterally orientated glenoid fossa), the resulting shoulder would seem to have poor functional capability due to the lateral set to the elbow. Perhaps then, the pectoral girdle/shoulder of early *H. erectus* was not configured as in modern humans. If not, the existence of a different configuration would indicate an unexpected transitional state in the evolution of the hominin shoulder.

A possible scenario for evolutionary transformation of the hominin shoulder from a roughly ape-like condition in early hominins (Fig. 7.4a) to something like that of modern humans (Fig. 7.4c), that could account for the morphology seen in early *H. erectus* is as follows: In response to growing dependence on tool-making and tool-using, it would seem likely that an important change from the ape-like early hominin condition would have been a drop in the position of the scapula on the thorax and a loss of the cranial orientation to the glenoid fossa, especially as reliance on use of the forelimb in overhead supporting postures decreased along with the frequency of arboreality. Since the reorientation of the glenoid had occurred at least by the time of early *H. erectus*, presumably the downward shift of the scapula had as well. Although the change in glenoid orientation involved morphological transformation of the scapula, if we imagine this reorientation and shift in scapular position as if they were brought about by a "glenoid-down" rotation of the scapula, the effect of a relatively short clavicle can be viewed as a constraint on the repositioning of the scapula on the thorax. The short

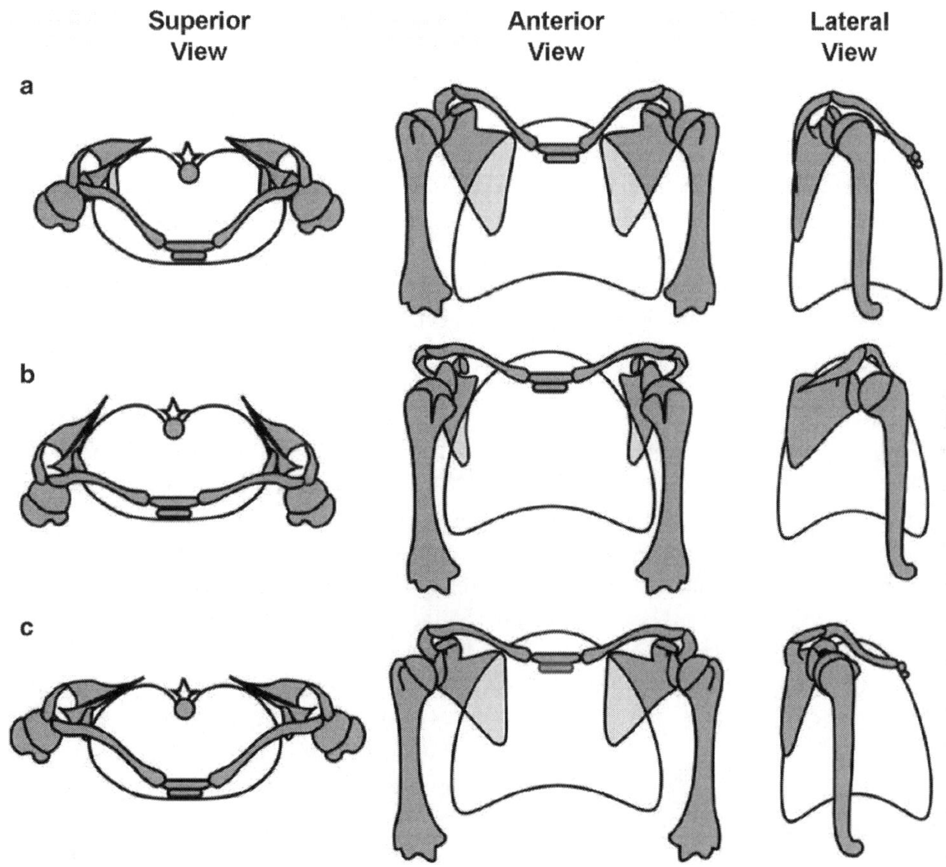

Fig. 7.4 Proposed course of hominin pectoral girdle evolution. (**a**) Superior, anterior and lateral schematic views of thorax showing pectoral girdle and shoulder of early hominin condition. Scapulae are dorsally positioned with a cranially facing glenoids, and clavicles are short and oriented obliquely resulting in a "hunched-shoulder" appearance. Humerus displays low to modest torsion. (**b**) Proposed transitional stage in hominin pectoral girdle evolution in which change from a cranially oriented glenoid fossa has been brought about in a way analogous to a downward rotation and translation of the scapula, constrained by a relatively short clavicle. Scapulae are more laterally positioned, and glenoid fossae face anteriorly. Parasagittal functioning of the elbow joint is maintained without major increases in humeral torsion. Such a configuration would explain the low degree of humeral torsion and relatively short clavicles seen in early *H. erectus* (KNM-WT 15000). (**c**) Pectoral girdle and shoulder of a modern human with elongated clavicles, dorsally positioned scapulae and laterally facing glenoid fossae. The humerus displays marked torsion to maintain a parasagittal plane for elbow function.

clavicle resulted in a forward as well as downward shift in position so that the scapula came to be more laterally positioned on the thorax (Fig. 7.4b). As a consequence, the glenoid fossa came to face anteriorly, and thus a humeral head that faced posteriorly, that is, had low to modest torsion, would produce an elbow that functioned in a parasagittal plane putting no restrictions on using the forelimbs for manipulation. This scenario therefore could explain the combination of pectoral girdle/shoulder features observed in early *H. erectus*.

Remarkably, an abnormality found occasionally in modern humans known as short or hypoplastic clavicle syndrome (Milgram, 1942; Guidera et al., 1991; Beals, 2000; Beals and Sauser, 2006) appears to closely mimic this configuration. Individuals typically present with significant forward displacement of the shoulders due to laterally positioned scapulae and resulting anteriorly directed glenoids. The vertebral borders of the scapulae are widely separated and are often prominent posteriorly. Other than their shortened length, the clavicles are normal in appearance. There are typically no other associated abnormalities or upper extremity dysfunction, and the chief complaint is abnormal posture. Unfortunately, the case studies describing short clavicle syndrome tend to be brief and offer little information beyond the appearance of the patient's shoulders. Although none report on the degree of humeral torsion in patients with this condition, Guidera et al. (1991) notes that the humeri are located anteriorly on CT scans. Given that humeral torsion is somewhat plastic developmentally (Krahl, 1947; Edelson, 1999, 2000), it would seem likely that these individuals have reduced humeral torsion to accommodate the anterior orientation of their glenoid fossae.

Indeed, the pronounced similarity between individuals displaying short clavicle syndrome and the shoulder configuration suggested here for early *H. erectus* leads one to wonder whether we are simply seeing examples of this syndrome in the fossil record. However, it would seem unlikely that both the Nariokotome boy and the Dmanisi hominins would display the identical pathological condition, suggesting that the pectoral girdle/shoulder configuration described above is truly characteristic of early *H. erectus*.

Scenarios for Evolution of Modern Human Shoulder Configuration

The observations summarized above indicate that a more protracted scapula with an anteriorly directed glenoid and low humeral torsion formed an intermediate stage between the more primitive condition of early hominins and the configuration of the pectoral girdle/shoulder in modern humans. What might the stimulus have been for this latter change, returning the scapula to a more dorsal position so that the glenoid fossa faced laterally (Fig. 7.4c), concomitantly requiring an increase in humeral torsion such as occurred in later hominin evolution? Such a shift in scapular position would dramatically increase the range of upper limb motion, particularly in the posterior direction. One potential factor stimulating such a change is throwing, which entails a significant component of posterior motion of the abducted arm during the cocking phase (e.g., Tullos and King, 1973; Atwater, 1979; Perry, 1983). As long as people have been attempting to explain the origins of upright posture and bipedalism, the throwing of objects for self-defense, hunting, etc., has been included as a significant factor contributing to the survival and success of the human lineage (e.g., Darlington, 1975; Calvin, 1983; Bingham, 1999; Dunsworth et al., 2003). The anterior position of the shoulder for early *H. erectus* postulated here would not have permitted the abducted arm posture that is an integral component of the form of overhand throwing with which we are familiar today. It is interesting to note in this context, that one incidental complaint of people with short clavicle syndrome is that they cannot throw well (Guidera et al., 1991; Beals, 2000). Effective throwing, therefore, could have an impetus for transformation of the pectoral girdle/shoulder complex from the proposed *H. erectus* condition to that of modern humans.

A second potential selective force for clavicular elongation leading to dorsal scapular positioning with laterally facing glenohumeral joints is running, which requires shoulder and upper body rotation to counteract the destabilizing torque created by the oppositely moving lower limbs. Although running ability to achieve higher speeds has obvious advantages, Carrier (1984) and more recently Bramble and Lieberman (2004) have argued that endurance running in particular may have played a major role in shaping the modern human body form. Indeed, Bramble and Lieberman (2004) suggest that endurance running may have been among the factors leading to the emergence of the genus *Homo*. However, the scenario for change in shoulder morphology presented here suggests that early *H. erectus* did not have particularly wide shoulders, and by inference, neither did earlier members of the genus. Although this would not have made running impossible for early *Homo*, the fact that their shoulders were narrow suggests that an effective upper body counter-rotation mechanism was not yet an important selective factor. As Bramble and Lieberman (2004) note, several of the changes in lower limb morphology seen in early *Homo* could also be explained as adaptations to long distance walking. However, running, whether for speed or endurance, could well have been an impetus for clavicular elongation to spread the shoulders apart in order to enhance the upper body counter-rotation mechanism at a somewhat later stage of human evolution.

Homo antecessor

Neanderthals as well as fossils attributed to early modern *Homo* display claviculohumeral ratios and degree of humeral torsion more comparable to those of modern humans. Therefore, at some point between early *H. erectus* and the appearance of these later fossil hominins, the clavicle underwent elongation to reposition the scapula more dorsally with a concomitant increase in humeral torsion. When might these changes in shoulder configuration have first appeared? The Early Pleistocene site of Gran Dolina, Sierra de Atapuerca, Spain has yielded a variety of postcranial remains attributed to *H. antecessor* (Carretero et al., 1999), including a complete adult clavicle (ATD 6-50), as well as one complete and one partial subadult clavicle. Although no humeri are known from the site with which to calculate a claviculohumeral index, the adult clavicle ATD 6-50 is absolutely quite long, falling at the upper fringes of size ranges for modern human samples (Carretero et al., 1999). It is possible, therefore, that *H. antecessor*, which has been proposed to represent the last common ancestor of *H. sapiens* and *H. neanderthalensis* (Bermudez de Castro et al., 1997), exhibits the clavicular elongation that is also seen in both later taxa. If this proves to be the case, then when a humerus is recovered from Gran Dolina, it should also display a degree of humeral torsion more comparable to that of later hominins.

Conclusions

The available fossil evidence documenting the evolution of the hominin shoulder and pectoral girdle indicates that despite making and using tools, early hominins retained much of the presumed ancestral condition in regard to shoulder morphology. The first major structural reorganization occurred in early *Homo erectus* with the loss of a cranial orientation to the glenoid fossa, probably accompanied by a caudal shift in scapular position. However, these changes appear to have been constrained by the retention of a relatively short clavicle so that the scapula not only shifted caudally, but also somewhat anteriorly. The transition in glenoid fossa orientation, therefore, was not from cranial to lateral, rather from cranial to anterior. This permitted parasagittal elbow function to be maintained with a humeral head that displayed only low to modest torsion, probably another retained primitive characteristic. Such a configuration with anteriorly positioned shoulder joints would have placed no limitations on manipulatory activities, but would not have permitted a very large range of posterior motion of an abducted upper limb.

The transformation of this configuration to something more like that of later hominins was accomplished largely by relative clavicle elongation. Such a change would spread the shoulders apart, pushing the scapula into a dorsal rather than lateral position on the rib cage, and the glenoid fossae would thereby come to face laterally. A significant increase in humeral torsion would also be necessary in order to maintain a parasagittally functioning elbow joint. Why such changes may have occurred is a matter of speculation, but a likely selective factor favoring these changes would be the advantages accrued from the dramatic increase in range of motion at the shoulder that would follow from this transformation.

References

Alemseged, A., Spoor, F., Kimbel, W.H., Bobe, R. Geraads, D., Reed, D., Wynn, J.G., 2006. A juvenile early hominin skeleton from Dikika, Ethiopia. Nature 443, 296–301.

Andrews, P., 1985. Family group systematics and evolution among catarrhine primates. In: Delson, E. (Ed.), Ancestors: The Hard Evidence. Alan R. Liss, New York, pp. 14–22.

Angel, J.L., Kelley, J.O., 1986. Description and comparison of the skeleton. In: Wendorf, F., Schild, R. (Eds.), The Wadi Kubbaniya Skeleton: A Late Paleolithic Burial from Southern Egypt. Southern Methodist University Press, Dallas, pp. 53–70.

Atwater, A.E., 1979. Biomechanics of overarm throwing movements and of throwing injuries. Exercise and Sport Sciences Reviews 7, 43–85.

Beals, R.K., 2000. The short clavicle syndrome. Journal of Pediatric Orthopedics 30, 389–391.

Beals, R.K., Sauser, D.D., 2006. Nontraumatic disorders of the clavicle. Journal of the American Academy Orthopedic Surgery 14, 205–214.

Bermúdez de Castro, J.M., Arsuaga, J.L., Carbonell, E., Rosas, A., Martínez, I., Mosquera, M., 1997. A hominid from the lower Pleistocene of Atapuerca, Spain: possible ancestor to Neandertals and modern humans. Science 276, 1392–1395.

Bingham, P.M., 1999. Human uniqueness: a general theory. Quarterly Review of Biology 74, 133–169.

Bramble, D.M., Lieberman, D.E., 2004. Endurance running and the evolution of *Homo*. Nature 432, 345–352.

Broom, R., Robinson, J.T., Schepers, G.W.H., 1950. Sterkfontein ape-men *Plesianthropus*. Transvaal Museum Memoir No. 4.

Campbell, B.G., 1966. Human Evolution. Heinemann Ed. Books, London.

Calvin, W.H., 1983. A stone's throw and its launch window: timing and precision and its implications for language and hominid brains. Journal of Theoretical Biology 104, 121–135.

Carretero, J.M., Arsuaga, J.L., Lorenzo, C. 1997. Clavicles, scapulae and humeri from the Sima de los Huesos site (Sierra de Atapuerca, Spain). Journal of Human Evolution 33, 357–408.

Carretero, J.M, Lorenzo, D., Arsuage, J.L., 1999. Axial and appendicular skeleton of *Homo antecessor*. Journal of Human Evolution 37, 459–499.

Carrier, D.R., 1984. The energetic paradox of human running and hominid evolution. Current Anthropology 25, 483–495.

Ciochon, R.L., 1983. Hominoid cladistics and the ancestry of modern apes and humans. In: Corruccini, R.S., Ciochon, R.L. (Eds), New Interpretations of Ape and Human Ancestry. Academic, New York, pp. 783–843.

Ciochon, R.L., Corruccini, R.S., 1976. Shoulder joint of Sterkfontein *Australopithecus*. South African Journal of Science 72, 80–82.

Churchill, S.E., 1994. Human upper body evolution in the Eurasian later Pleistocene. Ph.D. dissertation, University of New Mexico, Albuquerque.

Darlington, P.J., 1975. Group selection, altruism, reinforcement, and throwing in human evolution. Proceeding of the National Academy of Sciences 72, 3748–3752.

Day, M.H., 1978. Hominid postcranial material from Bed I, Olduvai Gorge. In: Isaac, G.L., McCown, E.R. (Eds), Human Origins: Louis Leakey and the East African Evidence. W.A. Benjamin, Menlo Park, pp. 363–374.

Dunsworth, H., Challis, J.H., Walker, A., 2003. Throwing and bipedalism: a new look at an old idea. Courier Forschungsinstitut Senckenberg 243, 105–110.

Edelson G., 1999. Variations in the retroversion of the humeral head. Journal of Shoulder and Elbow Surgery 8, 142–145.

Edelson, G., 2000. The development of humeral head retroversion. Journal of Shoulder and Elbow Surgery 9, 316–318.

Evans, F.G., Krahl, V.E., 1945. The torsion of the humerus: a phylogenetic study from fish to man. American Journal of Anatomy 76, 303–337.

Guidera, K.J., Grogan, D.P., Pugh, L., Ogden, J.A., 1991. Hypoplastic clavicles and lateral scapular redirection. Journal of Pediatric Orthopedics 11, 523–526.

Harrison, T., 1986. A reassessment of the phylogenetic relationships of *Oreopithecus bambolii* Gervias. Journal of Human Evolution 15, 541–583.

Harrison, T., 1987. The phylogenetic relationships of the early catarrhine primates: a review of the current evidence. Journal of Human Evolution 16, 41–80.

Inman, V.T., Saunders, J.B. de C.M., Abbott, L.C., 1944. Observations on the function of the shoulder joint. Journal of Bone and Joint Surgery 26B, 1–30.

Inouye, S.E., Shea, B.T., 1997. What's your angle? Size correction and bar-glenoid orientation in "Lucy" (A.L. 288–1). International Journal of Primatology 18, 629–650.

Jellema, L.M., Latimer, B., Walker, A., 1993. The rib cage. In: Leakey, R., Walker, A. (Eds), The Nariokotome *Homo erectus* Skeleton. Harvard University Press, Cambridge, pp. 294–325.

Johanson, D.C., Lovejoy, C.O., Kimbel, W.H., White, T.D., Ward, S.C., Bush, M.E., Latimer, B.M., Coppens, Y., 1982. Morphology of the Pliocene partial hominid skeleton (A.L. 288–1) from the Hadar Formation, Ethiopia. American Journal of Physical Anthropology 57, 403–451.

Jungers, W.L., 1994. Ape and hominid limb length. Nature 369, 194.

Jungers, W.L., Hartman, S.E., 1988. Relative growth of the locomotor skeleton in orang-utans and other large-bodied hominoids. In: Schwartz, J.E. (Ed.), Orang-utan Biology. Oxford University Press, Oxford, pp. 347–359.

Jungers, W.L., Falsetti, A.B., Wall, C.E., 1995. Shape, relative size, and size-adjustments in morphometrics. Yearbook of Physical Anthropology 38, 137–161.

Krahl, V.E., 1947. The torsion of the humerus: its localization, cause and duration in man. American Journal of Anatomy 80, 275–319.

Krahl, V.E., Evans, F.G., 1945. Humeral torsion in man. American Journal of Physical Anthropology 3, 229–252.

Larson, S.G., 1988. Subscapularis function in gibbons and chimpanzees: implications for interpretation of humeral head torsion in hominoids. American Journal of Physical Anthropology 76, 449–462.

Larson, S.G., 1996. Estimating humeral torsion on incomplete fossil anthropoid humeri. Journal of Human Evolution 31, 239–257.

Larson, S.G., 2007. Evolutionary transformation of the hominin shoulder. Evolutionary Anthropology 16, 172–187.

Leakey, R.E., Walker, A., Ward, C.V., Grausz, H.M., 1989. A partial skeleton of a gracile hominid from the upper Burgi member of the Koobi Fora formation, East Lake Turkana, Kenya. In: Giacobini, G. (Ed.), Hominidae: Proceedings of the 2nd International Congress of Human Paleontology. Jaka Books, Milano, pp. 167–173.

Le Gros Clark, W.E., 1959. The Antecedents of Man. Edinburgh University Press, Edinburgh.

Le Gros Clark, W.E., Thomas, D.P., 1951. Associated jaws and limb bones of *Limnopithecus macinnesi*. Fossil Mammals of Africa, No. 3. British Museum of Natural History, London.

Lordkipanidze, D., Jashashvili, T., Vekua, A., Ponce de León, M.S., Zollikofer, P.E., Rightmire, G.P., Pontzer, H., Ferring R., Oms, O., Tappen, M., Bukhsianidze, M., Agusti, J., Kahlke, R., Kiladze, G., Martinez-Navarro, B., Mouskhelishvili, A., Nioradze, M., Rook, L., 2007. Postcranial evidence form early *Homo* from Dmanisi, Georgia. Nature 449, 305–310.

Lovejoy, C.O., Johanson, D.C., Coppens, Y., 1982. Hominid upper limb bones recovered from the Hadar formation: 1974–1977 collections. American Journal of Physical Anthropology 57, 637–649.

Marquer, P., 1972. Novelle contribution a l'étude du squelette des pygmées occidentaux du Centre Africain comparé a celui des pygmées orientaux. Mémoires du Muséum National D'Histoire Naturelle, Nouvelle Série, Série A, Zoologie 72, 1–122.

Martin, L.B., 1986. Relationships among extant and extinct great apes and humans. In: Wood, B.A., Martin, L.B., Andrews, P. (Eds), Major Topics in Primate and Human Evolution. Cambridge University Press, Cambridge, pp. 161–187.

Martin, R., Saller, K., 1959. Lerbuch der Anthropologie. Gustav Fischer Verlag, Stuttgart.

McCown, T.D., Keith, A., 1939. The Stone Age of Mount Carmel. The Fossil Human Remains from the Levalloiso-Mousterian. Oxford University Press, Oxford.

Milgram, J.E., 1942. Congenital forward shoulders (quadrupedal type) – treatment by clavicular osteotomy. Bulletin of the Hospital for Joint Disease 3, 93–96.

Mivart, St.G., 1868. On the appendicular skeleton of the primates. Philosophical Transactions of the Royal Society of London 157, 299–426.

Mosimann, J.E., 1970. Size allometry: Size and shape variables with characterizations of the lognormal and generalized gamma distributions. Journal American of the Statistical Association 65, 930–945.

Napier, J.R., 1965. Comment on 'New Discoveries in Tanganyika: Their Bearing on Hominid Evolution by P.V. Tobias.' Current Anthropology 5, 402–403.

Ohman, J.C., 1986. The first rib of hominoids. American Journal of Physical Anthropology 70, 209–229.

Oxnard, C.E., 1968. A note on the fragmentary Sterkfontein scapula. American Journal of Physical Anthropology 28, 213–217.

Oxnard, C.E., 1969. A note on the Olduvai clavicular fragment. American Journal of Physical Anthropology 29, 429–432.

Partridge, T.C., Granger, D.E., Caffee, M.W., Clarke, R.J., 2003. Lower Pliocene hominid remains from Sterkfontein. Science 300, 607–612.

Perry, J., 1983. Anatomy and biomechanics of the shoulder in throwing, swimming, gymnastics, and tennis. Clinical Sports Medicine 2, 247–270.

Pickford, M., Johanson, D.C., Lovejoy, C.O., White, T.D., Aronson, J.L., 1983. A hominoid humeral fragment from the Pliocene of Kenya. American Journal of Physical Anthropology 60, 337–346.

Roberts, D., 1974. Structure and function of the primate scapula. In: Jenkins, F.A. (Ed.), Primate Locomotion. Academic, New York, pp. 171–200.

Robinson, J.T., 1972. Early Hominid Posture and Locomotion. University of Chicago Press, Chicago, IL.

Schmid, P., 1983. Eine Rekonstruktion des Skelettes von A.L. 288-1 (Hadar) und deren Konsequenzen. Folia Primatologica 40, 283–306.

Schultz, A.H., 1930. The skeleton of the trunk and limbs of higher primates. Human Biology 2, 303–438.

Schultz, A.H., 1937. Proportions, variability and asymmetries of the long bones of the limbs and the clavicles in man and apes. Human Biology 9, 281–328.

Sládek, V., Trinkuas, E., Hillson, S.W., Holliday, T.W., 2000. The People of the Pavlovian. Skeletal Catalogue and Osteometrics of the Gravettian Fossil Hominids from Dolní Věstonice and Pavlov. Academy of Sciences of the Czech Republic, Brno.

Smith, B.H., 1993. The physiological age of KNM-WT-15000. In: Leakey, R., Walker, A. (Eds), The Nariokotome *Homo erectus* Skeleton. Harvard University Press, Cambridge, MA, pp. 195–233.

Stern, J.T., Susman, R.L., 1983. The locomotor anatomy of *Australopithecus afarensis*. American Journal of Physical Anthropology 60, 279–317.

Trinkaus, E., 1983. The Shanidar Neandertals. Academic, New York.

Tullos, H.S., King, J.W., 1973. Throwing mechanisms in sports. Orthopedic Clinics of North America 4, 709–720.

Voisin, J.-L., 2006. Clavicle, a neglected bone: morphology and relation to arm movements and shoulder architecture in primates. Anatomical Record 288A, 944–953.

Vrba, E.S., 1979. A new study of the scapula of *Australopithecus africanus* from Sterkfontein. American Journal of Physical Anthropology 51, 117–130.

Walker, A., Leakey, R., 1993. The postcranial bones. In: Leakey, R., Walker, A. (Eds), The Nariokotome *Homo erectus* Skeleton. Harvard University Press, Cambridge, MA, pp. 95–160.

Zapfe, H., 1960. Die Primatenfunde aus der miozänen Spaltenfüllung von Neudorf an der March (Děvínská Nová Ves), Tschechoslowakei. Schweizerische Palaeontologische Abhandlungen 78, 1–293.

Chapter 8
Brains, Brawn, and the Evolution of Human Endurance Running Capabilities

Daniel E. Lieberman, Dennis M. Bramble, David A. Raichlen and John J. Shea

Keywords Endurance running • persistence hunting • thermoregulation • biomechanics • scavenging • habitat • *Homo habilis* • *Homo erectus*

The Evolutionary Question Posed by Human Running Capabilities

Theories about hominin evolution are often connected intimately with notions of what it is to be human. Such ideas have had a particularly strong influence on thinking about the definition and origin of the human genus (see Landau, 1993; Wood and Collard, 1999; Wood, 2009). Many, if not most scenarios for the evolution of the genus *Homo* emphasize the importance of quintessentially human traits such as large brains, tool-making, and complex cognition. Usually these derived features have been interpreted, explicitly or implicitly, as a suite of novel strategies that emphasize cognitive over athletic means of competing with the rest of nature ("red in tooth and claw"). Most animals compete with each other to a significant extent using athletic capabilities such as strength, power, agility and speed. Obviously, humans compare poorly with other mammals, including African apes, in these characteristics: we are weak, slow, and awkward creatures. Even though male chimpanzees weigh less than a typical adult modern human, they can produce much more force, can sprint more rapidly, and are obviously more agile during locomotion (Stedman et al., 2004). Yet, although no human alive could match a chimpanzee in hand-to-hand combat, our cognitive capacities are extraordinarily better developed. Accordingly, it seems reasonable to focus on evolutionary scenarios for the genus *Homo* that explain the triumph of brains over brawn.

Interestingly, the idea that humans are poor athletes is demonstrably wrong in one crucial respect. While humans have comparatively poor performance capabilities in terms of power and strength, we are unusually specialized endurance athletes, with surprisingly impressive aerobic performance capabilities. These capabilities are particularly remarkable for endurance running (ER), defined as running long-distances (>5 km) using aerobic metabolism. These capabilities, which have been reviewed in depth by Carrier (1984) and Bramble and Lieberman (2004), compare extremely well to other mammals, especially primates, in terms of several performance criteria such as speed and distance, especially in hot conditions.

D.E. Lieberman (✉)
Departments of Anthropology and Organismic & Evolutionary Biology, Harvard University, 11 Divinity Avenue, Cambridge, MA 02138, USA
e-mail: danlieb@fas.harvard.edu

D.M. Bramble
Department of Biology, University of Utah, Salt Lake City, UT 84112, USA
e-mail: Bramble@bioscience.utah.edu

D.A. Raichlen
Department of Anthropology, University of Arizona, 1009 E. South Campus Drive, Tucson, AZ 85721, USA
e-mail: raichlen@email.arizona.edu

J.J. Shea
Department of Anthropology and Turkana Basin Institute, Stony Brook University, Stony Brook, NY 11794-4364, USA
e-mail: John.Shea@sunysb.edu

Speed

Human ER speeds range from 2.3 to 6.5 m/s. While the latter are elite performance speeds for world-record holders, many amateurs without special training are easily capable of sustained running at 5 m/s. Such speeds are fast compared to the endurance speeds of specialized quadrupedal cursors. A dog of similar body mass to a human (65 kg) has a trot–gallop transition speed of 3.8 m/s, and can sustain a gallop at 7.8 m/s under ideal climatic conditions for only 10–15 min (Heglund and Taylor, 1988). Dogs and other quadrupedal cursors cannot gallop for long periods, especially when it is hot (see below). Thus, while a large dog can outrun a human over short distances of a kilometer or two, most fit humans can outrun any dog over longer distances. As detailed by Bramble and Lieberman (2004), humans also have remarkable endurance capabilities even in comparison to larger cursors such as ponies and horses. The latter offer a useful, extreme example because they have been bred intensively via artificial selection for running. Horses can easily outrun humans with a maximum gallop speed of 8.9 m/s for a 10 km race, but their sustainable galloping speed declines dramatically for runs longer than 10–15 min; in repeated runs over long distances, horses are constrained to about 5.8 m/s for approximately 20 km per day, above which they can sustain irreparable musculoskeletal damage (Minetti, 2003). By these standards, human ER capabilities are quite impressive, and explain why humans can sometimes best horses in long distance races such as marathons (e.g., http://www.man-v-horse.org.uk/).

Distance

Human ER capabilities are also comparable to the best quadrupedal cursors in terms of distance. Fit amateur humans can easily run 10 km or more a day, and are capable of far longer distances such as marathons and ultramarathons (although rarely on a daily basis!). Only a few other mammals, such as wolves, hunting dogs and hyenas, are known to habitually run long distances of 10–20 km a day (Pennycuick, 1979; Holekamp et al., 2000). These animals, notably, are all social carnivores in which natural selection has favored capabilities for running as a critical part of their hunting or scavenging strategy. Like humans, these cursorial specialists can also run distances greater than 10–20 km, but only when forced to do so by humans, and they are all restricted to a trotting gait, or, in the case of hyenas and wildebeest, a canter (a slow gallop just above the trot–gallop transition). Some dogs, for example, can be forced to trot as much as 100 km in cool conditions (e.g., with fans blowing air on them, or in the artic winter), but these feats are unnatural and often cause severe physiological distress (Dill et al., 1933; see below). Alaskan huskies are perhaps the extreme example of an animal specially bred for endurance: these dogs can run in packs up to 50 km in frigid conditions at a lope (a slow gallop), but for longer distances they must switch to a trot (Hinchcliff et al., 1998); in addition they cannot perform these feats in warm weather.

Environmental Context

While a few mammals can trot long distances, comparable to those that humans can run, they cannot run long distances *while galloping* in hot conditions without becoming hyperthermic. This major constraint derives from two aspects of mammalian biology. First, the thermogenic effects of exercise increase in proportion to the number and rate of cross-bridges that are activated during muscular contractions. In humans for example, running can generate as much as tenfold more heat than walking (Cheuvront and Haymes, 2001), and a sprinting cheetah generates so much heat that it must stop after approximately 1 km (Taylor and Rowntree, 1973). Second, the major mechanism by which most mammals cool themselves, panting, is problematic during galloping. Panting occurs via shallow breaths, about ten times the normal rate of respiration, in the dead space of the upper pharynx without any gas exchange occurring in the lungs (Schmidt-Nielson, 1990). Panting mammals, however, cannot satisfy their respiratory demands for oxygen during galloping, and the 1:1 coupling of locomotion with respiration that occurs during galloping is biomechanically incompatible with panting (Bramble and Jenkins, 1993; Entin et al., 1999).

Humans, however, have evolved a number of specialized modifications for effectively dissipating copious quantities of heat while running in hot, arid conditions. For one, humans do not have to couple respiration with stride (Bramble and Carrier, 1983). In addition, humans are considerably derived in terms of the number of eccrine sweat glands and the loss of almost all fur. Sweating is an effective means of cooling (evapotranspiration of 1 ml H_2O requires 580 cal of heat [Schmidt-Nielson, 1990]), but is ineffectual with fur, which traps air and moisture at the skin's surface, thereby considerably reducing convection (McArthur and Monteith, 1980). Therefore, other tropical cursorial mammals such as hyenas and hunting dogs that can run long distances are constrained to do so at night or during the dawn and dusk when the days are hot. Humans alone are capable of ER during midday heat. Human sweating, however, imposes high water demands, requiring as much as 1–2 l/h in well-conditioned athletes (Torii, 1995).

In short, humans are comparatively superb endurance athletes, particularly in hot, arid conditions that are conducive to heat-loss from sweating. In fact, humans appear to occupy a rare extreme in the general trade-off between aerobic and anaerobic capabilities (Wilson and James, 2004). Natural selection often favors speed over endurance because of the dynamics of predator–prey interactions: slower animals typically have lower fitness. Animals built for speed and power are rarely good at endurance and *vice versa*, in part because of muscle fiber composition. In most mammals, there is a predominance of Type IIb (fast-glycolytic) and Type IIa (fast oxidative) relative to Type I (slow oxidative) muscle fibers. The former fast-twitch fibers can produce several times more force but are anaerobic and fatigue quickly. Slow-twitch fibers have higher aerobic capacity, but produce less force. Most human leg muscles have about 50% of each type (McArdle et al., 1996), but can increase slow-twitch fiber content to about

80% through aerobic endurance training. They can also increase fast-twitch fiber content to between 70–80% through power training (Thayer et al., 2000). Such training effects for fast twitch fibers are more common in humans with a novel form of the ACTN3 gene that predisposes individuals to have a high fast twitch muscle fiber content (Yang et al., 2003). In general, quadrupedal cursors have higher percentages of fast-twitch fibers in hind limb extensor muscles than humans, with cheetahs having the highest known-values (Armstrong et al., 1982; Acosta and Roy, 1987; Williams et al., 1997).

Human endurance capabilities raise two questions. First, when did they evolve? Second, why did they evolve? Accordingly, we first review a few points about the evidence for ER capabilities in the genus *Homo* and its relationship to walking. We then consider some alternative hypotheses about the sort of conditions that might have led to selection for ER capabilities.

When Did ER Capabilities Evolve?

The derived ER capabilities of humans must have evolved sometime after the split of the human and chimpanzee lineages. Other primates rarely engage in any kind of running. Even patas monkeys (*Erythrocebus patas*), which have several typical cursorial specializations such as long, digitigrade limbs, sprint rarely and then only for short distances (Isbell, 1998). Importantly, running is also rare among chimpanzees; it comprises less than 1% of their locomotor repertoire (Hunt, 1992). Moreover, when chimpanzees run during hunting or chasing, they typically sprint rapidly for about 100 m, fatigue quickly, and then pant heavily while resting to cool down (R. Wrangham, personal communication).

Given that ER capabilities are derived in hominins, there are three alternative possibilities about their evolutionary origins. First, ER and walking capabilities might have coevolved with the origins of upright, habitual bipedalism. Second, ER capabilities might have evolved sometime around the morphological transition between *Australopithecus* and *Homo*. Third, ER capabilities may have evolved sometime more recently than earliest *Homo*, perhaps in *H. erectus*, *H. heidelbergensis* or *H. sapiens*. As emphasized by Bramble and Lieberman (2004), testing these hypotheses is a challenge because many of the physiological and anatomical features that improve ER performance do not fossilize. In addition, many features, such as long legs, that improve running performance capabilities also improve walking performance capabilities (e.g., Pontzer, 2005, 2007; Steudel-Numbers and Tilkens, 2004; Steudel-Numbers et al., 2007). However, several criteria may be useful for evaluating the extent to which morphological features evident in the fossil record may be adaptations for ER. The most useful of these derive from the biomechanical differences between running and walking.

Running Versus Walking

Running is biomechanically unlike walking in several crucial ways that can help specifically diagnose ER capabilities. Most importantly, walking is modeled as an inverted pendular gait in which the body's center of mass (COM) vaults over a relatively extended leg during stance. Potential energy is stored as the COM rises during the first half of stance; this energy is then released as kinetic energy as the COM falls during the second half of stance. During walking, kinetic and potential energy are thus out of phase. In contrast, kinetic and potential energy are in-phase during running, which saves energy in a completely different way via mass-spring mechanics. In this system, the COM falls during the first half of stance, storing elastic energy in collagen-rich tendons and ligaments in the leg; these structures then recoil during the second half of stance, as the COM rises, propelling the body into an aerial phase (see Alexander, 1991). Therefore, derived features in the human body relevant to mass-spring mechanics are evidence for selection for improving running, not walking, capabilities.

Another aspect of biomechanics in which running differs critically from walking is stabilization, primarily of the head and trunk. Walking is an inherently more stable gait than running, especially in upright, relatively stiff-legged bipeds such as humans. During walking, the human trunk is held upright above the hips, and the COM is rather stable with fluctuations of about 4–5 cm in the vertical and horizontal planes (Saunders et al., 1953). In contrast, running is somewhat like a controlled fall, in which the trunk and head are more flexed than during walking, each by approximately 10° in a typical runner (Thorstensson et al., 1984). In addition, ground reaction forces (GRFs) are more than twice (often as much as four times) as high in ER than walking (Keller et al., 1996). Since human bipeds have comparatively extended, stiff legs and upright axial columns compared to quadrupeds, the high GRFs generated at foot strike are transmitted as a rapid shock wave – the heel strike transient (HST) – up the legs, axial skeleton and into the head. GRFs in humans rise again more slowly after the HST, peaking at mid-stance when the COM reaches its nadir.

Maintaining stability is important to prevent a fall in all running animals, but is a special challenge for intrinsically unstable bipeds such as humans in which falls are more likely to cause serious injury. Running humans thus must stabilize the trunk and head in response to destabilizing forces at heel strike as well as at midstance. Trunk stabilization, which is needed to keep the body from falling over, is primarily achieved by contractions of the gluteus maximus (Lieberman et al., 2006). Head stabilization is more complex but no less important, because of the need to stabilize gaze via the vestibulo-ocular reflexes (VORs) which sense angular accelerations of the head and adjust eye movements to stabilize images on the retina. Because the head is not balanced, forces that are generated during running have the tendency to cause rapid pitching. These movements are problematic if they exceed 200°/s, the threshold above which the VORs decrease in performance, causing significant losses of balance and visual acuity (Gauthier et al., 1984; Maas et al., 1989; Cromwell et al., 2001). Other quadrupedal and bipedal cursors have somewhat horizontally oriented necks and cantilevered heads, which enable them to stabilize the head through flexion and extension of the neck. Humans, however, must stabilize the head by other means, because our necks are vertical and emerge from near the center of the cranial base (see below).

Finally, running and walking differ in the intensity of the thermoregulatory and mechanical demands they impose. As noted above, GRFs, hence joint reaction forces (JRFs), are several times higher during running than walking. In addition, running generates as much as an order of magnitude more heat than walking. It follows that adaptations for thermoregulation are considerably more limiting for running than walking. Both walking and running in hot, midday, arid conditions would have benefited from derived human sweating capabilities (Wheeler, 1991), but it is reasonable to conclude that running would not be possible without them. In addition, the lack of adaptations such as sweating and hair loss in other African mammals, all of whom cannot run for long in hot conditions, lends extra support to the hypothesis that ER was a factor that led to their evolution and/or elaboration in humans (eccrine glands are a derived feature of catarrhines, but are vastly more numerous in humans than other primates [Jablonski, 2006]).

Evidence for Skeletal Features That Improve ER Performance

Based on these criteria, several lines of fossil evidence suggest that ER capabilities first emerged in the genus *Homo*. These features are discussed at length in Bramble and Lieberman (2004), but a few that are illustrated in Fig. 8.1 merit brief mention here. First, while there are some indications in the skeleton of morphological specializations related to the mass-spring mechanics of running, features related to stabilization are more prevalent. In terms of trunk stabilization, the cranial portion of the gluteus maximus, which plays a critical role in running but not walking, has a considerably expanded origin in *H. erectus* relative to *Australopithecus* (Rose, 1984; Lieberman et al., 2006). The gluteus maximus also acts in concert with the erector spinae to stabilize the trunk, and the sacroiliac trough in which the latter originates may be considerably expanded in *Homo* compared to *Australopithecus* (see Lovejoy, 1988). Even more concrete evidence of derived mechanisms for stabilization relevant only to running is in the head. As shown by Spoor et al.

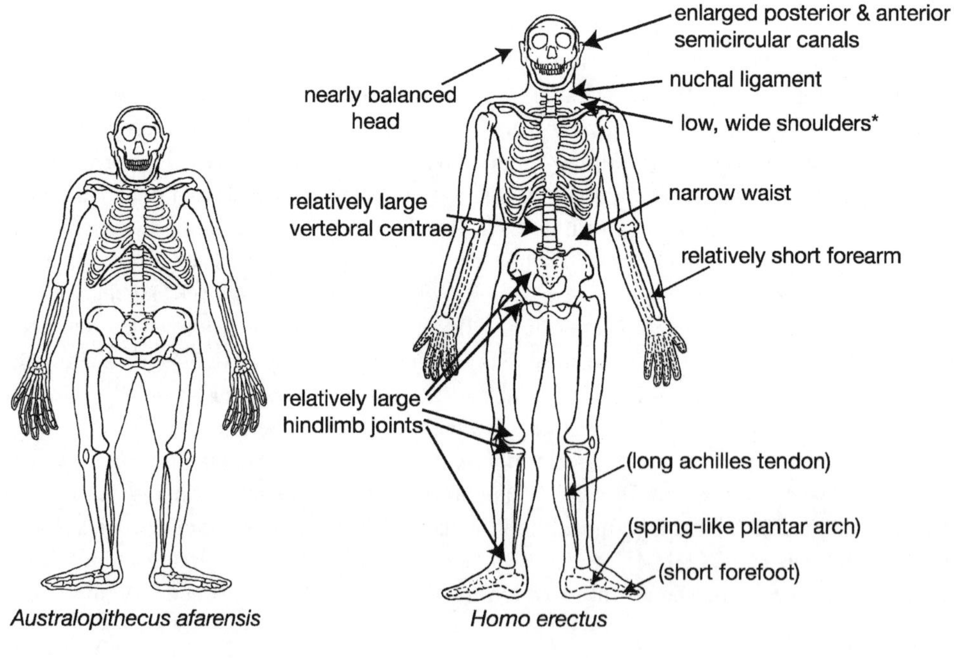

Fig. 8.1 Illustration of basic body shape differences between *A. afarensis* (*left*) and *H. erectus* (*right*) highlighting features discussed in the text that are derived in *H. erectus* and which would improve endurance running performance. Features in parentheses are as yet unknown (in the foot) or hypothetical reconstructions (e.g., Achilles tendon length). Note that shoulder position (indicated with an *) in *H. erectus* is unresolved (Modified from Bramble and Lieberman, 2004).

(1994), the diameters of the anterior and posterior semicircular canals relative to body mass, which influence their sensitivity to head pitching accelerations, are first expanded in early *Homo* compared to *Australopithecus* and *Pan*. The vestibular system is fully formed prior to birth (Jeffery and Spoor, 2004), and is not significantly challenged during walking. It is difficult to think of any human activity other than running that would have selected for increased sensitivity to head pitching.

The anatomical relationships between the shoulder and the head comprise another set of derived features of *Homo* that are absent in *Australopithecus* and which have key roles in head stabilization during running. During walking, the head is stabilized in minor ways by inertia, the viscoelastic properties of the ligaments and muscles that connect the head to the axial skeleton, and by contractions of the head extensors (Hirasaki et al., 1999; Winter et al., 1990). During running, however, the heelstrike transient imparts such a rapid and substantial pitching impulse to the head that it needs to be stabilized almost instantly to avoid vestibular overload. Humans do so by a novel mechanism (a mass-damping system), in which the long axis accelerations of the arm counter pitching accelerations of the head via an out-of-phase elastic linkage (Bramble et al., 2009). A critical component of this system is an almost complete decoupling of the head and shoulder so they can act as linked masses. In chimpanzees, the shoulder and head are tightly connected by a massive trapezius, the rhomboideus, and the atlantoclavicularis (Aiello and Dean, 1990). These connections have all been lost in humans with the exception of the cleidocranial portion of the trapezius (CCT). This muscular strap between the shoulder and midline occiput interdigitates with another novel feature in humans, the nuchal ligament (NL). This tendon-like structure originates on the midline of the occiput and connects with the upper trapezius as well as a deeper fascial septum that attaches to the cervical spines (Mercer and Bogduk, 2003). A NL is present in other cursors such as canids, equids and bovids, as well as in a few species with massive heads (Dimery et al., 1985; Bianchi, 1989). In running, but not walking, the CCT fires before HS on the stance side arm, linking the mass of this arm with the head in the midsagittal plane via the NL. Critically, evidence for this linkage is first present in the fossil record of early *Homo* (all *H. erectus* skulls as well as KNM-ER 1813) because the NL leaves a trace on the skull in the form of a sharp, everted, median nuchal line that that is not present in *Australopithecus* or *Pan*. While apes and australopiths sometimes have a rounded ridge in the midline of the occipital, this ridge lacks the everted contour indicative of a NL.

Other derived changes in *Homo* relevant to stabilization during running but not walking may be evident in the shape of the waist, thorax and neck. Leg swing during the aerial phase of running causes substantial angular momentum, which, unchecked, would cause the body to rotate around a vertical axis before heel strike. Humans counteract this momentum not only by swinging the arms in opposition to the legs, but also by rotating the thorax independently of the hips and head (Hinrichs, 1990). Such rotations, which are neither important nor particularly marked during walking, are made possible by two zones of separation: a relatively narrow, tall waist; and a relatively tall neck with low, wide shoulders. Although the waist in *Australopithecus* was probably as tall as in *Homo*, it was relatively wider as judged by the greater bi-iliac breadth of the australopithecine pelvis (Lovejoy, 1988; Schmid, 1983). A narrow waist in *Homo* may reflect smaller guts (Aiello and Wheeler, 1995), but it would also have improved running performance by reducing resistance between the pelvis and ribcage, and decreasing inertial moments for thorax rotation.

The second rotational zone of separation, between the upper thorax and neck, is harder to assess in early *Homo*. Whether the thorax of *Australopithecus* was funnel-shaped, as in apes, or barrel-shaped, as in humans, is debated (Schmid, 1991; Ward, 2002), but most evidence suggests that a barrel-shaped upper thorax is first present in the KNM-WT 15000 skeleton (Jellema et al., 1993). The primitive condition of a narrow upper thorax in combination with more muscular connections between the shoulder and head would have no effect on walking performance capabilities. They are useful for helping generate torque in the shoulder for orthograde climbing (Larson, 1993), but would decrease the ability to stabilize the head during running. It is interesting to speculate that selection for running capabilities may have come at the expense of adaptations for climbing, explaining why *Homo* is the first non-arboreal primate. However, Larson et al. (2007; see also Larson, 2009) have suggested that the KNM-WT 15000 shoulder was somewhat narrow with a relatively short clavicle and a glenoid fossa that faced anteriorly in order to accommodate a low degree of humeral torsion. A humerus from Dmanisi (D2700) also has a low degree of humeral torsion (Lordkipanidze et al., 2007). It is hard to interpret these data in part because both KNM-WT 15000 and D2700 are juveniles. The clavicle (which grows intramembranously) is the last bone in the human body to attain adult size, and both skeletons have clavicles that fall in the range of humans at equivalent ontogenetic stages (Scheuer and Black, 2000). Without better reconstructions of the upper thorax itself, it is difficult to assess the relative breadth and position of the shoulders in these specimens, one of which (KNM-WT 15000) has evidence for axial pathologies that may have affected upper thoracic anatomy. Regardless, low humeral torsion in *H. erectus* would have compromised its ability to throw effectively (Larson et al., 2007; Larson, 2009), raising questions about how the species was able to hunt (see below).

As noted above, it is much harder to document elastic structures in the skeleton relevant to the mass-spring form of energy exchange used in running but not walking. The most important anatomical components of the system are extensive tendons, especially the Achilles, which are substantially longer in humans relative to chimpanzees or gorillas. The size of the Achilles tendon insertion in the Hadar calcanei (Susman et al., 1984) suggests that they had an ape-like configuration, but such inferences must remain speculative without evidence of some relationship between tendon length and tendon insertion morphology. A more promising anatomical region for evidence of mass-spring anatomy that requires further study is the foot, especially the plantar arch. Some form of arch is useful in bipedal walking in order to act as a windlass to stiffen the foot for effective toe-off (Kappel-Bargas et al., 1998), but in running the arch functions quite differently as a spring, storing and releasing approximately 17% of the energy generated during each impact of the foot with the ground (Ker et al., 1987). Although australopithecines clearly had some form of plantar arch, there are several indications that the arch had a different configuration in *Homo*. In particular, the navicular in apes and australopithecines has a relatively expanded medial tuberosity, suggesting that it was a weight-bearing element (Harcourt-Smith, 2002). In addition, the first hominin fossil with a close-packed calcaneo-cuboid joint (as evident by an expanded medial flange on the proximal cuboid) is OH 8, a specimen attributed to early *Homo* (Lewis, 1989; Susman, 2008). Together, these novel features – along with an unequivocally adducted big toe and a relatively shorter forefoot (see Susman et al., 1984; Aiello and Dean, 1990) – hint that elastic storage mechanisms in the foot necessary for running may be derived features of the genus *Homo*.

Finally, it is important to note that there are more than a dozen other derived skeletal features of the genus *Homo,* particularly in *H. erectus*, which improve performance for *both* walking and running (summarized in Bramble and Lieberman, 2004). Given that hominins were habitual bipeds for at least 4 million years before the origin of *H. erectus* with little evidence for any major change in postcranial anatomy (reviewed in Ward, 2002), it is difficult to imagine that selection for walking alone was responsible for the derived features of *Homo*. The most likely scenario is that *H. erectus* was the first hominin with a substantially expanded diurnal day range made possible by *both* walking *and* running. Indeed, both gaits are important ways to travel long distances, and one can expect that hominins would have walked rather than run whenever possible (see below). Thus, the extent to which selection acted on running versus walking is impossible to assess, as both would have been important. That said, it is worthwhile noting that the considerably more extreme thermoregulatory and mechanical demands of running might have imposed a greater selective benefit on performance capabilities in running than walking. In addition, many ancestral features of australopithecines that improve climbing performance, such as long forearms and heavily muscled shoulders, do not conflict with the biomechanical demands of walking, but may impede the ability to stabilize the head. Selection for running capabilities may thus have selected against arboreal capabilities in *Homo*.

Put together, there is much evidence that *H. erectus* but not *Australopithecus* was capable of ER. However, this inference does not imply that *H. erectus* was necessarily as good as modern humans or even later archaic *Homo* at ER. Some modern human features that improve ER performance may have evolved since early *H. erectus*. In addition, there are some hints that *H. habilis* may have possessed some ER capabilities, but the evidence is sparse and equivocal (see Bramble and Lieberman, 2004). While it is possible that ER capabilities had evolved by the time of *H. habilis*, it is premature to be definitive, and there are theoretical reasons to hypothesize that such capabilities, if they existed, were not as developed as in *H. erectus*. Natural selection tends to take advantage of existing variations in the context of some fitness benefit. Thus, it is unlikely that selection would have favored the evolution of ER-related features if hominins had not already been engaged to some extent in a form of ER. One potential scenario is that early *Homo* during the Oldowan started to scavenge and/or hunt to a limited extent. At some point, hominins that were better at ER for various reasons (longer legs, larger anterior and posterior semicircular canals, and so on) had a slight fitness benefit, leading to the evolutionary changes that we observe in *H. erectus*.

Why Did Endurance Running Capabilities Evolve?

Given the above evidence that ER capabilities are derived in the genus *Homo*, and that they were probably present to some extent in *H. erectus*, the final question to address is why these capabilities evolved in the first place. Answering this question, however, is a challenge because it is obvious that humans today – including contemporary hunter–gatherers – no longer need to practice ER (although it remains a potentially useful component of some hunter–gatherer subsistence strategies). Thus, answers need to be sought primarily in past rather than present behaviors. Nevertheless, ethnographic studies of recent humans provide several lines of evidence which suggest that ER would have significantly improved performance in scavenging and/or hunting activities prior to the invention of sophisticated projectile technology such as the bow and arrow. In order to explore these hypotheses, we first outline several alternative ways in which ER may have been useful for scavenging versus hunting during the Early

Stone Age, and we evaluate the recent ethnographic and paleoenvironmental evidence relevant to both forms of meat procurement.

Endurance Running and Scavenging

The debate about scavenging in human evolution is long and contentious, largely because it is difficult to prove that the animal bones found in early archaeological sites were procured by scavenging or hunting (for reviews, see Bunn, 2001; Dominguez-Rodrigo, 2002). Another point of contention has been the challenge of evaluating how reliably or effectively early *Homo* would have been able to scavenge in various habitats. Regardless of the extent to which scavenging occurred, the most likely source of scavenged carcasses would have been lion kills, because lions, unlike hyenas, do not consume all their prey, but instead leave behind marrow, brains and sometimes flesh (Blumenschine, 1987, 1988). Leopard and sabertooth tiger kills might have been additional possible sources of edible animal tissue (Cavallo and Blumenschine, 1989; Marean, 1989), but it is unclear how common such carcasses would have been, and how much of the carcass sabertooths would have consumed (Van Valkenburgh, 2001). In any event, early *Homo* might have used two general strategies to scavenge from lion kills. One possibility is that hominins scavenged opportunistically when they happened to come across carcasses in the course of their daily foraging activities. Alternatively, or additionally, hominins might have sought out scavenging opportunities strategically by searching for carcasses through long range cues, the most common of which is seeing vultures circling in the air from a distance.

Apart from whether hominins were scavenging opportunistically or strategically, to do so effectively they would have faced two considerable challenges, both of which are relevant to ER. First, carcasses are comparatively rare and ephemeral resources, largely because of hyenas, which are impressively efficient at finding kills. According to Cooper (1991), hyenas in Kruger Park typically arrive at lion kill sites within 30 min of a kill, even at night. Given that a large percentage of kills occur at night, it is probable that only a fraction of kills, notably those made during the day, were available for scavenging by diurnal hominins. In addition, it is often argued that hominins in environments such as the Serengeti would have been most effective at scavenging in riparian habitats where the density of hyenas is lower and scavengeable carcasses survive for longer (Blumenschine, 1986, 1987). In wetter, less seasonally arid environments (e.g., the Parc National des Virunga), carcasses might have been more available in more open habitats, but they still would have been rare and rapidly consumed (Tappen, 2001).

The second serious challenge that hominins would have faced while scavenging is competition. To become scavengers (or hunters), they would have entered the carnivore guild, which means competing with other carnivores. In fact, most interspecific interactions between carnivores occur in the context of competition for a kill (Van Valkenburgh, 2001). Carnivores compete through a combination of strength, speed, stealth, and cooperation, and the risk of mortality associated with these interactions is quite high. Human hunters are no exception to this competition: a high percentage of scavenging opportunities observed among modern hunter–gatherers are classified as "competition" or "power" scavenging in which groups of foragers drive off lions or hyenas from a kill using weapons (O'Connell et al., 1988; Potts, 1988; Bunn and Ezzo, 1993). According to O'Connell and colleagues (1988), 85% of the total carcass weight that the Hadza scavenged was acquired by driving off or killing the initial predator (mostly lions). Since it is probable that early *Homo*, like modern humans, was neither strong nor powerful, but also lacked the sophisticated weapons available to modern foragers, it is debatable to what extent they would have been able engage in competition scavenging. It may strain credulity to imagine hominins successfully driving off a pack of lions or hyenas armed only with stones and sharpened sticks, but Hadza foragers seem to be able to do just that with relatively simple weapons. It must be remembered, however, that the Hadza's armature includes projectile weapons, and the carnivores in question have undergone thousands (perhaps millions) of years of natural selection for avoiding encounters with groups of well-armed humans.

The combination of ephemerality and competition has led many researchers to suggest that scavenged meat was not a commonly available resource for early hominins (e.g., Bunn, 2001; Tappen, 2001; Dominguez-Rodrigo, 2002). However, it is possible that ER provided an additional means to improve access to this potentially very valuable resource. In particular, hominins during the day in open habitats would be able to identify scavenging opportunities by seeing vultures in the distance, often many kilometers away. If they just walked to the kill site, it is likely that little meat would be left to scavenge, and/or there would be considerable competition with hyenas. But, as demonstrated above, early *Homo* might have been able to run the few kilometers necessary to get to the kill *before* other scavengers. Since hyenas face the same thermoregulatory constraints as other non-human mammals for running long distances in extreme heat (they run primarily at night and during the dawn or dusk), hominins would have had a competitive advantage over hyenas for getting to diurnally available carcasses, particularly in the dry season. Whether and to what extent hominins did this is debatable, but modern ethnographic evidence provides some support for this potential strategy. As reported by O'Connell et al. (1988: 357), when Hadza believe they have a scavenging opportunity, they "abandon other activities and move quickly to the spot, *often at a run* [emphasis added]." In another example, a !Kung bushwoman, Nisa (Shostak, 1981: 93), recounts an instance in which she used ER to secure quickly an opportunistically discovered carcass before it is lost to other scavengers:

> I remember another time, when I was the first to notice a dead wildebeest, one recently killed by lions, lying in the bush. Mother and I had gone gathering and were walking along, she in one direction and I a short distance away. That's when I saw the wildebeest... She stayed with the animal while I ran back, but we had gone deep into the mongongo groves and soon I got tired. I stopped to rest. Then I got up and started to run again, following along on our tracks, ran and rested and then ran until I finally got back to the village. It was hot and everyone was resting in the shade... My father and my older brother and everyone in the village followed me [back to the wildebeest]. When we arrived, they skinned the animal, cut the meat into strips, and carried it on branches back to the village.

In short, ER might have opened a new niche for scavenging that was previously unavailable.

Endurance Running and Hunting

Another key, perhaps even more important role for ER in *H. erectus* and possibly earlier *Homo* may have been during hunting. As noted above, a wide array of evidence suggests that hominins were actively hunting, at least by the time that *H. erectus* appears circa 1.9 Ma (for reviews see Potts, 1988; Bunn, 2001; Dominguez-Rodrigo, 2002). The evidence for hunting includes a large proportion of bones with cut-marks indicative of flesh removal from regions of shafts that would not have had flesh had they been scavenged. In addition, many of these bones are from medium- to large-sized mammals. One question that arises from these findings is how early humans managed to kill their prey? Humans not only lack the natural weaponry of cursorial predators such as claws and fangs, but also cannot run fast enough to capture most prey by sprinting. The fastest human sprinters can run approximately 10 m/s for about 20–30s; in contrast, most African mammals that were apparently hunted by *Homo* can run approximately twice as fast for several minutes (Garland, 1983). Thus, most scenarios of early human hunting posit, not unreasonably, that humans managed to hunt only with the aid of various forms of technology.

The reliance of human hunters on technology poses an interesting quandary relevant to ER, because the extremely limited, simple technology of the Early Stone Age (ESA) has led some researchers to doubt that early *Homo* was capable of hunting (e.g., Binford, 1984; Brain, 1981). Stone tools, namely Acheulian handaxes and spheroids are viewed by some researchers as possible hunting weapons (O'Brien, 1981; Clark, 1955), but the evidence that they were specifically designed for such tasks is weak or equivocal (Shea, 2006b). Handaxes perform poorly as thrown projectiles (Whittaker and McCall, 2001), and it is a myth that spheroids (putative bola stones) are found at ESA sites in groups of two or three (cf., Cole, 1963: 148). More plausible ESA weapons might have included sharpened wooden spears, such as those recovered from Middle Pleistocene contexts (ca. 400 ka BP) at Schöningen, Germany, although it is unlikely that early *Homo* spears would have been as sophisticated as the Schöningen example (Theime, 1997). Importantly, even if we assume that ESA hunters made spears, there is no evidence that they made stone-tipped or bone-tipped spears, which are capable of inflicting serious damage from a distance. The most effective Paleolithic technologies for hunting, the bow and arrow and the spear thrower (atlatl), were not invented until quite recently, probably after the origin of modern *H. sapiens* (Shea, 2006a). In this crucial respect, modern hunter–gatherers such as the Hadza and the Bushmen, who have bows and arrows as well as other weapons such as poison and tipped spears, are not particularly useful analogues for how early *Homo* would have hunted (see Lieberman et al., 2007). Moreover, as noted above, Larson (Larson et al., 2007; Larson, 2009) has suggested that *H. erectus* lacked a modern shoulder configuration, which would have compromised the species' ability to throw projectiles effectively.

In spite of the deficiencies of the ethnographic record, studies of recent hunters suggest that the lack of any sophisticated projectile technology during the ESA would have presented early hominin hunters with several significant challenges, especially prey encounter, and risk of injury. According to analyses by Binford (1984) and Churchill (1993), hunters typically employ five general strategies to kill prey: (1) disadvantaging, in which prey are first immobilized by mechanisms such as traps, mud, water, and hunting dogs, and then killed; (2) ambushing, in which hunters hide (often behind a blind) until prey come close enough to kill; (3) approach, in which hunters stalk free moving animals until they are within weapon range; (4) encounter, in which hunters kill prey that happen to be within range as they encounter them by chance; and (5) pursuit, in which hunters chase an animal until it is within range or collapses from exhaustion. In a review of ethnographic and ethnohistoric literature from 96 recent human groups, Churchill (1993) has shown that the bow and arrow and atlatl are by far the most common weapons used to hunt for most of these strategies, and the Hadza and Bushmen are no exception. Spears, which might have been available (albeit in crude form) to ESA hunters, are rarely used in ambush, approach or encounter hunting, but instead are used primarily to dispatch disadvantaged prey that have been immobilized or incapacitated. In addition, modern hunters not only use stone- or bone-tipped spears that ESA hunters did not have, but also usually use them after they have disadvantaged their prey using dogs or other recent technologies (e.g., boats, snares, nets) that were probably also unavailable to ESA hunters.

There are two reasons that hunters use spears primarily to kill only disadvantaged prey. First, the killing range of spears is very limited. Experiments with replicas of the Schöningen spears by trained athletes suggest they may be effective out to 15 m (Rieder, 2003), but the controlled conditions of an athletic field do not precisely replicate the conditions of hunting large dangerous mammals at close quarters and/or in dense vegetation or on uneven terrain. The mean distance from which ethnographic throwing spears are cast at their targets is only 7.8 ± 2.2 m ($n = 14$) (Churchill, 1993). Moreover, it is important to note that the ability of ESA hunters to kill with spears would have been considerably less than observed in modern hunters because ESA spears, if they existed, lacked stone or bone points. These points greatly increase the effectiveness of the spear because they are much sharper than a wooden tip, enabling the spear to penetrate hair and hide. In addition, the major way by which spears disable or kill prey is from causing hemorrhaging of an animal's internal organs, or by laming the animal. Thus, thrown wooden spears have a much lower, possibly negligible probability of mortally wounding or disabling an animal. The differential lethality of wooden vs. stone- or bone-tipped weapons is intuitively obvious, but imprecisely quantifiable. Some measure of support for this hypothesis can be seen in the strong association in ethnographic contexts between the use of stone projectile points and the hunting of large dangerous mammals and warfare. In a nutshell, people use stone and bone-tipped armatures to improve penetration and to minimize the chances that their target can either recover or retaliate. The main cost of this strategy lies with the considerable time and effort needed to haft stone or bone armatures (or their modern metal counterparts) to wooden shafts. Such weapons are frequently among the most complex subsistence aids used by recent hunter–gatherers (Oswalt, 1976). Simple wooden spears, on the other hand, can be made quickly, repaired easily, and unlike bone- or stone-tipped weapons, they retain considerable functional versatility. Moreover, given their short effective range, hunters using such simple spears are unlikely to miss their targets. In contrast, the effective ranges of the atlatl and the bow and arrow are approximately 40 and 26 m, respectively (Churchill, 1993). Moreover, these weapons have a much greater chance of causing internal bleeding and death, and are thus much more effective. The countervailing cost of such projectile weapons are that, as noted, they require considerable time and energy to build and maintain, and using them requires the learning and practice of specialized skills (Blurton-Jones and Marlowe, 2002). The bow and arrow and atlatl have completely changed human hunting practices since their invention in the Late Pleistocene (Cattelain, 1997; Shea, 2006a).

The other reason of relevance that hunters use spears mostly to kill disadvantaged large prey is to minimize risk to themselves. It is possible to kill small animals, such as gazelles or duikers, at close range by stabbing or clubbing them, but getting within a few meters of any medium- to large-sized animal is clearly very risky. Understandably, we have no data on injury rates for humans who try to kill such animals at close quarters with ESA technology because rational humans apparently will not attempt such feats on large, non-disadvantaged animals. But it is reasonable to assume that such behaviors would be extremely hazardous. It is doubtful that any reader of this paper would be willing to try to sneak up on a wildebeest or kudu and kill it with a sharpened wooden stick: one well-aimed kick or impact with the animal's horns could cause serious, potentially fatal injury! American rodeo athletes, who regularly interact at close quarters with large, dangerous mammals, frequently incur injuries, such as broken legs, that would have killed or disabled a Pleistocene hominin (Berger and Trinkaus, 1995). It follows that ESA hunters would have faced significant and considerable challenges in their efforts to kill prey using untipped spears without some reliable method of disadvantaging their prey. Put differently, evidence that ESA hunters appear to have been able to hunt medium- to large-sized mammals such as wildebeest, zebra, waterbuck and various other antelopes (e.g., Bunn and Kroll, 1986; Potts, 1988), strongly suggests that they were somehow able to get close enough to their prey to kill them with crude, non-projectile weapons without serious risk of injury. Given the absence of dogs, nets, and other technologies typically employed by recent hunter–gatherers to disadvantage large animals, the most likely method by which this occurred was persistence hunting (PH), a form of pursuit hunting in which humans use ER during the midday heat to drive animals into hyperthermia and exhaustion so they can be easily killed. Although ethnographic evidence indicates that PH is practiced relatively rarely by recent hunter–gatherers, PH is not only a low-risk method by which ESA hunters could become effective predators, but is also surprisingly low in terms of energetic cost.

PH has been observed among a variety of recent human groups, all in tropical, arid habitats. Among others, PH has been documented for the Kalahari Bushmen (Schapera, 1930; Marshall, 1958; Washburn, 1960; Liebenberg, 1990, 2006), the Tarahumara of Northern Mexico (Bennett and Zingg, 1935; Balke and Snow, 1965; Groom, 1971; Pennington, 1963), the Navajo and Paiutes of the American Southwest (Nabokov, 1981), and Australian aborigines (McCarthy, 1957). In all these cases, PH has three basic characteristics that make it an effective, albeit time-consuming and intensive method of hunting for a poorly-armed human. First, PH primarily occurs during the day when it is hot - often during the hottest time of the year and the day. In the Kalahari, for example, most persistence hunts occur in temperatures of 39–42°C

(Liebenberg, 2006). Second, once hunters spot a target prey animal, they chase the animal at a run, preferably between the prey's preferred trot and gallop speeds. This speed is significant because, as noted above, most mammals cannot gallop long distances, but instead quickly become hyperthermic because of their inability to thermoregulate fast enough via panting. In addition, and in contrast to humans (and kangaroos), most prey have a U-shaped cost of transport (COT, the energy per unit body weight to go a given distance) and thus prefer to trot and gallop only at those speeds that minimize cost (Hoyt and Taylor, 1981). Running at an intermediate speed therefore elevates the animal's cost, hastening its rate of fatigue. When chased at such speeds, prey typically gallop away from the hunter, and try to cool down and rest while the human catches up (Carrier, 1984; Liebenberg, 1990; Heinrich, 2002). Since the animal usually cannot lose heat fast enough during this interval, core body temperature in the prey rises until the animal suffers from heat stroke and exhaustion. Even a kangaroo, which is capable of sweating and has a speed-independent COT, reaches lethal core body temperatures after 1 to 2 h of running (Dawson et al., 1974).

The third key characteristic of PH is the need to track the animal. As noted by Liebenberg (1990, 2006), tracking is a considerable skill that requires the hunter to be able to distinguish tracks in the ground, but also to think like the animal. Tracking is often done while walking, but the faster the hunter can track his/her prey, the quicker the prey becomes hyperthermic. When the cognitive capacity necessary for tracking first evolved is impossible to document, but it seems reasonable to hypothesize that tracking abilities were present in *H. erectus* given its relatively larger brain not to mention its ability to make symmetrical tools that required some mental template (Wynn, 2002). Tracking is also a skill that has to be taught and learned in the field and on the go. In the context of an ER-based hunting strategy, juveniles and/or adolescents would have to have been able to keep up with adults while learning tracking skills.

Viewed from the perspective of ESA hunting technologies and hominin ER capabilities, PH via ER has several key advantages. First, this method of hunting is low risk, and comparatively easy for any human capable of ER and who has the ability to track animals. Second, PH has a relatively high success rate. Approximately 50% of the persistence hunts documented by Liebenberg in the Kalahari were successful, leading to an approximately 70% higher yield of meat per day than hunting using a bow and arrow (Liebenberg, 2006). Third, PH has a surprisingly low metabolic cost. Although PH has frequently been discounted as an unlikely strategy for hunting because the metabolic cost of human running is about 50% higher than an average quadrupedal mammal's after adjusting for body mass, closer inspection of the evidence reveals that the actual cost of ER is not that high, particularly compared to the potential pay-off. The COT of ER in humans is approximately 0.211 O_2/kg/km (Margaria et al., 1963; Cavagna and Kaneko, 1977), about 30–40% higher than the minimal cost of O_2 (0.161 O_2/kg/km) consumed during walking. In addition, while the COT for walking is U-shaped, with an optimal speed (about 1.3 m/s or 5 km/h), the COT for running is independent of speed in humans. In other words, a running human consumes the same amount of energy per unit distance running at a slow jog (3 m/s) or a competitive pace (6 m/s). Assuming an average conversion rate of 4.8 kcal/l O_2, then running 15 km at any ER speed costs approximately 980 Kcal, whereas walking the same distance at an optimal speed costs 750 Kcal. Put in everyday terms, running 15 km to kill a large antelope requires fewer calories than the 1,040 Kcal consumed from a Big Mac® and a medium-sized french fries at McDonald's (http://www.mcdonalds.com/app_controller.nutrition.index1.html)! Since a large antelope weighs more than 200 kg and contains several orders of magnitude more calories than McDonald's can manage to pack into one of its meals, one can easily appreciate that the pay-off is clearly worthwhile, even if the chances of success are only 50%.

A final, possibly important advantage of PH is that it does not require any sophisticated technology other than the simplest weaponry such as a spear or club. Hunting is generally a male activity in recent hunter–gatherer societies, but older children and women (the latter unaccompanied by children or infants) who were good at ER would also have been effective persistence hunters with little risk. Again, Nisa provides an excellent example of this point (Shostak, 1981: 101–102):

> Another day, when I was already fairly big, I went with some of my friends and with my younger brother away from the village and into the bush. While we were walking, I saw the tracks of a baby kudu in the sand. I called out "Hey, Everyone! Come here! Come look at these kudu tracks." The others came over and we all looked at them.
>
> We started to follow the tracks and walked and walked and after a while, we saw a little kudu lying quietly in the grass, dead asleep. I jumped up and tried to grab it. It cried out "Ehnnn... ehnnn..." I hadn't really caught it well and it freed itself and ran away. We all ran, chasing after it, and we ran and ran. But I ran so fast that they all dropped behind and I was alone, chasing it, running as fast as I could. Finally, I was able to grab it, I jumped on it and killed it... I gave the animal to my cousin and he carried it. On the way back, one of the other girls spotted a small steenbok and she and her older brother ran after it. They chased it and finally her brother killed it. That day we brought a lot of meat back to the village and everyone had plenty to eat.

Despite the many advantages of PH, it also has some disadvantages that probably account for its comparative rarity among modern hunter–gatherers with dogs, bows and arrows and other such recently invented (or domesticated) technologies. First, PH is clearly more demanding metabolically and physically than other methods of hunting. It is difficult to imagine why any recent human since the invention of the bow and arrow would regularly engage in PH if other, less grueling forms of obtaining meat were available. PH is also not an option for old or infirm individuals. That said, peak ER performance, as judged by marathon times, is achieved by humans in their 30s; individuals in their 40s typically run long distances such as marathons within 10–20% of their peak performance time (Noakes, 2003). Another cost of PH is that humans require considerable quantities of water in order to thermoregulate adequately during these feats. According to Liebenberg (2006), Kalahari Bushmen always precede a persistence hunt by drinking as much water as they possibly can; carrying water in a gourd or some other form of bottle also improves a hunter's chances. Finally, long distance ER requires dietary sources of salt, which is lost at high rates in sweating, as well as high concentrations of glycogen and triglycerides that can be stored in both the muscle and liver and hydrolyzed into free fatty acids (for review, see Coyle, 2000). Although "carbohydrate-loading" increases these stores, the capability to store glycogen and free fatty acids is highly labile in response to training and does not require diets that are abnormally high in simple carbohydrates (Tsintzas and Williams, 1998).

Another requirement, although not a disadvantage, of PH is that, like other forms of hunting, it requires a cooperative social system in which individuals share food. When an unsuccessful hunter returns to camp, he or she still needs to consume enough calories to pay not only for normal metabolic costs but also for the additional costs of running (at most 50% more than walking). This can only be accomplished by social networks based on food-sharing, and division of labor (Isaac, 1978).

Habitats and Endurance Running

A final consideration relevant to the evolution of ER is habitat. ER, whether for scavenging or for persistence hunting, is obviously an activity suited primarily to relatively open habitats, especially short grass savannas such as the Serengeti ecosystem, as well as more open, arid habitats such as the semi-desert Kalahari or the scrubland of the Turkana Basin. ER is also possible in lacustrine and open woodland zones that lack dense ground vegetation, but is not practicable in tall grass savannas, dense woodland, forest, or marshes and swamps. Although open habitats are a prerequisite for ER, we do not suggest that hominins capable of ER lived exclusively in such zones. Like humans today, early *Homo* almost certainly exploited a wide range of environments. In addition, PH and/or scavenging were probably seasonal behaviors. Thus, our hypothesis is that open habitats in combination with ER capabilities may have provided an important new niche for diurnal scavenging and/or hunting that was one component of their food procurement strategy.

When the open, semi-arid grassland habitats that are now so common in East Africa originated has been the subject of much debate. One theory is that open savannas emerged rapidly during the Pliocene, sometime between 2.8 and 2.5 Ma, as demonstrated by the evolution and prevalence of grazing bovid species such as alcelaphines (e.g., wildebeest, hartebeest and topi) and antelopes (e.g., gazelle) (Vrba, 1995). However, it seems more probable that the process of aridification that occurred prior to the Pleistocene was complex, highly variable, and not as radical as originally suggested (Behrensmeyer et al., 1997; Bobe et al., 2002). Xeric habitats were present prior to 1.8 Ma, for example at Laetoli, which was a dry savanna circa 3.2 Ma (Hay, 1987), but such habitats were probably not widespread until about 1.8 Ma (for review, see Potts, 1988). Thus, regardless of whether earliest *Homo* had evolved ER capabilities (see above), the degree to which habitats conducive to ER were prevalent prior to 1.8 Ma, just after *H. erectus* first appears in the fossil record, is unresolved.

Several lines of evidence suggest that *H. erectus* was probably the first hominin species regularly able to exploit open, hot and arid savanna environments conducive to ER. One source of evidence is the body of form of *H. erectus* itself, whose long limbs and narrow thorax is ideally suited to thermoregulating in the midday sun (Ruff, 1991). More importantly, paleontological and geological evidence from a variety of sites, including the Turkana Basin and Olduvai Gorge, indicate that grasslands were present in the environments in which *H. erectus* lived. In the Turkana Basin, for example, a major environmental change occurred around 1.9 Ma when a lake formed in the central portion of the basin in place of the meandering Omo River, and there was a coincident expansion of open habitats (Feibel et al., 1991; Rogers et al., 1994). At this time, oxygen isotopes from soil samples record a major increase in the percentage of C4 grasses, and faunal assemblages indicate an increased percentage of open habitat grazers (Feibel et al., 1991; Cerling, 1992). By 1.7 Ma, the lake in the Turkana basin had gone, but the region continued to host a diverse range of environments, in which the marginal zones of the basin had "open woodland along ephemeral drainages, giving place laterally to scrub, thicket and dry grassland" (Feibel et al., 1991: 334). These latter habitats would probably have been ideal for PH during dry seasons, and have been present ever since, including around 1.5 Ma when the Nariokotome boy lived (Feibel and Brown, 1993; Harris and Leakey, 1993).

Olduvai Gorge presents a similar picture. According to Cerling and Hay (1986), Lower Bed I of Olduvai was a wet, marshy habitat, but by the top of Bed I (approximately 1.8 Ma), the environment was more open and arid. At the top of Bed II, which is dated to approximately 1.7 Ma, there is a prevalence of dry vegetation and open habitats (Cerling and Hay, 1986). For example, oxygen isotope analyses of soils indicate that C4 vegetation went from between 20–40% to between 60–80% around 1.8 Ma, for a phase that lasted at least 50,000 years (Hay, 1976). Although conditions at Olduvai and elsewhere certainly fluctuated considerably throughout the Pleistocene (see Potts, 1998), it is reasonable to conclude that within the general region of the Gorge there was an abundance of open habitats after about 1.75 Ma that would have been conducive either to scavenging or hunting by ER.

A relationship between ER and open-country adaptations by *H. erectus* may also be involved in the marked shift in the quality of evidence for hominin dispersal into temperate Eurasia after 0.9 Ma. Prior to this time, evidence for hominin occupation is sparse, but afterwards there is clear and consistent evidence of such occupation (Dennell, 2003). The onset of Middle Pleistocene glaciations after 0.9 Ma, and the increasingly open-steppic landscapes throughout much of Eurasia may have made this region a more hospitable venue for *H. erectus*' ER-based hunting adaptation.

Conclusion

In conclusion, humans have a surprisingly impressive ability to run long distances at relatively high speeds and in extremely hot conditions compared to other specialized cursors. In many respects, these capabilities can also exceed those of the few other mammals – all social carnivores – known to engage in ER. In addition, human ER capabilities are all the more special because other primates generally eschew running other than occasional sprinting, and they lack endurance capabilities.

If humans are so good at ER, then why have these capabilities received so little attention in the history of research on human evolution? There have been countless articles and numerous books on the evolution of bipedalism in hominins, yet, with the exceptions of Carrier (1984) and Bramble and Lieberman (2004), none have considered running in any depth (see also Bortz, 1985; Heinrich, 2002). There are three major reasons for this lack of attention. First, what is out of sight is often out of mind: humans no longer need to run very much, and do so now primarily for pleasure or health. Second, we consistently underrate our abilities as athletes, primarily because we tend to focus on aspects of athleticism related to speed and power in which humans are pathetic compared to most mammals. The idea that brains have triumphed over brawn is so deeply engrained that it rarely receives much consideration.

Finally, students of the fossil record of human evolution have, understandably, focused on the origins of walking. There is substantial evidence that the earliest hominins were bipedal (Haile-Selassie, 2001; Galik et al., 2004; Zollikofer et al., 2005; Richmond and Jungers, 2008), and that walking was a key part of the transition that set early hominins off on a strikingly different evolutionary trajectory than chimpanzees (Darwin, 1871). There has been much debate over the extent to which early hominins were arboreal and whether these capabilities compromised their ability to walk optimally (Lovejoy, 1988; Stern, 2000; Ward, 2002), but few doubt that australopithecines were capable, habitual bipeds. However, it is important to note that the biomechanics of running and walking are substantially different, especially for a biped. In addition, the physiological demands of ER are quite different from those of sprinting or walking. Thus, evidence for walking capabilities in early hominins is not necessarily evidence for ER capability. Instead, a diverse array of comparative functional morphological and physiological evidence (Carrier, 1984; Bramble and Lieberman, 2004) suggests that human ER capabilities are not a byproduct of selection on bipedal walking alone. It follows that human ER capabilities demand some explanations for when and why they evolved.

Unfortunately, we cannot pinpoint precisely when ER first evolved. As outlined above and by Bramble and Lieberman (2004), the majority of the fossil evidence points to *H. erectus* as the first endurance runner. But, as also noted, we cannot rule out the possibility that *H. habilis* had some ER capabilities, nor can we rule out the hypothesis that later hominins had better performance capabilities than early *H. erectus*. More definitive answers require more evidence and more research. That said, the available evidence suggests that *Australopithecus* lacked many, if not most, of the derived features of *Homo* that improve ER performance. Some of these derived features, such as relatively large anterior and posterior semicircular canals, and the nuchal ligament, are specific to running. Other derived features, such as long legs, would have benefited both running and walking. Thus it is reasonable to speculate that selection for ER occurred in the context of selection for *both* walking and running long distances.

If there has been any skepticism about the ER hypothesis, then it has been with regard to why ER evolved (see, for example, comments in Carrier, 1984; also Pickering and Bunn, 2007). Because modern humans, including recent hunter–gatherers, no longer require ER in their daily lives, it is hard for many scholars to imagine how ER would have been selected for in the distant past. However, such reliance on the ethnographic present – what Wobst (1978) has referred to as the "tyranny of ethnography" – is problematic since recent inventions (such as the bow and arrow and the domestication of the dog) have substantially changed human hunting strategies in precisely those aspects that relate to ER. Critically, these innovations allow humans to hunt and kill animals from a distance without getting close to large prey. But for most of the history of the genus *Homo*, it appears that hominins have been able to kill large, prime age, adult prey that would have posed serious risks to any hunter armed solely with an untipped spear. ER, however, would have changed that equation by allowing hunters in the hot, arid and open habitats that have existed in Africa since at least 1.9 Ma, to run their prey into exhaustion, thereby disadvantaging them sufficiently to be slain with minimal risk and a high probability of success. While ER-based persistence hunting would have required the cognitive skills to track an animal combined with abundant access to water, the energetic costs are surprisingly low in comparison to walking, and well worthwhile in terms of the payoff. Like other methods of hunting, ER and PH would also have required social groups with food-sharing.

Although the extent to which scavenging was an important behavior among early hominins is still debated, it is likely that scavenging played some component of early *Homo* subsistence strategies, just as it now does among the Hadza and Bushmen. Since carcasses are an evanescent resource in which early access improves the chance of getting something to eat and minimizing competition with other carnivores, then it would have benefited from ER capabilities.

In short, there is a compelling case to be made that ER would have given early *Homo* the ability to create a new niche within the carnivore guild: that of a diurnal predator within the increasingly open habitats in Africa by 1.9 Ma. In particular, ER would have provided ESA hunters with various means of getting meat at comparatively low risk and low cost. Observations that ER is rare among modern hunter–gatherers who possess weapons (such as the bow and arrow and atlatl) are not disproof of the hypothesis. Instead, the persistent, albeit rare, use of ER in scavenging and persistence hunting by modern hunter–gatherers such as the Bushmen, the Tarahumara and others are testaments to the importance of running in hunting in general, and the effectiveness of persistence hunting in particular, despite the invention of technologies that have made these athletic feats obsolete.

Finally, it is fun to conclude by speculating on a possible scenario for the evolution of ER in the genus *Homo*. Natural selection works by tinkering (Jacob, 1977). That is, selection can work only by taking advantage of small-scale heritable variations that somehow improve performance within a particular fitness context. One can well imagine circumstances in which the earliest members of the genus *Homo* or perhaps australopithecines began to scavenge or possibly hunt a little. In such a context, individuals with variations such as larger anterior and posterior semicircular canals, longer legs, narrower waists, more sweat glands, and so on might have enjoyed some fitness benefit because their improved performance in long distance running and/or walking that helped them acquire more meat. Over time – depending on factors such as the strength of selection, how much variation was available, and population size – modern ER capabilities, along with a modern-shaped body evolved, probably first in *H. erectus*. These capabilities apparently enabled *H. erectus* to kill medium- to large-sized animals in the hot, open habitats of Africa in the Early Pleistocene without any weaponry more sophisticated than a sharpened wooden stick. After the ESA, more sophisticated projectile technologies evolved (e.g., stone and bone tipped spears, bows and arrows, spear throwers and nets) that gave hunters other, less grueling options to bring home the bacon. As a result, persistence hunting has become less important. In addition, many hominins started to move out of Africa into temperate zones where PH was no longer possible. But the traces of our ancestry persist in a body well-suited to ER, a behavior that nowadays serves primarily as a means of relaxation and a way to stay healthy.

References

Acosta, L. Jr, Roy, R.R., 1987. Fiber-type composition of selected hindlimb muscles of a primate (cynomolgus monkey). Anatomical Record 218, 136–141.

Aiello, L., Dean, M.C., 1990. An Introduction to Human Evolutionary Anatomy. Academic, London.

Aiello, L., Wheeler, P., 1995. The expensive tissue hypothesis: the brain and digestive system in human and primate evolution. Current Anthropology 36, 199–221.

Alexander, R.M., 1991. Elastic mechanisms in primate locomotion. Zeitschrift fur Morphologie und Anthropologie 78, 315–320.

Armstrong, R.B., Saubert, C.W. 4th, Seeherman, H.J., Taylor, C.R., 1982. Distribution of fiber types in locomotory muscles of dogs. American Journal of Anatomy 163, 87–98.

Balke, B., Snow, C., 1965. Anthropological and physiological observations on Tarahumara endurance runners. American Journal of Physical Anthropology 23, 293–301.

Behrensmeyer, A.K., Todd, N.E., Potts, R., McBrinn, G.E., 1997. Late Pliocene faunal turnover in the Turkana Basin, Kenya and Ethiopia. Science 278, 1589–1594.

Bennett, W.C., Zingg, R.M. 1935. The Tarahumara: an Indian Tribe of Northern Mexico. University of Chicago Press, Chicago, IL.

Berger, T.D., Trinkaus, E., 1995. Patterns of trauma among the Neandertals. Journal of Archaeological Science 22, 841–852.

Bianchi, M., 1989. The thickness, shape and arrangement of the elastic fibres within the nuchal ligament from various animal species. Anatomischer Anzeiger 169, 53–66.

Binford, L., 1984. Bones of contention: a reply to Glynn Isaac. American Antiquity 49, 164–167.

Blumenschine, R.J., 1986. Early Hominid Scavenging Opportunities: Implications of Carcass Availability in the Serengeti and Ngorongoro Ecosystems. British Archaeological Reports International Series, 283, Oxford

Blumenschine, R.J., 1987. Characteristics of an early hominid scavenging niche. Current Anthropology 28, 383–407

Blumenschine, R.J., 1988. An experimental model of the timing of hominid and carnivore influence on archaeological bone assemblages. Journal of Archaeological Science 15, 483–502.

Blurton-Jones, N.G., Marlowe, F.W., 2002. Selection for delayed maturity: does it take 20 years to learn to hunt and gather? Human Nature 13, 199–238.

Bobe, R., Behrensmeyer, A.K., Chapman, R.E., 2002. Faunal change, environmental variability and late Pliocene hominin evolution. Journal of Human Evolution 42, 475–497.

Bortz, W.M., 1985. Physical exercise as an evolutionary force. Journal of Human Evolution 14, 145–155

Brain, C.K., 1981. The Hunters or the Hunted? An Introduction to African Cave Taphonomy. University of Chicago Press, Chicago, IL.

Bramble, D.M., Carrier, D.R., 1983. Running and breathing in mammals. Science 219, 251–256.

Bramble, D.M., Jenkins, F.A.J. Jr., 1993. Mammalian locomotor-respiratory integration: implications for diaphragmatic and pulmonary design. Science 262, 235–340.

Bramble, D.M., Lieberman, D.E., 2004. Endurance running and the evolution of *Homo*. Nature 432, 345–352.

Bramble, D.M, Lieberman, D.E., Raichlen, D.A., Whitcome, K., 2009. Passive head stabilization in running humans. Science (in review).

Bunn, H.T. 2001. Hunting, power scavenging, and butchering by Hadza foragers and by Plio-Pleistocene *Homo*. In: Stanford, C.B., Bunn, H.T. (Eds), Meat-Eating and Human Evolution. Oxford University Press, Oxford, pp. 199–218.

Bunn, H.T., Ezzo, J.A., 1993. Hunting and scavenging by Plio-Pleistocene hominids: nutritional constraints, archaeological patterns, and behavioural implications. Journal of Archaeological Science 20, 365–398

Bunn, H.T., Kroll, E.M., 1986. Systematic butchery by Plio-Pleistocene hominids at Olduvai Gorge, Tanzania. Current Anthropology 27, 431–452.

Cattelain, P., 1997. Hunting during the Upper Paleolithic: bow, spearthrower, or both? In: Knecht, H. (Ed.) Projectile Technology. Plenum, New York, pp. 213–240.

Cavagna, G.A., Kaneko, M., 1977. Mechanical work and efficiency in level walking and running. Journal of Physiology 268, 467–481.

Cavallo, J.A., Blumenschine, R.J., 1989. Tree-stored leopard kills: expanding the hominid scavenging niche. Journal of Human Evolution 18, 393–399

Carrier D.R., 1984. The energetic paraodox of human running and hominid evolution. Current Anthropology 24, 483–495.

Cerling, T.E., 1992. Development of grasslands and savannas in East Africa during the Neogene. Palaeogeography, Palaeoclimatology, Palaeoecology 97, 241–247.

Cerling, T.E., Hay, R.L., 1986. An isotopic study of paleosol carbonates from Olduvai Gorge. Quaternary Research 25, 63–78.

Cheuvront, S.N., Haymes, E.M., 2001. Ad libitum fluid intakes and thermoregulatory responses of female distance runners in three environments. Journal of Sports Sciences 19, 845–854

Churchill, S.E., 1993. Weapon technology, prey size selection and hunting methods in modern hunter-gatherers: implications for hunting in the Palaeolithic and Mesolithic. In: Peterkin, G.L., Bricker, H.M. Mellars, P.A. (Eds), Hunting and Animal Exploitation in the Later Palaeolithic and Mesolithic of Eurasia. Archeological Papers of the American Anthropological Association No. 4, pp. 11–24.

Clark, J.D., 1955. The stone ball, its associations and use by prehistoric Man in Africa. In: L. Balout (Ed.), Actes du Congrès Pan-African de Prèhistoire, IIe session, Alger, 1952. Arts et Métiers, Paris, pp. 403–407.

Cole, S., 1963. The Prehistory of East Africa. Mentor, New York.

Cooper, S.M., 1991. Optimal hunting group size: the need for lions to defend their kill against loss to spotted hyenas. African Journal of Ecology 29, 130–136.

Coyle, E.F., 2000. Physical activity as a metabolic stressor. American Journal of Clinical Nutrition 72, 512S-20S.

Cromwell, R.L., Newton, R.A., Carlton, L.G., 2001. Horizontal plane head stabilization during locomotor tasks. Journal of Motor Behavior 33, 49–58

Darwin, C., 1871. The Descent of Man. John Murray, London.

Dawson, T.J., Robertshaw, D., Taylor, C.R., 1974. Sweating in the kangaroo: a cooling mechanism during exercise, but not in the heat. American Journal of Physiology 227, 494–498.

Dennell, R., 2003. Dispersal and colonisation, long and short chronology: how continuous is the Early Pleistocene record for hominids outside East Africa. Journal of Human Evolution 45, 421–440.

Dill, D.B., Bock, A.V. Edwards, H.T., 1933. Mechanism for dissipating heat in man and dog. American Journal of Physiology 104, 36–43.

Dimery, N.J., Alexander, R.M., Deyst, K.A., 1985. Mechanics of the ligamentum nuchae of some artiodactyls. Journal of Zoology, London 206, 341–351.

Domínguez-Rodrigo, M., 2002. Hunting and scavenging by early humans: the state of the debate. Journal of World Prehistory 16, 1–54.

Entin, P.L., Robertshaw D., Rawson R.E., 1999. Effect of locomotor respiratory coupling on respiratory evaporative heat loss in the sheep. Journal of Applied Physiology 87, 1887–1893.

Feibel, C.S., Brown, F.H., 1993. Microstratigraphy and environments. In: Walker, A., Leakey, R.E.F. (Eds.) The Nariokotome *Homo erectus* skeleton. Harvard University Press, Cambridge, MA, pp. 21–39.

Feibel, C.S., Harris, J.M., Brown, F.H., 1991. Palaeoenvironmental context for the late Neogene of the Turkana Basin. In: Harris, J.M. (Ed.): Koobi Fora Research Project, Vol. 3. Clarendon, Oxford, pp. 321–370.

Galik, K., Senut, B., Pickford, M., Gommery, D., Treil, J., Kuperavage, A.J., Eckhardt, R.B., 2004. External and internal morphology of the BAR 1002′00 *Orrorin tugenensis* femur. Science 305, 1450–1453.

Garland, T., 1983. The relation between maximal running speed and body mass in terrestrial mammals. Journal of Zoology 199, 1557–1570.

Gauthier, G.B., Piron, J.P., Roll, J.P., Marchetti, E., Martin, B., 1984. High-frequency vestibulo-ocular reflex activation through forced head rotation in man. Aviation, Space and Environmental Medicine 55, 1–7.

Groom, D., 1971. Cardiovascular observations on Tarahumara Indian runners – the modern Spartans. American Heart Journal 81, 304–314.

Haile-Selassie, J., 2001. Late Miocene hominids from the Middle Awash, Ethiopia. Nature 412, 178–181.

Harcourt-Smith, W.E.H., 2002. Form and function in the hominoid tarsal skeleton. Ph.D. thesis, University College London, London.

Harris, J.M, Leakey, M., 1993. The faunal context. In: Walker, A., Leakey, R.E.F. (Eds), The Nariokotome *Homo erectus* skeleton. Harvard University Press, Cambridge, MA, pp. 54–60.

Hay, R.L., 1987 Geology of the Laetoli area. In: Leakey, M.D., Harris, J.M. (Eds), Laetoli: A Pliocene Site in Northern Tanzania. Clarendon, Oxford, pp. 23–47.

Hay, R.L., 1976. Geology of the Olduvai Gorge. University of California Press, Berkeley.

Heglund, N.C., Taylor, C.R,. 1988. Speed, stride frequency and energy cost per stride. How do they change with body size and gait? Journal of Experimental Biology 138, 301–318.

Heinrich, B., 2002. Why We Run: a Natural History. Harper Collins, New York.

Hinchcliff, K.W., Shaw, L.C., Vukich, N.S., Schmidt, K.E., 1998. Effect of distance traveled and speed of racing on body weight and serum enzyme activity of sled dogs competing in a long-distance race. Journal of the American Veterinary Medicine Association 213, 639–44.

Hinrichs, R.N., 1990. Upper extremity function in distance running. In: Cavanagh, P.R (Ed.) Biomechanics of Distance Running. Human Kinetics Books, Champaign, IL, pp. 107–133.

Hirasaki, E., Moore, S.T., Raphan, T., Cohen, B., 1999. Effects of walking velocity on vertical head and body movements during locomotion. Experimental Brain Research 127, 117–30

Holekamp, K.E., Boydston, E.E., Smale, E., 2000. Group travel in social carnivores. In: Boinski, S., Garber, P. (Eds), On the Move: How and Why Animals Travel in Groups. University of Chicago Press, Chicago, IL, pp, 587–627.

Hoyt, D.F., Taylor, C.R., 1981. Gait and the energetics of locomotion in horses. Nature 292, 239–240.

Hunt, K.D., 1992. Positional behavior of *Pan troglodytes* in the Mahale Mountains and Gombe Stream National Parks, Tanzania. American Journal of Physical Anthropology 87, 83–105.

Isaac, G.L., 1978. Food sharing and human evolution: archaeological evidence from the Plio-Pleistocene of East Africa. Journal of Anthropological Research 34, 311–325.

Isbell, L.A., Pruetz, J.D., Lewis, M., Young, T.P., 1998. Locomotor activity differences between sympatric patas monkeys (*Erythrocebus patas*) and vervet monkeys (*Cercopithecus aethiops*): implications for the evolution of long hindlimb length in *Homo*. American Journal of Physical Anthropology 105, 199–207.

Jablonski, N., 2006. Skin: a Natural History. University of California Press, Berkeley.

Jacob, F., 1977. Evolution and tinkering. Science 196, 1161–1166.

Jeffery, N., Spoor, F., 2004. Prenatal growth and development of the modern human labyrinth. Journal of Anatomy 204, 71–92.

Jellema, L.M., Latimer, B., Walker, A., 1993. The rib cage. In: Walker, A., Leakey, R. E.F. (Eds), The Nariokotome *Homo erectus* Skeleton. Harvard University Press, Cambridge, MA, pp 294–325.

Kappel-Bargas, A., Woolf, R.D., Cornwall, M.W., McPoil, T.G., 1998. The windlass mechanism during normal walking and passive first metatarsalphalangeal joint extension. Clinical Biomechanics 13, 190–194.

Keller, T.S., Weisberger A.M., Ray J.L., Hasan S.S., Shiavi R.G., Spengler D.M., 1996. Relationship between vertical ground reaction force and speed during walking, slow jogging, and running. Clinical Biomechanics 11, 253–259.

Ker, R.F., Bennett, M.B., Bibby, S.R., Kester, R.C., Alexander, R. McN., 1987. The spring in the arch of the human foot. Nature 325, 147–149.

Landau, M., 1993. Narratives in Human Evolution. Yale University Press, New Haven.

Larson, S.G., 1993. Functional morphology of the shoulder in primates. In: Gebo, D.L. (Ed.), Postcranial Adaptation in Nonhuman Primates. Northern Illinois University Press, Dekalb, pp. 45–69.

Larson, S.G., 2009. Evolution of the hominin shoulder: early *Homo*. In: Grine, F.E., Fleagle, J.G., Leakey, R.E. (Eds), The First Humans: Origin and Early Evolution of the Genus *Homo*. Springer, Dordrecht, pp. 65–75.

Larson, S.G., Jungers, W.L., Morwood, M., Sutikna, T., Jatmiko, Saptomo, E.W., Due, R.A., Djubiantono, T., 2007. *Homo fbr esiensis* and the evolution of the hominin shoulder. Journal of Human Evolution 53, 718–731.

Lewis, O.J., 1989. Functional Morphology of the Evolving Hand and Foot. Oxford University Press, Oxford.

Liebenberg, L., 1990. The Art of Tracking: The Origin of Science. David Philip Publishers, Claremont (South Africa).

Liebenberg, L., 2006. Persistence Hunting by Modern Hunter-Gatherers. Current Anthropology 47, 1017–1025.

Lieberman, D.E., Raichlen, D.A, Pontzer, H., Bramble, D.M., Cutright-Smith, E., 2006. The human gluteus maximus and its role in running. Journal of Experimental Biology 209, 2143–2155

Lieberman, D.E., Bramble, D.M., Raichlen, D.A., Shea, J.J., 2007. Endurance running and the tyranny of ethnography: a reply to Pickering and Bunn. Journal of Human Evolution 53, 439–442.

Lordkipanidze, D., Jashashvili, T., Vekua, A., Ponce de León, M.S., Zollikofer, C.P., Rightmire, G.P., Pontzer, H., Ferring, R., Oms, O., Tappen, M., Bukhsianidze, M., Agusti, J., Kahlke, R., Kiladze, G., Martinez-Navarro, B., Mouskhelishvili, A., Nioradze, M., Rook, L. 2007. Postcranial evidence from early *Homo* from Dmanisi, Georgia. Nature 449, 305–310

Lovejoy, C.O., 1988. The evolution of human walking. Scientific American 259, 118–125.

Maas, E.F., Huebner, W.P., Seidman, S.H., Leigh, R.J., 1989. Behavior of human horizontal vestibulo-ocular reflex in response to high-acceleration stimuli. Brain Research 499, 153–156.

Marean, C.W., 1989. Sabertooth cats and their relevance for early hominid diet and evolution. Journal of Human Evolution 18, 559–582.

Margaria, R., Cerretlli, P., Aghemo, P., Sassi, G., 1963. Energy cost of running. Journal of Applied Physiology 18, 367–370.

Marshall, J., 1958. Man as hunter. Natural History 72, 291–309, 376–395.

McCarthy, F.D., 1957. Australian Aborigines: Their Life and Culture. Colorgravure Publications, Melbourne.

McArdle, W.D., Katch, F.I., Katch,V.L., 1996. Exercise Physiology: Energy, Nutrition and Human Performance. 4th edition. Lea and Febiger, Philadelphia, PA.

McArthur, A.J., Monteith, J.L., 1980. Air movement and heat loss from sheep. I. Boundary layer insulation of a model sheep, with and without fleece. Proceedings of the Royal Society of London, Part B, Biological Sciences 209, 187–208.

Mercer, S.R., Bogduk, N., 2003. Clinical anatomy of ligamentum nuchae. Clinical Anatomy 16, 484–493.

Minetti, A.E., 2003. Physiology: efficiency of equine express postal systems. Nature 426, 785–786.

Nabokov, P., 1981. Indian Running: Native American History and Tradition. Ancient City Press, Santa Fe, NM.

Noakes, T., 2003. The Lore of Running. 4th Edition. Human Kinetics, Champaign, IL.

O'Brien, E.M., 1981. The projectile capabilities of an Acheulean handaxe from Olorgesailie. Current Anthropology 22, 76–79.

O'Connell, J.F., Hawkes, K., Blurton-Jones, N.G., 1988. Hadza scavenging: implications for Plio-Pleistocene hominid subsistence. Current Anthropology 29, 356–363.

Oswalt, W., 1976. An Anthropological Analysis of Food-Getting Technology. Wiley, New York.

Pennington, Campbell W., 1963. The Tarahumara of Mexico, Their Environment and Material Culture. University of Utah Press, Salt Lake City.

Pennycuick, C.J., 1979. Movements of the migratory wildebeest population in the Seregeti. In: Sinclair, A.R.E., Norton-Griffiths, M. (Eds), Serengeti: Dynamics of an Ecosystem. University of Chicago Press, Chicago, IL, pp. 164–184.

Pickering T.R., Bunn, H.T., 2007. The endurance running hypothesis and hunting and scavenging in savanna-woodlands. Journal of Human Evolution 53, 438–442.

Pontzer, H., 2005 A new model predicting locomotor cost from limb length via force production. Journal of Experimental Biology 208, 1513–24.

Pontzer, H., 2007. Limb length and the scaling of locomotor cost in terrestrial animals. Journal of Experimental Biology 210, 1752–1761.

Potts, R., 1988. Environmental hypotheses of human evolution. Yearbook of Physical Anthropology 41, 93–136.

Richmond, B.G., Jungers, W.L., 2008. *Orrorin tugenensis* femoral morphology and the evolution of hominin bipedalism. Science 319, 1662–1665.

Rieder, H., 2003. Der Große Wurf der frühen Jäger: Nachbau altsteinzeitlicher Speere. Biologie im Unserer Zeit 33, 156–160.

Rogers, M.J., Feibel, C.S., Harris, J.W.K., 1994. Changing patterns of land use by Plio-Pleistocene hominids in the Lake Turkana Basin. Journal of Human Evolution 27, 139–158.

Rose, M.D., 1984. A hominine hip bone, KNM-ER 3228, from East Lake Turkana, Kenya. American Journal of Physical Anthropology 63, 371–378.

Ruff, C.B., 1991. Climate, body size and body shape in hominid evolution. Journal of Human Evolution 21, 81–105.

Saunders J.B.D.M., Inman, V.T., Bresler, B., 1953. The major determinants in normal and pathological gait. Journal of Bone and Joint Surgery 35A, 543–555.

Schapera, I., 1930. The Khoisan Peoples of South Africa: Bushmen and Hottentots. Routledge and Kegan Paul, London.

Scheuer, L., Black, L., 2000. Developmental Juvenile Osteology. Academic Press, London.

Schmid, P., 1983. Eine Rekonstruktion des skelettes von A.L. 288–1 (Hadar) und deren Konsequenzen. Folia Primatologica 40, 283–306.

Schmid, P., 1991. The trunk of the australopithecines. In: Coppens, Y., Senut, B. (Eds), Origine(s) de la Bipédie chez les Hominidés. Presses du CNRS, Paris, pp. 225–234.

Schmidt-Nielsen, K., 1990. Animal Physiology: Adaptation and Environment. 4th edition. Cambridge University Press, Cambridge.

Shea, J.J., 2006a. The origins of lithic projectile point technology: evidence from Africa, the Levant, and Europe. Journal of Archaeological Science 33, 823–846.

Shea, J.J., 2006b. What stone tools can (and can't) tell us about early hominin diets. In: Ungar, P. (Ed.), Early Hominin Diets: The Known, the Unknown and the Unknowable. Oxford University Press, Oxford, pp. 321–351.

Stedman, H.H., Kozyak, B.W., Nelson, A., Thesier, D.M., Su, L.T., Low, D.W., Bridges, C.R., Shrager, J.B., Minugh-Purvis, N., Mitchell, M.A., 2004. Myosin gene mutation correlates with anatomical changes in the human lineage. Nature 428, 415–418.

Shostak, M., 1981. Nisa: The Life and Words of a !Kung Woman. Harvard University Press, Cambridge, MA.

Spoor, F., Wood, B., Zonneveld, F., 1994. Implications of early hominid labyrinthine morphology for evolution of human bipedal locomotion. Nature 369, 645–648.

Stern, J.T., 2000. Climbing to the top: a personal memoir of *Australopithecus afarensis*. Evolutionary Anthropology 9, 113–133.

Steudel-Numbers, K.L., Tilkens, M.J., 2004. The effect of lower limb length on the energetic cost of locomotion: implications for fossil hominins. Journal of Human Evolution 47, 95–109.

Steudel-Numbers, K.L., Weaver, T.D., Wall-Scheffler, C.M., 2007. The evolution of human running: effects of changes in lower-limb length on locomotor economy. Journal of Human Evolution 53, 191–196.

Susman, R.L., 2008. New evidence bearing on the status of *Homo habilis* from Olduvai Gorge. American Journal of Physical Anthropology 135 (S46), 203.

Susman, R.L., Stern. J.T., Jungers, W.L,. 1984. Arboreality and bipedality in the Hadar hominids. Folia Primatologica 43, 113–156.

Tappen, M., 2001. Deconstructing the Serengeti. In: Stanford, C.B., Bunn, H.T. (Eds), Meat-Eating and Human Evolution. Oxford University Press, Oxford, pp. 13–32.

Taylor, C.R., Rowntree, V.J., 1973. Temperature regulation and heat balance in running cheetahs: a strategy for sprinters? American Journal of Physiology 224, 848–851.

Thayer, R., Collins, J., Noble, E.G., Taylor, A.W., 2000. A decade of aerobic endurance training: histological evidence for fibre type transformation. Journal of Sports Medicine and Physical Fitness 40, 284–289.

Theime, H., 1997. Lower Paleolithic hunting spears from Germany. Nature 385, 807–810.

Thorstensson, A., Nilsson, J., Carlson, H., Zomlefer, M.R., 1984. Trunk movements in human locomotion. Acta Physiologica Scandinavica 121, 9–22.

Torii, M., 1995. Maximal sweating rate in humans. Journal of Human Ergology (Tokyo) 24, 132–152.

Tsintzas, K., Williams, C., 1998. Human muscle glycogen metabolism during exercise. Effect of carbohydrate supplementation. Sports Medicine 25, 7–23.

Van Valkenburgh, B,. 2001. The dog-eat-dog world of carnivores: a review of past and present carnivore community dynamics In: Stanford, C.B., Bunn, H.T. (Eds), Meat-Eating and Human Evolution. Oxford University Press, Oxford, pp. 101–121

Vrba, E.S., 1995. The fossil record of African antelopes (Mammalia, Bovidae) in relation to human evolution and paleoclimate. In: Vrba, E.S., Denton, G.H., Partridge, T.C., Burckle, L.H. (Eds), Paleoclimate and Evolution, With Emphasis on Human Origins. Yale University Press, New Haven, pp. 385–424.

Ward, C.V., 2002. Interpreting the posture and locomotion of *Australopithecus afarensis*: where do we stand? Yearbook of Physical Anthropolology 35, 185–215.

Washburn, S.L., 1960. Tools and human evolution. Scientific American 203, 169–181.

Wheeler, P.E., 1991. The thermoregulatory advantages of hominid bipedalism in open equatorial environments: the contribution of increased convective heat loss and cutaneous evaporative cooling. Journal of Human Evolution 21, 107–115.

Whittaker, J.C., McCall, G., 2001. Handaxe-hurling hominids: an unlikely story. Current Anthropology 42, 566–572.

Williams, T.M., Dobson, G.P., Mathieu-Costello, O., Morsbach, D., Worley, M.B., Phillips, J.A., 1997. Skeletal muscle histology and biochemistry of an elite sprinter, the African cheetah. Journal of Comparative Physiology, Part B 167, 527–535.

Wilson R.S., James R.S., 2004. Constraints on muscular performance: trade-offs between power output and fatigue resistance. Proceedings of the Royal Society, Biological Science 271 Supplement 4, S222–225.

Winter, D.A., Ruder, K.G., MacKinnon, C.D., 1990. Control of balance of upper body in gait. In: Winters, J.M., Woo, S.L.Y. (Eds), Multiple Muscle Systems: Biomechanics and Movement Organization. Springer, New York, pp. 534–541.

Wobst, M., 1978. The archaeo-ethnology of hunter-gatherers or the tyranny of the ethnographic record in archaeology. American Antiquity 43, 303–309.

Wood, B.A., 2009. Where does the genus *Homo* begin, and how would we know? In: Grine, F.E., Fleagle, J.G., Leakey, R.E. (Eds), The First Humans: Origin and Early Evolution of the Genus *Homo*. Springer, Dordrecht, pp. 17–28.

Wood, B.A., Collard, M., 1999. The human genus. Science, 284, 65–71.

Wynn, T., 2002. Archaeology and cognitive evolution. Behavioral and Brain Science 25, 389–402

Yang, N., MacArthur, D.G., Gulbin, J.P., Hahn, A.G., Beggs, A.H., Easteal, S., North, K. 2003., ACTN3 genotype is associated with human elite athletic performance. American Journal of Human Genetics 73, 627–631.

Zollikofer, C.P., Ponce de Leon, M.S., Lieberman, D.E., Guy, F., Pilbeam, D., Likius, A., Mackaye, H.T., Vignaud, P., Brunet, M., 2005. Virtual cranial reconstruction *of Sahelanthropus tchadensis*. Nature 434, 755–759.

Chapter 9
Interlimb Proportions in Humans and Fossil Hominins: Variability and Scaling

William L. Jungers

Keywords *Australopithecus* • *Homo erectus* • *Homo floresiensis* • humerofemoral index • intermembral index • interlimb proportions • scaling

Introduction

Our understanding of the evolutionary trajectory of hominin limb proportions and overall body shape is severely hampered by the paucity of fossil individuals with associated elements from early in the human career (Walker, 1973; McHenry and Coffing, 2000; Richmond et al., 2002; Reno et al., 2005). It is now apparent that hind-limb elongation and modern human interlimb proportions emerged at least by the time of early *Homo erectus* (=*ergaster*) as represented by the fossils from Dmanisi at almost 1.8 million years ago (Lordkipanidze et al., 2007) and the "Nariokotome boy", KNM-WT 15000, at approximately 1.6 million years ago (Ruff, 1993; Ruff and Walker, 1993). Interlimb proportions based on complete bone lengths in australopithecines are poorly known, and "Lucy" (A.L. 288-1, *Australopithecus afarensis*) still represents our best example of the "primitive" hominin condition (Johanson and Edey, 1981). With a humerofemoral index (100 × humerus length/femur length) near 85, *A. afarensis* is "intermediate" between African apes and modern humans (Jungers, 1982, 1991; Jungers and Stern, 1983; Richmond et al., 2002). This difference from humans is driven by a relatively short femur (but not as short as in apes), not by long arms (Jungers, 1994). It is also important to note that interlimb proportionality based on lengths need not correspond to estimates of interlimb size and shape based on diaphyseal and articular dimensions (McHenry, 1978; McHenry and Berger, 1998; Green et al., 2007). Long bone lengths, especially of the hind-limb elements, are most relevant to questions about the evolution of locomotor efficiency and the kinematics of bipedalism (e.g., Jungers, 1982; Bramble and Lieberman, 2004; Pontzer, 2005; Steudel-Numbers, 2006).

BOU-VP-12/1 is another skeleton of a later australopith (possibly *A. garhi*), but its humerus and femur are both quite damaged and incomplete (Asfaw et al., 1999). Estimated and reconstructed lengths of these long bones are controversial and exhibit large confidence intervals, whether based on regression or by eye (Richmond et al., 2002; Haeusler and McHenry, 2004; Reno et al., 2005). Although possible, it seems premature to conclude with any confidence that its humerofemoral proportions were already "human-like" (Reno et al., 2005). Similarly, *Homo habilis* as represented by OH 62 has essentially indeterminate interlimb proportions, probably lying somewhere between those of gorillas and modern humans (Korey, 1990; Reno et al., 2005; but see Haeusler and McHenry, 2004). It appears risky at this time to develop complex evolutionary scenarios about limb length proportions predicated on BOU-VP-12/1 and/or OH 62.

The recent discovery of a partial skeleton of a new species of small-bodied hominin in Indonesia dated to the Late Pleistocene (*Homo floresiensis*) is significant and relevant to this discussion (Brown et al., 2004; Morwood et al., 2005; Argue et al., 2006). The type specimen (LB 1) has a humerus and femur very similar in lengths to those observed in A.L. 288-1, and exhibits a humerofemoral index of almost 87; as in *A. afarensis*, this difference from modern humans derives primarily from an absolutely and relatively short femur (280 mm). Wolpoff (1983) argued that the limb proportions seen in A.L. 288-1 were simply the allometric result of small body size; this argument assumed that the humerofemoral index necessarily increases as adult human body size decreases. Although Wolpoff's size-required model was vigorously contested (Jungers and Stern, 1983; Jungers, 1991; see also Haeusler and McHenry, 2004), this allometric argument was redeveloped by Franciscus and Holliday (1992) and again by Lovejoy (1993). Lovejoy (1993) extrapolated the bivariate relationship between femur length and the humerofemoral index seen in Causasian males and contended that *A. afarensis* interlimb proportions were reasonably well-predicted. However, the relevant regression equation, the prediction error, and the

W.L. Jungers (✉)
Department of Anatomical Sciences,
Stony Brook University Medical Center,
Stony Brook, NY 11794-8081, USA
e-mail: William.Jungers@sunysb.edu

correlation coefficient were not provided by Lovejoy. Vancata (1996) also ascribed to this "allometric explanation" (*sic*; allometry is a description, not an explanation) for the limb proportions of *A. afarensis*, as did Holliday and Franciscus (2001: 82, so-called "allometric consequences of small body size"). Richards (2006) has recently invoked essentially the same argument to explain away the australopithecine limb proportions of *H. floresiensis*; i.e., they all supposedly have a high humerofemoral index simply because they are small-bodied. Again, the underlying assumption is that there exists a negative and necessary correlation between the humerofemoral index and body size ("negative allometry", sensu Mosimann and James, 1979). I revisit these issues here and explicitly test some of these allometric hypotheses, while also offering new data bearing on intraspecific variation in interlimb proportions in modern humans.

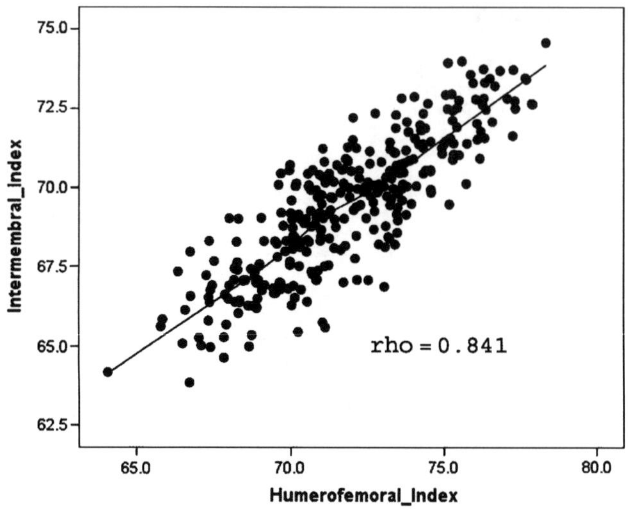

Fig. 9.1 Bivariate plot of the intermembral index against the humerofemoral index in a diverse ethnic sample of 314 modern humans. The two indices are highly correlated but not isometric because the intercept is significantly different than 0. Spearman's rank order correlation (rho) is 0.841. A nonparametric Loess line (tension = 0.5) is fit to the data.

Human Variation in Interlimb Proportions

The humerofemoral index (defined above) is highly correlated with another interlimb metric, the intermembral index (100 × [humerus + radius]/[femur + tibia]). Figure 9.1 is a bivariate plot of these two indices in a human skeletal sample of 314 individuals drawn from a diversity of ethnicities, climates and body sizes: African Pygmies, Andaman Islanders, Khoesan, Zulu, African Americans, Sami and Inuit. The parametric and rank order correlations are highly significant ($p<0.001$) and similar in magnitude at 0.848 and 0.841, respectively. One can therefore predict the intermembral index from the humerofemoral index, but the relationship is not an isometric one. The regression in raw data space is

$$\textit{Intermembral Index} = 0.676 \times \textit{Humerofemoral Index} + 20.851 \quad (9.1)$$

and the 95% confidence interval on the intercept does not include 0 (cf. Jungers et al., 1995).

Table 9.1 provides the descriptive statistics for the humerofemoral index in these samples, for the total sample, as well as additional data on small-bodied "Asian Negritos" from Martin and Saller (1959), and three fossils (A.L. 288-1, LB 1, and KNM-WT 15000). Figure 9.2 presents these data in a box-and-whiskers graphical format. A great deal of variation is apparent (although not as much as imagined by Eckhardt, 2000), and there are many significant differences among groups that are unrelated to overall body size. A one-way analysis-of-variance is highly significant ($p<0.001$), and Games-and-Howell posthoc comparisons (employed due to heterogeneous variances) reveal the following:

Table 9.1 Descriptive statistics for the humerofemoral index in humans and fossil hominins

Group	N	Mean	Standard deviation	Lower 95%	Upper 95%	Minimum	Maximum
African Pygmies	24	73.7	1.5	73.1	74.4	70.2	76.3
Andaman Islanders	31	69.9	2.2	69.1	70.7	65.9	74.5
Khoesan	27	69.9	2.8	68.8	71.1	64.1	76.3
Zulu	47	70.5	1.8	70.0	71.0	66.7	75.9
African Americans	43	70.5	2.4	69.7	71.2	65.8	77.3
Sami	57	74.6	2.0	74.1	75.2	70.3	78.3
Inuit	85	72.1	2.2	71.7	72.6	66.6	77.3
Total Human Sample	314	71.8	2.7	71.5	72.1	64.1	78.3
Asian "Negritos"[a]	15 averages	70.5	1.7	69.5	71.4	67.6	73.3
AL 288-1	1	85.4	–	–	–	–	–
LB 1	1	86.8	–	–	–	–	–
KNM-WT15000	2 (juvenile & adult)	75.5	–	–	–	73.8 (juvenile)	77.2 (adult)

[a]Means from Martin and Saller (1959), after Schebesta (1952). The "15 averages" are used as individual data points to create a mean of means, standard deviation of means and the range of means for these Asian groups of small stature.

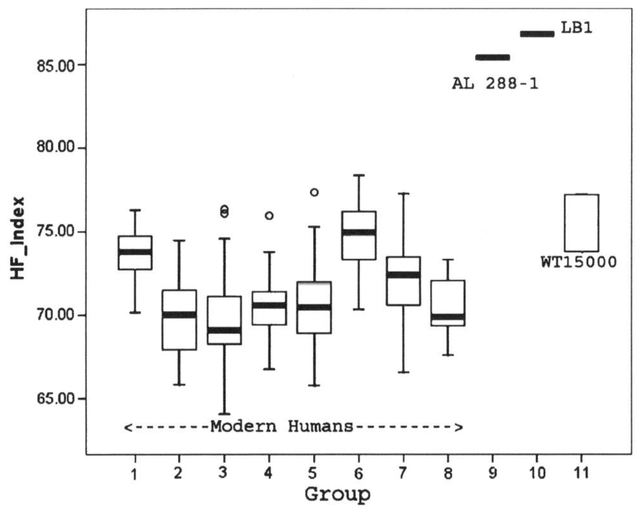

Fig. 9.2 Box-and-whiskers plots of the humerofemoral index (HF Index) in modern humans and fossil hominins. The samples are numbered as follows: 1 – African Pygmies, 2 – Andaman Islanders, 3 – Khoesan, 4 – Zulu, 5 – African Americans, 6 – Sami, 7 – Inuit, 8 – "Asian Negritos", 9 – AL 288-1, 10 – LB 1, 11 – KNM-WT15000. Sample sizes are in Table 9.1. The bold horizontal line indicates the median value, and the "box" encompasses 50% of the data points; the "whiskers" encompass the remainder of the sample except for extreme outliers (individual circles). Both A.L. 288-1 and LB 1 fall far above all modern humans, but KNM-WT-15000 is decidedly more human-like (note that the box for KNM-WT 15000 reflects the range from juvenile to estimated adult proportions).

- African Pygmies are significantly different from (greater than) all other groups except the Sami.
- Andaman Islanders are significantly different from (less than) African Pygmies, Sami and Inuit.
- Khoesan are significantly different from (less than) African Pygmies, Sami and Inuit.
- Zulu are significantly different from (lower than) African Pygmies, Sami and Inuit.
- African Americans are significantly different from (lower than) African Pygmies, Sami and Inuit.
- Sami are significantly different from (greater than) all groups except the African Pygmies.
- Inuit are significantly different from all other groups.

The Sami and African Pygmies have the highest humerofemoral indices, and the Andaman Islander and Khoesan have the lowest. The mean of averages for small-bodied "Asian Negritos" (Aeta, Semang, and Senoi) is below my total sample average of 71.8 but very similar to that seen in Zulu and African Americans (70.5). KNM-WT 15000 (as a juvenile) overlaps with several modern human groups, but the estimated adult index (using the average of 11 year-old and 12 year-old models [Ruff and Walker, 1993]) is appreciably higher; however, it can still be matched by individuals from the Sami, Inuit and African American samples. Both A.L. 288-1 and LB1 fall far above all of the human samples, *including the small-bodied groups* (African Pygmies, Andaman Islanders, and Khoesan) that almost certainly contain some very small individuals comparable in adult body mass to both fossils.

It is obvious that the individual fossils represent point estimates of the species mean for the humerofemoral index, but we have no idea where any of them really fall with respect to the variation seen in the populations from which they are drawn. It is statistically possible to calculate confidence intervals for samples of N = 1, but these intervals are enormous, biologically absurd and "unlikely to be useful" (Smith, 2005). Using a human-driven model, Korey (1990) estimated the standard deviation for A.L. 288-1 at roughly 2.8; 2 standard deviations below a value of 85.4 is still outside the range observed in my heterogeneous human sample (as well as any of the many values presented in Martin and Saller (1959) for a wide variety of modern human groups). Moreover, the value of 85.4 used here is perhaps too conservative. Recent attempts at 3-D digital reconstruction of femur length in A.L. 288-1 suggest that 281 mm could well be too long (Sylvester et al., 2007). Regardless, I am unaware of any normal humans with a humerofemoral index approaching the values seen in either A.L. 288-1 or LB 1, and that includes skeletons drawn from the smallest people on earth. These observations alone render the allometric hypotheses described above as highly suspect. Direct tests described below lead to unequivocal rejection of such size-required scenarios.

Scaling of the Humerofemoral and Intermembral Indices

The relationship between body mass and humerus length is remarkably consistent among modern humans and African apes (Jungers, 1994). Humerus length divided by the cube root of body mass does not distinguish among gorillas, chimpanzees, bonobos, African Pygmies and larger-bodied humans; rephrased in a slightly different manner, the interspecific log–log scaling relationship between these two variables is highly significant and isometric. This perhaps surprising result indicates that humerus length itself can be used as a valuable surrogate for body mass or overall body size. I take advantage of this empirical finding to examine the scaling relationship between the humerofemoral and intermembral indices and humerus length (much like Lovejoy [1993] did with femur length) in order to evaluate the hypothesis that these interlimb indices are necessarily inversely correlated with "size" in humans.

Figure 9.3 is a bivariate plot of the humerofemoral index on humerus length in the total sample of 314 modern humans. Figure 9.4 is a similar plot of the intermembral index on humerus length. In both plots, a nonparametric Loess line (tension = 0.5) is provided to guide the eye through a scatter

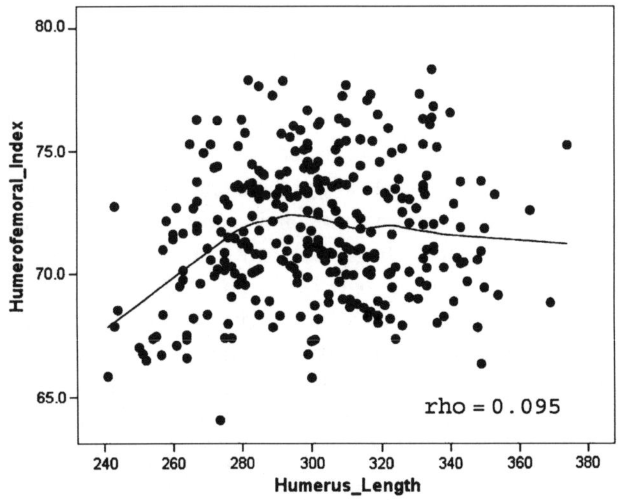

Fig. 9.3 Bivariate plot of the humerofemoral index against humerus length (used here as a size surrogate, in mm). A Loess line is provided to guide the eye through the scatter of points. The relationship is not statistically significant as judged by a Spearman rank order correlation (rho). This indicates that that the humerofemoral index does not increase as size decreases (i.e., the relationship is isometric).

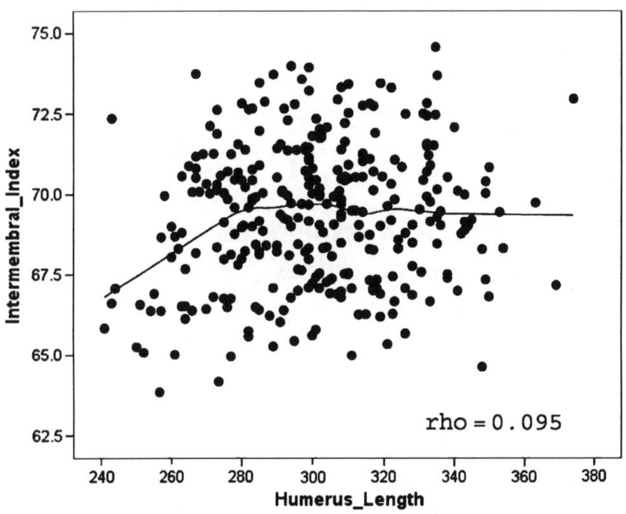

Fig. 9.4 Bivariate plot of the intermembral index against humerus length (in mm). The same conventions apply here as in Fig. 9.3. The intermembral index does not change in a predictable manner with size.

of points that are statistically uncorrelated (Spearman's rho = 0.095 in both). This lack of correlation between a shape variable (the index) and a size variable (humerus length) is what Mosimann and James (1979) define as "isometry." That is, shape fails to change in a predictable fashion with size. The Loess fit might suggest that both indices actually increase with humerus length initially and then change very little if at all afterwards. The important take-home message from this finding is that there is no increase in either index as size decreases. The predicted negative allometry between each index and size simply does not obtain for the total sample.

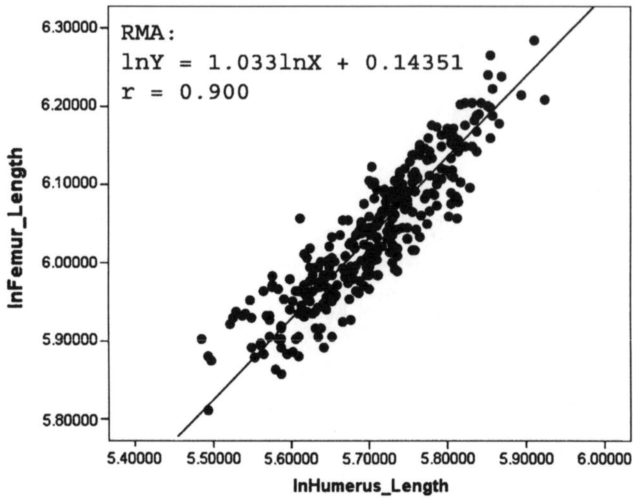

Fig. 9.5 Bivariate ln–ln plot of femur length (mm) on humerus length (mm). The parametric correlation coefficient is 0.9, and the reduced major axis (RMA) slope is 1.033. The 95% confidence interval includes the value of 1.0, indicating thereby an isometric scaling relationship (as was demonstrated in a different way in Fig. 9.3).

For those readers more familiar with the log–log regression approach to scaling, it is easy to demonstrate the equivalency of the two methods. Figure 9.5 is a bivariate plot of ln(femur length) on ln(humerus length) for the total sample. The correlation is 0.9 and the reduced major axis estimate of the slope is 1.033. The full equation is

$$ln(FL) = 1.033 \times ln(HL) + 0.14351 \qquad (9.2)$$

Permutation-based 95% confidence intervals on the slope are 0.981 and 1.086. In other words, the relationship is best described again as isometric, just as it was using the Mosimann and James methodology. The expectation of negative allometry derived from the size-required hypotheses is firmly rejected regardless of how one approaches the problem.

One might still wonder if the predicted negative scaling relationships might exist within more homogeneous human subsamples, especially among the smallest modern humans. I evaluate this possibility by calculating correlations between each index and humerus length within each of the seven ethnic samples, and the results are summarized in Table 9.2. In most groups, the correlations are nonsignificant (4/7 for the humerofemoral index, 5/7 for the intermembral index), indicating an essentially isometric relationship in which the index does not change in a predictable direction. Equally important, when there is a significant correlation, it is in the *wrong direction* for the allometric scenarios that predict higher indices in smaller humans. For example, both indices in the Andaman Islanders and Khoesan (among some of the smallest people on earth) are significantly positive; i.e., the indices increase with size rather than decrease. Accordingly, if one were to predict humerofemoral indices for LB1 and AL288-1

Table 9.2 Rank order correlations (Spearman's rho) between humerus length (mm) and interlimb indices

Group	Humerofemoral index and humerus length (rho)	Intermembral index and humerus length (rho)
African Pygmies	0.073	0.143
Andaman Islanders	0.613*	0.674*
Khoe-san	0.746*	0.658*
Zulu	0.141	0.196
African Americans	0.194	0.223
Sami	0.330**	0.244
Inuit	0.062	0.019
Total sample	0.095	0.095

* $p < 0.01$, **$p < 0.05$

from the Andaman Islanders and Khoesan, the relevant equations would be:

*Humerofemoral Index = 0.090 × Humerus Length + 45.7 (**Andaman Islanders**)* (9.3)

*Humerofemoral Index = 0.089 × Humerus Length + 44.0 (**Khoesan**)* (9.4)

The Andaman Islander equation predicts an index of 67.6 for LB1 and 67.2 for A.L. 288-1; the Khoesan equation predicts 65.6 and 65.3, respectively. These values are obviously poor estimates of the known values (Table 9.1).

Conclusions

The hydra-like predictions that small-bodied fossil hominins like A.L. 288-1 and LB 1 necessarily have unusual limb proportions simply because they are extrapolated examples of size-required allometries in modern humans are demonstrably wrong. No small-bodied human remotely approaches the high humerofemoral index found in both of these diminutive fossils. No metric evidence can be found across or within modern humans (including very small-bodied ethnic groups) that serves to corroborate the repeated inferences that the humerofemoral index has to scale negatively due to some unspecified and unknown biological constraint; i.e., that it *must* increase as size decreases. Little fossil hominins are not small people in terms of their limb proportions and likely locomotor performance (Sylvester et al., 2008).

The "intermediate" humerofemoral index of A.L. 288-1 most probably reflects a modest degree of hind-limb elongation from the ancestral condition related to selection for increased stride length and enhanced locomotor economy in a facultative biped (Jungers, 1982, 1991). It should not be interpreted as evidence of a long upper limb (Jungers, 1994). Nevertheless, such an index might well provide some mechanical advantage in climbing (Jungers, 1982; Stanley, 1992; Argue et al., 2008) when compared to that capacity in modern humans (but see Devine, 1985). Similarly, the high humerofemoral index of *Homo floresiensis* is not due to long upper extremities, but again derives from relatively short hind limbs (contra Morwood et al., 2005; see Jungers et al., 2008). There are documented primitive features in the skeleton of *H. floresiensis* (Tocheri et al., 2007; Larson et al., 2007), and one might speculate that the humerofemoral index is also a primitive retention. However, this would probably rule out *Homo erectus* as a possible ancestor in view of the human-like limb proportions seen in KNM-WT 15000, unless what we see in the "hobbit" represents a reversed convergence back to an earlier body design (i.e., homoplasy). Clearly, LB 1 and modern humans are not geometrically similar, and analytical models that assume they are should be re-evaluated (e.g., Blaszczyk and Vaughan, 2007).

The emergence of truly human-like limb proportions in early African *H. erectus* (= *H. ergaster*) is now firmly linked biomechanically to enhanced locomotor performance in a striding, obligate biped (Steudel, 1994; Bramble and Lieberman, 2004; Steudel-Numbers and Tilkens, 2004; Pontzer, 2005, 2007a, b; Steudel-Numbers, 2006; Lieberman et al., 2009). The australopithecine version of bipedality was a smashing success if it is judged by its duration in the fossil record (Richmond and Jungers, 2008), but the hominin locomotor skeleton was fundamentally reorganized by 1.8–1.6 million years ago in early *Homo erectus/ergaster*, and elongated hind limbs are clearly part of this new package. We currently lack an adequate fossil record to assess confidently whether or not some of these proportional changes were also evident earlier in *H. habilis* or perhaps in a later species of *Australopithecus*.

Acknowledgments It was my pleasure to participate in the Human Evolution Workshop at Stony Brook University, and I thank Fred Grine, John Fleagle and Richard Leakey for the invitation. I sincerely thank Ozzie Pearson and Fred Grine for sharing unpublished data. Thanks also are due to the many museum curators who graciously provided access to skeletal material in their care. I also am very grateful to Mike Morwood and Tony Djubiantono for inviting me to work on the fossils of *Homo floresiensis*. Brigitte Demes offered valuable comments on an earlier draft of this paper. This research was supported in part by grants from the Wenner-Gren Foundation, the Leakey Foundation and the National Geographic Society.

References

Argue, D., Donlon, D., Groves, C., Wright, R., 2006. *Homo floresiensis:* microcephalic, pygmoid, *Australopithecus*, or *Homo*? Journal of Human Evolution 51, 360–374.

Asfaw, B., White, T., Lovejoy, O., Latimer, B., Simpson, S., Suwa, G., 1999. *Australopithecus garhi*: a new species of early hominid from Ethiopia. Science 284, 629–635.

Blaszczyk, M.B., Vaughan, C.L., 2007. Re-interpreting the evidence for bipedality in *Homo floresiensis*. South African Journal of Science 103, 409–414.

Bramble, D.M., Lieberman, D.E., 2004. Endurance running and the evolution of *Homo*. Nature 424, 345–352.

Brown, P., Stikna, T., Morwood, M.J., Soejono, R.P., Jatmiko, Wayhu Saptomo, E., Awe Due, R., 2004. A new small-bodied hominin from the late Pleistocene of Flores, Indonesia. Nature 431, 1055–1061.

Devine, J., 1985. The versatility of human locomotion. American Anthropologist 87, 550–570.

Eckhardt, R.B., 2000. Human Paleobiology. Cambridge University Press, Cambridge.

Franciscus, R.G., Holliday, T.W., 1992. Hindlimb skeletal allometry in Plio-Pleistocene hominids with special reference to AL 288–1 ("Lucy"). Bulletin de la Memoires Society Anthropologie, Paris 4, 5–20.

Green, D.J., Gordon, A.D., Richmond, B.G., 2007. Limb-size proportions in *Australopithecus afarensis* and *Australopithecus africanus*. Journal of Human Evolution 52, 187–200.

Haeusler, M., McHenry, H.M., 2004. Body proportions of *Homo habilis* reviewed. Journal of Human Evolution 46, 433–465.

Holliday, T.W., Franciscus, R.G., 2001. Multivariate assessment of body shape in A.L. 288–1 ("Lucy"). American Journal of Physical Anthropology 32 (supplement), 82.

Johanson, D.C., Edey, M., 1981. Lucy: The Beginnings of Humankind. Simon & Schuster, New York.

Jungers, W.L., 1982. Lucy's limbs: skeletal allometry and locomotion in *Australopithecus afarensis*. Nature 297, 676–678.

Jungers, W.L., 1991. A pygmy perspective on body size and shape in *Australopithecus afarensis* (AL 288-1, "Lucy"). In: Senut, B., Coppens, Y. (Eds), Origine(s) de la Bipedie chez les Hominides. Cahiers de Paleoanthropologie, Paris, pp. 215–224.

Jungers, W.L., 1994. Ape and hominid limb length. Nature 369, 194.

Jungers, W.L., Stern, J.T., 1983. Body proportions, skeletal allometry and locomotion in Hadar hominids: a reply to Wolpoff. Journal of Human Evolution 12, 673–684.

Jungers, W.L., Falsetti, A.B., Wall, C.E., 1995. Shape, relative size, and size-adjustments in morphometrics. Yearbook of Physical Anthropology 38, 137–161.

Jungers, W.L., Sutikna T., Jatmiko, Saptomo, E.W., Awe Due, R., Djubiantono, T., Morwood, M., 2008. Body size and shape in *Homo floresiensis* and pygmy humans. African Genesis: A Symposium on Hominid Evolution in Africa, p. 14.

Korey, K.A., 1990. Deconstructing reconstruction: the OH 62 humerofemoral index. American Journal of Physical Anthropology 83: 25–33.

Larson, S., Jungers, W.L., Morwood, M. J., Sutikna, T., Jatmiko, Saptomo, E.W., Due, RA., Djubiantono, T., 2007. *Homo floresiensis* and the evolution of the hominin shoulder. Journal of Human Evolution 53: 718–731.

Lieberman, D.E., Bramble, D.M., Raichlen, D.A., Shea, J.J., 2009. Brains, brawn and the evolution of endurance running capabilities. In: Grine, F.E., Fleagle, J.G., Leakey, R.E. (Eds), The First Humans: Origin of the Genus *Homo*. Springer, Dordrecht, pp. 77–92.

Lordkipanidze D., Jashashvili T., Vekua A., Ponce de León M.S., Zollikofer C.P.E., Rightmire G.P., Pontzer H., Ferring R., Oms O., Tappen M., Bukhsianidze M., Agusti J., Kahlke R., Kiladze G., Martinez-Navarro B., Mouskhelishvili A., Nioradze M., Rook L., 2007. Postcranial evidence from early *Homo* from Dmanisi, Georgia. Nature 449, 305–310.

Lovejoy, C.O., 1993. Modeling human origins: are we sexy because we're smart, or smart because we're sexy? In: Rasmussen, D.T. (Ed), The Origin and Evolution of Humans and Humanness. Jones & Bartlett, Boston, MA, pp. 1–28.

Martin, R., Saller, K., 1959. Lehrbuch der Anthropologie. Gustav Fischer Verlag, Stuttgart.

McHenry, H.M., 1978. Fore- and hindlimb proportions in Plio-Pleistocene hominids. American Journal of Physical Anthropology 49, 15–22.

McHenry, H.M., Berger, L.R., 1998. Limb lengths in *Australopithecus* and the origin of the genus *Homo*. South African Journal of Science 94, 447–450.

McHenry, H.M., Coffing, K.E., 2000. *Australopithecus* to *Homo*: transformations in body and mind. Annual Review of Anthropology 29, 125–146.

Morwood, M.J., Brown, P., Jatmiko, Sutikna, T., Wahuy Saptomo, E., Westaway, K.E., Awe Due, R., Roberts, R.G., Maeda, T., Wasisto, S., Djubiantono, T., 2005. Further evidence for small-bodied hominins from the Late Pleistocene of Flores, Indonesia. Nature 437, 1012–1017.

Mosimann, J.E., James, F.C., 1979. New statistical methods for allometry with application to Florida red-winged blackbirds. Evolution 33, 444–459.

Pontzer, H., 2005. A new model predicting locomotor cost from limb length via force production. Journal of Experimental Biology 208, 1513–1524.

Pontzer, H., 2007a. Predicting the energy cost of terrestrial locomotion: a test of the LiMb model in humans and quadrupeds. Journal of Experimental Biology 210, 484–494.

Pontzer, H., 2007b. Effective limb length and the scaling of locomotor cost in terrestrial animals. Journal of Experimental Biology 210, 1752–1761.

Reno, P. L., DeGusta, D., Serrat, M. A., Meindl, R.S., White, T.D., Eckhardt, R.B., Kuperavage, A.J., Galik, K, Lovejoy, C.O., 2005. Plio-Pleistocene hominid limb proportions. Current Anthropology 46, 575–588.

Richards, G.D., 2006. Genetic, physiologic and ecogeographic factors contributing to variation in *Homo sapiens*: *Homo floresiensis* reconsidered. Journal of Evolutionary Biology 19, 1744–1767.

Richmond, B.G., Aiello, L.C., Wood, B.A., 2002. Early hominin limb proportions. Journal of Human Evolution 43, 529–548.

Richmond, B.G., Jungers, W.L., 2008. *Orrorin tugenensis* femoral morphology and the evolution of hominin bipedalism. Science 319, 1662–1665.

Ruff, C.B., 1993. Climatic adaptation and hominid evolution: the thermoregulatory imperative. Evolutionary Anthropology 2, 53–60.

Ruff, C.B., Walker, A.C., 1993. Body size and body shape. In: Walker, A.C., Leakey, R.E. (Ed), The Nariokotome *Homo erectus* Skeleton. Harvard University Press, Cambridge, MA, pp. 234–265.

Schebesta, P., 1952. Die Negrito Asiens. St. Gabriel Verlag, Vienna.

Smith, R.J., 2005. Species recognition in paleoanthropology: implications of small sample sizes. In: Lieberman, D.E., Smith, R.J., Kelley, J. (Eds), Interpreting the Past: Essays on Human, Primate, and Mammal Evolution. Brill Academic Publishers, Boston, MA, pp. 207–219.

Stanley, S.M., 1992. An ecological theory for the origin of *Homo*. Paleobiology 18, 237–257.

Steudel, K., 1994. Locomotor energetics and hominid evolution. Evolutionary Anthropology 3, 42–48.

Steudel-Numbers, K., 2006. Energetics in *Homo erectus* and other early hominins: the consequences of increased lower limb length. Journal of Human Evolution 51, 445–453.

Steudel-Numbers, K., Tilkens, M., 2004. The effect of lower limb length on the energetic cost of locomotion: implications for fossil hominids. Journal of Human Evolution 47, 95–109.

Sylvester, A.D., Merkl, B.C., Mahfouz, M.R., 2007. Reconstructing the AL 288–1 femur using three-dimensional computer models. American Journal of Physical Anthropology 44 (supplement), 228–229.

Sylvester, A.D., Kramer, P.A., Jungers, W.L., 2008. Modern humans are not (quite) isometric. American Journal of Physical Anthropology 137, 371–383.

Tocheri, M., Jungers, W., Larson, S., Orr, C., Sutikna, T., Jatmiko, Saptomo, Due, E.R., Djubiantono, T., Morwood, M., 2007. Morphological affinities of the wrist of *Homo floresiensis*. Abstracts of the 2007 Annual Palaeoanthropology Society Meeting, p. 28. Available at *http://www.paleoanthro.org/meeting.htm*

Vancata, V., 1996. Major patterns of early hominid evolution: body size, proportions, encephalization and sexual dimorphism. Anthropologie 34, 11–25.

Walker, A., 1973. New *Australopithecus* femora from East Rudolph, Kenya. Journal of Human Evolution 2, 545–555.

Wolpoff, M.H., 1983. Lucy's lower limbs: long enough for Lucy to be fully bipedal? Nature 304, 59–61.

Part IV
Perspectives on Development, Diet and Behavior

Chapter 10
Growth and Development of the Nariokotome Youth, KNM-WT 15000

M. Christopher Dean and B. Holly Smith

Keywords Growth • development • maturation • dentition • *Homo erectus* • East Africa

Introduction

A substantial number of the large, slow-growing fauna on several continents were famously lost in the Pleistocene extinctions. At present, many slow-growing, long-lived mammals are threatened with extinction all over the world. Rhinos, elephants, chimpanzees, orangutans, and gorillas all make an enormous investment in their single offspring, taking years to raise one to independence. Recovery from population downturns is always difficult and sometimes impossible. Large investments in a single offspring, in theory, produce high quality offspring that may be larger, smarter, or more social, but slow maturers risk dying before reproducing. Of all these species, humans are the slowest to begin reproducing.

In recent years it has become clear that human growth and development, like human life history in general, was not present in australopiths (see Kuykendall, 2003 for a recent review). Evidence suggests, on the contrary, that much of what makes our life history unique took shape during the evolution of the genus *Homo*. Much of our evidence comes from the study of incremental records of daily growth preserved in teeth (see Dean, 2006), and some from correlative studies of life history in living primates (Smith, 1989, 1991). Given time and judicious use of some partially destructive analytical techniques, it seems likely that we will eventually be able to retrieve a time scale of dental maturation for each species within the Homininae and thus be able to reconstruct the general pace of maturation for each. In this effort, even single broken or sectioned teeth can make a contribution to knowledge. Even so, the occasional rare fossil of a more complete juvenile offers the chance to attempt a more complete reconstruction of maturation, or to ask questions that go beyond time scale alone. The juvenile male *Homo erectus* skeleton of Nariokotome (KNM-WT 15000) is one of these.

Indeed, the Nariokotome juvenile preserves more information about the species than any other individual find to date. It documents the appearance of a long-legged form with a barrel shaped chest, near-human proportions of the limbs, and with a brain well enlarged with respect to previous hominins (Brown et al., 1985; Walker and Leakey, 1993). Since the publication of the monograph on Nariokotome (Walker and Leakey, 1993) and the first extended study of his maturation (Smith, 1993), conflicting opinions have appeared about his growth or development: one fits Nariokotome as an ordinary human (e.g., Clegg and Aiello, 1999); a second sees both sides of the issue (S.L. Smith, 2004), whereas a third finds evidence of a much more primitive condition (Dean et al., 2001). For these reasons alone it is worth revisiting the issue of maturation of the Nariokotome youth and what he can tell us about growth and development in early *Homo erectus*.

Maturation of the Teeth and Skeleton of KMN-WT 15000: Basic Observations

To review briefly, the Nariokotome juvenile is judged to be male on the basis of sciatic notch morphology, skull robusticity, and overall size of the skeleton (Ruff and Walker, 1993). KNM-WT 15000 certainly died at early adolescence, before reaching adult size and proportion of the axial and appendicular skeletons; further he had most likely initiated, but not completed, physical and behavioral sexual maturation (Table 9.7 in Smith, 1993). Major ossification centers of long bone epiphyses had appeared and most remained unfused; of note, the triradiate cartilage still separated primary elements of the innominate. Epiphyses of the distal humerus had begun to fuse (see below, and Smith, 1993;

M.C. Dean (✉)
Department of Cell and Developmental Biology, University College London, Anatomy Building, Gower Street, London WC1E 6 BT, UK
e-mail: ucgacrd@ucl.ac.uk

B.H. Smith
Museum of Anthropology, The University of Michigan, Ann Arbor, MI 48109-1107, USA
e-mail: bhsmith@umich.edu

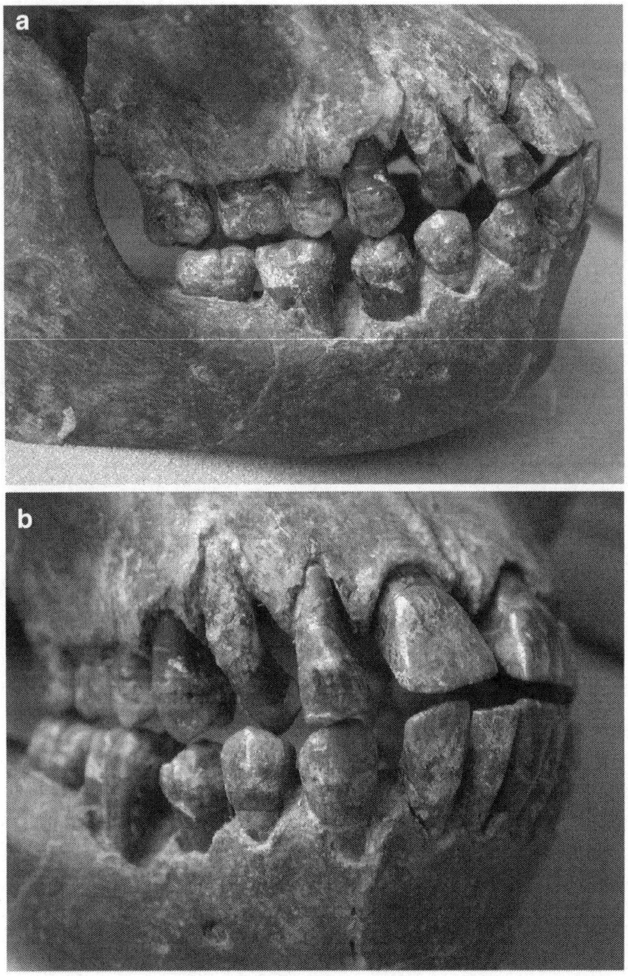

Fig. 10.1 The teeth and jaws of KNM-WT 15000. **a**), the recently erupted second permanent molars, premolars and lower canine are viewed laterally. The upper third molar region is broken away on this side (although the unerupted crown is preserved on the opposite side). **b**), the oblique view shows permanent upper and lower incisors; the permanent canine is erupted in the mandible, but the deciduous canine was still present in the maxilla.

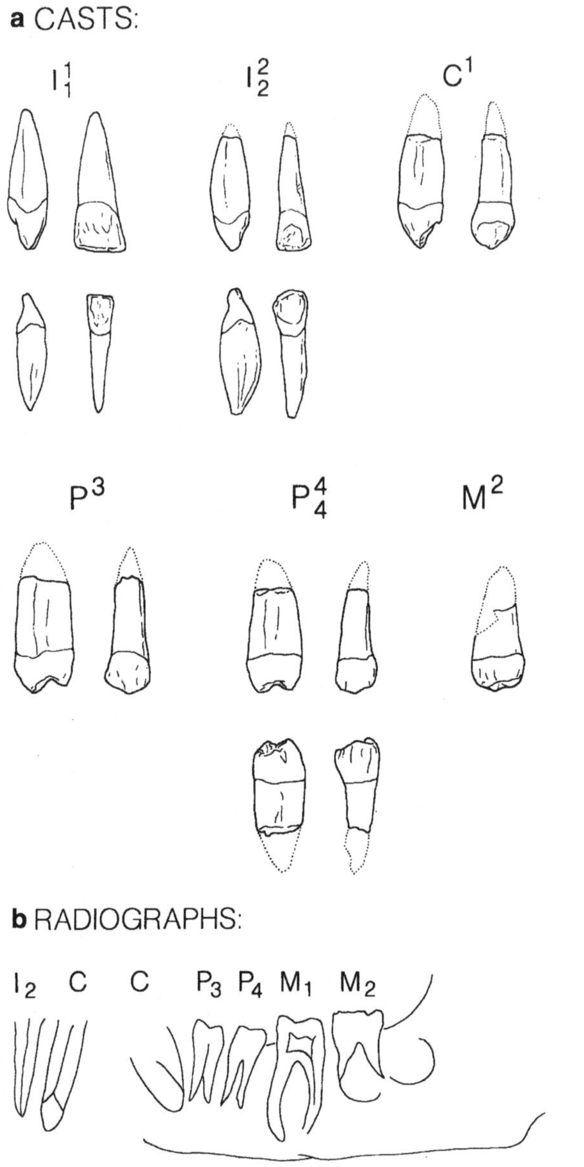

Fig. 10.2 The isolated teeth of KNM-WT 15000 drawn from sharp casts (**a**) and from radiographs (**b**). Dotted outlines project remaining growth. The drawing of the second radiograph shows it was taken in oblique view, which elongates the true tooth lengths (Reproduced from Smith, 1993. With permission of the editors and Harvard University Press).

Walker and Leakey, 1993). In the dentition, 26 permanent teeth had emerged, all but third molars and upper canines (Figs. 10.1 and 10.2).

The two deciduous upper canines remained in place. While the right upper third molar was never recovered, the upper left third molar crown can be observed through a window in the bone. There is no sign of either of the lower M3s on the best available radiographs (Brown and Walker, 1993); these had apparently failed to form. Wear on teeth suggests that upper and lower second molars had been in functional occlusion for a short time, and that the lower canines were probably the last teeth to erupt before death. Roots of most teeth (I^2, I_2, C, P3, P4 and M2) were incompletely formed at death, and the upper third molar crown was unerupted and probably incomplete.

Adolescence

Mammalian life stages can be defined in any number of ways, based on behavior and survival (Goodall, 1986), maturation of hard tissues (Schultz, 1956), growth curves (Bogin, 1990), hormone profiles (Molinari and Gasser, 2004), or a combination of the above (Bogin and Smith, 2000). Molinari and Gasser (2004), for example, key three phases of modern human growth to three phases of hormone activity: Infancy, beginning before birth and ending before 1.5–2 years, steered by thyroid hormones; childhood, in which the addition and powerful influence of growth hormones dominate growth, even into initial stages of puberty; and the pubertal phase, in which sex hormones initiate maturation of non-skeletal secondary sexual characters as well as the pubertal growth spurt and simultaneously lead to eventual termination in growth by triggering the closure of epiphyses. "Adolescence" is by most definitions the stage that follows the onset of puberty and lasts until adulthood, which is a period of relative physiological homeostasis (e.g., Bogin and Smith, 2000). Both humans and great apes typically experience puberty, the onset of sexual maturation, years before full somatic maturation, a gap which creates a true and extended adolescence – something which is not the case for all primates (see Smith, 1992). Although humans and great apes share many characteristics of their life cycles, humans mature on a much slower time scale, reaching marker events of hard tissue maturation at ca. 1.8 times the age typical for chimpanzees (Table 10.1).

Maturation of the skeleton and dentition (see below) makes it clear that the Nariokotome youth was somewhere in the adolescent *stage* of growth and development at death, but what is less clear, however, is whether the pubertal growth spurt itself existed at his time, 1.5 million years earlier in the evolution of the genus *Homo* (see Smith, 1993; Tardieu, 1998; Antón and Leigh, 2003). A number of studies have found that the human combination of a protracted slow phase of growth in late childhood followed by a peak or spurt in *both* weight and linear dimensions appears to be unique even among higher primates (Bogin, 1999; Smith, 1993; see Bogin and Smith, 2000; Hamada and Udono, 2002).

Table 10.1 Age of attainment of maturation events in chimpanzees and humans, and the similar ratio between them (Kerley, 1966; Scheuer and Black, 2000; Liversidge, 2003: 86; see also Smith, 1993, 2000)

	Age of attainment (year)		
Maturation event	Chimp ♂	Human ♂	Ratio (H/C)
Emergence of M_1	3 1/3	5.7	1.7
Emergence of M_2	6 1/2	11.4	1.8
Rise in ♂ serum testosterone	7 1/2	12–13	1.7
Elbow begins ossifying	7 1/2	12–13	1.7
Tri-radial plate pelvis ossifies	8	14–17	1.9
Emergence of M_3	10 1/3	18–20	1.8
Shoulder ossification	13 1/2	20	1.5
Life span	55+	100+	1.8

Skeletal Age

Like many other primates, humans ossify the appendicular skeletal joints in sequence from elbow, hip, ankle, knee, wrist, to shoulder (Schultz, 1956), although great apes and humans delay the start of this fusion sequence until puberty or after (Figure 9.2 in Smith, 1993). We can place Nariokotome firmly past puberty and into adolescence by evidence that the process of elbow joint ossification had begun (Fig. 10.3), uniting some of the four elements of the distal humerus epiphysis, and by evidence that his shoulders had broadened (see below).

He is also kept to the younger side of adolescence by the lack of fusion of the remaining major long bone epiphyses. In the human elbow, individual centers of ossification of the trochlea, capitulum and lateral epicondyle fuse together at about age 12 in boys; fusion of the whole composite distal epiphysis to the humeral shaft starts about age 12.5; final fusion of the fourth element, the medial epicondyle, lags until approximately age 15 (all ages for boys, see Scheuer and Black, 2000). For Nariokotome, Walker and Leakey (1993) note that the medial (trochlear) half of the distal

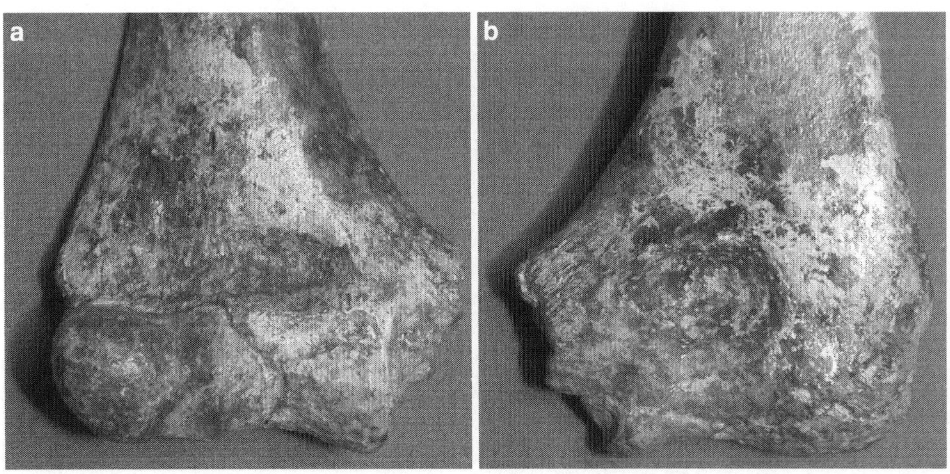

Fig. 10.3 Anterior (*left*) and posterior (*right*) views of the right distal humerus of KNM-WT 15000. The trochlea, capitulum and lateral epicondylar centers were fused together at the time of death. This composite distal humeral epiphysis had also fused with the shaft along the full length of the trochlea anteriorly (but incompletely posteriorly). The medial epicondylar epiphysis was lost and therefore unfused at the time of death. A fracture, seen most easily in the anterior view (left image), runs obliquely across the trochlea indicating those portions of the lateral trochlea and capitulum found separately but fused together. The total preserved bi-epicondylar width is 52.0 mm (Walker and Leakey, 1993).

humeral epiphysis was fused to the shaft anteriorly but not posteriorly or laterally (the capitulum was fractured off with a portion of the trochlea and found separately). Thus, fusion of the three centers to the shaft had begun and the (missing) medial epicondyle, the fourth element of the growing distal humeral complex, had not fused. The state of his elbow thus places Nariokotome (in human calibration) at >12.5 but <15, but nearer the lower limit because, as Scheuer and Black (2000) note, once the elements of the epiphysis coalesce, the composite epiphysis will not long remain separate from the shaft. Lastly, Ruff and Walker (1993) cite Flecker (1932/1933) and Modi (1957; as cited in Krogman, 1962) who gave a median age or mode of about 13–14 years for the start of fusion of the whole distal humeral epiphysis. Thus, in all, 13 or 13.5 years seems to be a good compromise estimate of skeletal age in this individual using a modern human schedule.

Although most studies of bone age are from the Northern Hemisphere, some data can be found for African children. Agossou-Voyeme et al. (2005) document hand-wrist ossification by age among black children from Benin. Benin boys were generally delayed in hand bone development by 1–2.5 years compared to European, North American, and Asian boys; carpal bone maturation was even more delayed, by 3.5–3.8 years. Authors commented that upper socio-economic levels in Benin resembled average children in British studies, suggesting a nutritional effect. Thus, the Benin study suggests that a strictly African standard for elbow ossification would find Nariokotome's bone age advanced to beyond 13 years, unless comparison was restricted to only the topmost socioeconomic level.

Body Proportion and Stature

On first sight, the long limbs of Nariokotome set him apart from earlier hominins. As Ruff and Walker (1993) have shown, the distal segments of his limbs are particularly long. Both the crural (tibia/femur length × 100) and brachial (radius/humerus length × 100) indices for Nariokotome indicate an individual with tropical/subtropical limb proportions. His crural index (88 as compared, for example, with 84.5 in the Denver growth sample of Ruff, 2007) is particularly extreme, falling ~1 S.D. above the mean values for five out of six living tropical populations (Ruff and Walker, 1993). On this basis Nariokotome resembles living Nilotic peoples such as the Dinka, Nuer and Shilluk in body proportion, East Africans living today in a hot dry climate. Because the relevant data are not available for Nilotes, Ruff and Walker used data from Ugandan and South African Bantu to estimate stature at death of Nariokotome and his stature at adulthood (using Allbrook's, 1961 Ugandan study and Lundy's South African Bantu study, see Lundy and Feldesman, 1987). For present stature, predictions from equations regressing femur and or tibia lengths on stature clustered around 160 cm (5′3″), with 95% confidence intervals of ±5–7 cm expected if stature were predicted for a modern Bantu adult (Ruff and Walker, 1993: Table 11.6; see also Ruff, 2007). More recently, Ruff (2007) has recalculated stature predictions for KNM-WT 15000 based on a sample of 20 individuals from the Denver Growth Study. Based on femur length, the new stature estimate for KNM-WT 15000 was 161 cm while based on tibia length (which is relatively long in comparison with the Denver sample) stature was estimated at 160 cm. Ruff (2007) points out that 95% confidence intervals are still about ±5–7 cm, nonetheless, these estimates are remarkably close to those reported previously by Ruff and Walker (1993).

Estimating stature at death depends on chronological age estimate in only one point, that is, as Ruff and Walker (1993) argue, that an early adolescent has limb/stature proportions similar to adults. In a case as complete as the Nariokotome skeleton, a stature prediction can also be checked for overall reasonableness. The second estimation, however, that of stature had Nariokotome lived to maturity, depends much more on age assignment. Ruff and Walker (1993) stress that their analyses assume that Nariokotome was about 12 years of age (striking a midpoint between dental and skeletal age estimates) and that he would have followed a modern human-like growth curve with an adolescent growth spurt yet to come. With these assumptions, and after an in depth study, they settled on a predicted adult stature of 185 cm (6′ 1″). Yet, in reading their section on adult stature prediction, it is clear that Ruff and Walker found it difficult to wrest a sensible prediction from the data. The method they settled on was a double extrapolation: first they estimated remaining tibial growth for a 12 year old; second they estimated adult stature from the predicted adult tibial length.[1] Notably, simpler one step extrapolations of juvenile to adult stature produced estimates that were considerably larger than 185 cm, from 189.5–197 cm (6′2 1/2″ – 6′5 1/2″).

Body Mass

Although body mass is often estimated using a biomechanical approach with regressions of either long bone articular surfaces or diaphyseal cross-sections (Ruff, 2007), for the earlier study Ruff and Walker (1993) adopted a morphometric approach that used body size and shape measurements for Nariokotome. Using bi-iliac crest distance and stature to predict body mass at death gave 48 kg (~105 lbs), with a range of 47.7–50.7 kg based on a range of seven reference samples. A similar prediction for body mass at adulthood using the

[1] The 95% confidence interval of the prediction (±4 cm, or 5′10″ to 6′4″) estimates error for a case of direct measurement of an adult Bantu tibia; clearly many more sources of variation come in to play in the Nariokotome stature prediction.

same reference samples gave a best estimate of 68 kg (150 lbs), (range 67.1–68.5 kg). For comparison, Ruff and Walker cite Roberts and Bainbridge (1963) for average height and weight of adult African males from the Shilluk and Dinka populations as 181.5 cm and 58.2 kg. While the predicted adult stature for Nariokotome is remarkably close to that in these modern groups, the adult body mass estimate for Nariokotome is some 10 kg greater than the mean. Ruff and Walker (1993) also point out that even if had they used a smaller present stature for their calculation, the result would only have increased juvenile bi-iliac/stature ratio and thus the adult body femoral head breadth known for 20 individuals from the Denver Growth Study. Body mass estimates for Nariokotome ranged from 50.0–52.8 kg rising slightly as age estimates rise above 11–12 years. Ruff (2007) points out that the 95% confidence intervals are large (±13–18 kg) but there is remarkable correspondence between estimates based on the mechanical relationship between body weight and articular size and previous ones based on body proportions.

Another sign of the advanced size and maturity of Nariokotome can be seen in the clavicle. Walker and Leakey (1993) give clavicle length as 130.5 (right) and 130.4 (left) for Nariokotome. According to Walker and Leakey (1993) a known adult *Homo erectus* clavicle (KNM-ER 808) would only have been a little larger than those of Nariokotome. Even though the aged sample size in Black and Scheuer (1996) is very small, such a length is not approached until 15–16 years of age and already falls within the adult range for modern humans. Although the aged sample in Black and Scheuer (1996) is mainly of European origin, Terry (1932) noted no significant differences in adult clavicle length between races. Thus, the Nariokotome youth's shoulders had broadened nearly to an adult size. Interestingly, Tardieu (1998) also previously estimated Nariokotome to be 15 years old based on the relationship between femoral length and the shape of the distal femoral epiphysis.

Stature and Body Mass

Ruff and Walker (1993) note Nariokotome was a large individual; indeed Nariokotome emerges as large even compared to all nine known adult early *Homo* specimens (Table 11.15, KNM-ER 1472; KNM-ER 1481; KNM-ER 3728; KNM-ER 736; KNM-ER 1808; KNM-ER 737; OH 34; OH 28 and KNM-WT 15000). The mean stature of these nine adult specimens is 162 cm (5′ 4″) and the mean body mass estimate 54 kg (119 lbs), both of which, however, include estimates for Nariokotome – and are not in fact far different from the estimates of Nariokotome as a juvenile of 12 years of age. Without Nariokotome in the adult sample, the mean of the remaining eight early *Homo* adults would be 159.4 cm and 52 kg, with both values, incidentally, identical to those given for modern living African adults (Ruff and Walker, 1993). Nariokotome is then 99.6% of the mean estimated stature for the other eight adult early *Homo* specimens and 96% of their estimated body mass. Tanner (1962), for example, provides data that show modern human adolescents do not typically reach this percentage of adult stature until 17–18 years of age or 16–17 years when advanced.

The question that emerges is how likely these heights and weights for Nariokotome are for a modern human today? Gray et al. (2004) provide some comparative data for growth in height and weight of modern Turkana and Karimojong children, both East African Rift Valley populations. For these children, growth in weight is far below U.S. children in the National Child Health Statistics norms, with these African means hovering close to the U.S. 5th percentile. In stature, however, African children eventually neared the NCHS 50th percentile, with the data suggesting that adolescent growth spurts occurred later and that growth extended to a later age than in NCHS youth. Gray et al. (2004) showed that for African boys in their study, a stature of 160 cm and a weight of ~48 kg (predicted size at death for Nariokotome) was typical of 16–18 year old boys. A number of older African studies concur: MacKay and Martin (1952; Tables XII and XIII) found that Bantu boys with a stature of 162 cm and a weight of 48 kg (~105 lbs) were typically 17 years of age; at this stature and weight, a 13 year old Bantu boy would be more than 2 S.D. above the mean.

Similarly, children from various rural and urban localities in Ethiopia also appear to track European standards at around the 3rd–10th percentile (Clegg et al., 1972), although these authors cite one early study of Ugandan Baganda children (Welbourn, 1956) that met the 50th percentile of Tanner and Whitehouse's (1982) standards for UK children. Even the few data available for true 'hyper-tropical' Nilotes (Roberts, 1960) find them similar to other African populations at juvenile ages. Thus repeatedly, the stature and body mass predictions for Nariokotome seem more typical of modern 15–18 year old African children.

If, indeed, Nariokotome was as much as 12 years of age, as used in most evaluations of his size, the only good comparison for him is found in comparatively well fed, more rapidly growing European populations.[2] Even here, however, his size nears the 97th percentile (compare MacKay and Martin, 1952; Hass and Campirano, 2006). The body mass estimate for Nariokotome, however, is even more remarkable than the estimate for stature at the time of death. Nariokotome is estimated to have been ~48 kg which is ~70% of the estimate for adult

[2] By way of example, Molinari and Gasser (2004) show that Count Montbeillard's son was ~160 cm at 13 years of age; his peak height velocity occurred at just over 14 years of age, and his final adult height was ~185 cm at age 17.5 years.

body mass (68 kg) in this individual. Turkana and Karimojong boys today (Gray et al., 2004) appear not to achieve 70% of their adult body mass until 16–17 years of age. Twelve year old children tend to weigh closer to 35 than 48 kg and be nearer 55% of their adult body mass (Tanner, 1962).

Among Western populations, especially of European origin, children who are 160 cm tall at 12 years of age tend to be the tallest in their class at school and tend to grow into tall adults. Cases approaching the size reconstructed for Nariokotome can certainly be found in well-fed Western populations: Shelley Smith (2004: 113), for example, describing growth in 40 Canadian boys aged 10–15, found one early-maturing 12 year old boy with stature of 158.5 cm and weight of 45.2 kg. Such cases may become more common as secular trends continue in Western society. Ruff (2007) found the individual who came closest to matching KNM-WT 15000 in age and skeletal dimensions was a 12.5 years old female with a body mass of 48.9 kg. As Cameron (2002) points out, early maturing children today are typically heavier than late maturing children and experience a bigger growth spurt earlier on (although this growth pattern does not disqualify a child from attaining median stature). Nariokotome might then be interpreted in a modern human context as resembling a well-fed very early maturing modern child. But by any standard, Nariokotome is large, the size expected for a 15+ year old African adolescent or among that of the tallest of well-fed 12 year olds in North America.

Dental Development in Nariokotome

To evaluate the whole individual means that skeletal maturation, body mass and tooth development, anything we can study in the fossil record, must fit together. The dentition of the Nariokotome youth is so complete (Fig. 10.1) that we can make in depth comparisons to see if we can make sense of the youth as either an early or late maturing modern human.

Tooth Emergence and Root Formation

A study by Grøn (1962) documented the relationship between gingival emergence, skeletal age and chronological age in 847 white children from the lower socioeconomic population of the Greater Boston area. The Nariokotome youth died shortly after an active phase of tooth eruption in which second molars, second premolars and lower canines all emerged into functional occlusion. Relative wear suggests an emergence order of M_2 P_4 C, not unheard of in modern humans, but the reverse of the sequence of the Boston children. The mean ages of emergence in these teeth in the Bostonian children were, respectively 11.96, 11.04, and 10.91 years (see Table 10.2). However, the earliest ages that gingival emergence was observed in these teeth were respectively, 8.89, 8.49 and 8.33 years. Those children with a young skeletal age (9–10 years) tended to erupt second molars with less root formed than those of an older skeletal age (13–14 years). Compared with these white Bostonian children, Nariokotome, with a skeletal age of ~13 years and an M_2 with two or three occlusal wear facets, would be expected to have a full root length with an open apex formed at gingival emergence. At best, however, roots of the Nariokotome M_2 are three fourths complete (Fig. 10.2) with between 9–10 mm of root on the maxillary M^2 (Dean et al., 2001). In short, the formation stage of the second molar in Nariokotome does not seem typical of what one might expect in a modern human child with a skeletal age of ~13 years.

Mean Dental Age

Smith (1993) assessed the stages of individual tooth formation in Nariokotome (see also Brown and Walker, 1993) and compared them with those defined by Moorrees et al. (1963) and Anderson et al. (1976). The average dental age for mandibular permanent teeth that are still forming is 10.3 years when scored against the standards of Moorrees et al. (1963) (higher if one estimates a value for M_3); the average of all ten immature mandibular and maxillary teeth is 10.6 years adding in values from Anderson et al. (1976). The addition of more recent data confirms the former assessment: Liversidge et al. (2006) analysed the raw data for 4,480 girls and 4,522 boys from eight countries, re-calculating age of attainment for each group (using the slightly different stages defined by Demirjian et al., 1973). In her worldwide sample, boys matching Nariokotome mandibular teeth within a formation stage were, on average, 10.24 years of age (see Table 10.2). It is worth pointing out that the only individual tooth suggesting a dental age of 12 or more is the upper third molar, a tooth often neglected in contemporary studies (but see Liversidge and Townsend, 2005; Liversidge, 2008a, b).

Although we have more data on Europeans for tooth formation worldwide, it is also clear that changing wholly to a black African standard of tooth formation would only increase the discrepancies Nariokotome displays. From a sample of more than 800 black children from Johannesburg, Pretoria and Cape Town, South Africa (Liversidge, H., pers. comm., 2008) the mean dental age for the same stages of tooth development in Nariokotome is 10.1 years, just slightly younger than when the large combined worldwide sample is used (Table 10.2).

Skeletal Age Versus Dental Age

Conservatively, skeletal age in Nariokotome was more than 2.5 years greater than his mean dental age (SA – DA = 13 – 10.2 = 2.8 years); the discrepancy only rises if strictly African standards

Table 10.2 Dental age of KNM-WT 15000 by human standards compared to ages of children in similar developmental stages in other samples

Tooth	Stage	N	Min	Mean	Max	S.D.	S.E.
KNM-WT 15000 (Smith, 1993)							
UI1	Ac	1		>10.60			
UI2	Rc-A 1/2	"		10.10			
UM3	Cr 3/4	"		12.30			
LI2	Ac	"		>9.90			
LC	R 3/4	"		10.20			
LP3	R 1/2–3/4	"		10.00			
LP4	R 1/2–3/4	"		10.50			
LM2	R 1/2	"		10.50			
Dental age (mandibular teeth)				**10.30**			
Dental age (all)				**10.60**			
Worldwide sample-boys (Liversidge et al., 2006)							
LI2	G	688	8.76	8.83	12.98	1.100	0.040
LC	F	1,069	7.05	9.78	14.88	1.220	0.040
LP3	F	930	8.56	10.29	14.38	1.240	0.040
LP4	F	958	5.56	10.98	15.98	1.430	0.050
LM2	F	575	5.56	11.34	14.99	1.180	0.050
Age in stage				**10.24**			
Modern Africans-boys (Liversidge, H., pers. comm., 2008)							
UI1	Rc	24	7.10	8.54	10.50	0.902	0.184
UI2	R 3/4	41	5.74	8.74	11.50	1.299	0.203
UM3	C 3/4	15	6.5	11.09	13.50	2.056	0.531
LI2	Rc	16	5.74	7.81	10.00	0.994	0.249
LC	R 3/4	76	7.50	10.16	13.92	1.276	0.146
LP3	R 3/4	66	7.50	10.67	13.50	1.260	0.154
LP4	R 3/4	85	8.50	11.17	14.20	1.260	0.137
LM2	R 3/4	48	9.66	11.92	14.20	1.139	0.164
Age in stage				**10.01**			
Boston (Grøn, 1962)							
LI1	Gingival emergence	41	4.85	6.45	7.81	0.450	
LI2	"	52	5.89	7.37	9.21	0.320	
LC	"	50	8.33	10.91	13.90	1.180	
LP3	"	51	8.18	10.60	14.45	1.140	
LP4	"	50	8.49	11.04	13.36	1.130	
LM1	"	30	4.68	6.35	7.40	0.690	
LM2	"	58	8.89	11.96	15.09	1.160	

are applied (14.0 – 10.1 = 3.9 years). It is firmly established that skeletal and dental development are separate processes with only moderate correlations (see Lewis, 1991), yet, when children are assessed by experts, there is a characteristic degree of difference between the two. Fairly extensive data exist on the contrast between dental and skeletal ages for normal American children (Lewis, 1991; S.L. Smith, 2004) and there is a growing data base on children with endocrine disorders (Garn et al., 1965; Vallejo-Bolaños et al., 1999) and disease (Holderbaum et al., 2005).

The most comprehensive study of skeletal versus dental age comes from Lewis (1991), who described 694 Ohio children presenting (but not yet treated) as orthodontic patients. In two thirds of cases, skeletal age and dental ages differed by 1 year or less; 95% of cases differed by ±2 years or less. Extremes extended to as much as 3 years, but a lag of skeletal age was more common than the reverse. Lewis presented a detailed distribution broken down in months, reproduced here in Table 10.3 for the 320 boys. Only one of 320 boys nears Nariokotome in advancement of skeletal age.

Shelly Smith (2004) also explored variation in dental, skeletal and chronological age using data from Demirjian's well known study of growth of Canadian children, and her data are shown for further comparison. Matched for sex and grossly similar tooth formation stages to Nariokotome, discrepancy between skeletal and dental maturation ran from −1.8 to +1.6 years (−21.6 to +19.6 months) in 13 normal boys. The resulting distribution of SA – DA (skeletal age minus dental age) largely mirrors that of Lewis, although it is less extreme. Enlarging her search, Smith tracked all 40 boys in the study from ages 10–15, looking for a skeletal age advanced more than 2 years over dental age. Of 221 records over the 6 years, only four records (1.8%) showed >2 year

Table 10.3 Distribution of discrepancy between skeletal age (SA) and dental age (DA) in North American boys, children with endocrine disorders, and the Nariokotome *Homo erectus* youth KNM-WT 15000

	Percent of cases (nearest whole %)					
SA – DA mos	American boys[a]	Canadian boys[b]	Hypopituitary[c]	Sexual precocity[c]	KNM-WT 15000	Direction
–36 or more	3		20			
–30 to –35.9	0		20			
–24 to –24.9	5					
–18 to –23.9	8	15	40			
–12 to –17.9	15	15	20			
–6 to –11.9	17	8				
0 to –5.9	19	23				Skeleton lags ↑
0–5.9	16	15				Skeleton advances ↓
6–11.9	6	15				
12–17.9	6			20		
18–23.9	2	8		20		
24–29.9	1					
30 or more	1			60	X	
N	320	13	5	5	1	
Chronological age range (year)	6–15	10–13	~8–13	~7–10		

[a] Ohio boys presenting for orthodontic treatment (Lewis, 1991).
[b] Canadian boys from Demirjian's study of normal growth; this subset was matched for tooth formation grossly similar to Nariokotome by Shelley Smith (2004).
[c] Male and female patients with endocrine disorders affecting growth (Garn et al., 1965).

advance of skeletal age, and all four were 14 or 15 years old, where dental age prediction begins to tail off in accuracy as most teeth reach maturity. Although Smith interpreted her study as a cautionary one about variation, she also clearly recognized that Nariokotome was atypical.

We do know that truly eye-catching discrepancies between SA and DA can occur in endocrine disorders. Classic data from patients with endocrine disorders presented by Garn et al. (1965) are reproduced in Table 10.3 for children of comparable age range. In hypopituitary patients, Garn et al. showed that the skeletal development lags behind dental development; these cases make up a distribution at one end of the extremes of Table 10.3. The discrete pile up of 3% of Lewis's cases in the extreme skeletal delay category also suggests some of these orthodontic patients had an underlying growth deficiency. The opposite condition characterizes children with sexual precocity, when sex hormones are released years too early. In this case, skeletal maturation is accelerated – in 60% of the patients beyond 30 months in advance of dental development. The most extreme case was a 9 year old who had nearly closed all epiphyses (SA – DA = 7.3 years!). Comparing the data sets in Table 10.3, it is clear that for a child like Nariokotome, with skeletal age advanced by 34+ months over the dentition, a pediatrician would be justified in sending the case to an endocrinologist.

As is typical when experts gather the data, all the studies in Table 10.3 (see also Vallejo-Bolaños et al., 1999) found that dental age was the more accurate predictor of chronological age. Although both skeletal and dental development can be delayed by undernutrition or advanced by supernutrition, the dentition is much more resistant to environmental effects than is the skeleton (see Smith, 1991). Thus, chronic undernutrition, disease, or growth hormone deficiencies hit the skeleton harder, producing a lag of skeletal to dental age (SA – DA = a negative value in Table 10.3). Undernutrition likely contributes to the repeated finding that skeletal age in African children is delayed with respect to European children, quite markedly so in children under 10 years of age (Mackay and Martin, 1952; Clegg et al., 1972). When we turn to prehistory and cemetery samples, a lag of skeletal age behind dental age should be much more common than the reverse, judging from other evidence of growth faltering (e.g., Humphrey, 2003). Nariokotome, of course, shows just the opposite: his skeletal age is *greater* than his dental age, something more commonly found in obese or sexually precocious children today (Garn et al., 1967), a direction of difference which is particularly unexpected.

The hundreds of cases described in Table 10.3 show Nariokotome outside 99% limits for normal children and well into the distribution of growth disorders. One study, however, gives results at odds with the literature: Clegg and Aiello (1999) presented data from historical burials at Spitalfields for ten children they claimed had widely disparate dental and skeletal ages (ranging between –3.3 to + 3.5 years at least), a study sometimes cited as evidence that Nariokotome is not so unusual (Antón and Leigh, 2003; S.L. Smith, 2004; Ruff, 2007). Several aspects of the data, however, isolate this study: skeletal age was a better predictor of chronological age than dental age and 80% of cases were graded as advanced in skeletal age, as in obese children or sexual precocity, despite the fact that the Spitalfields mortuary is famous for small body size and late growth (Molleson and Cox, 1993). Certainly, most of the Clegg and Aiello subjects were too old for the problem at hand: indeed six of the ten were ages 14–18, with nearly every tooth mature except third molars, a poor comparison to Nariokotome, a much younger adolescent with ten immature teeth. For all these teenagers, systematic undergrading of mature teeth by Clegg and Aiello built up large apparent errors. But in any case, fatal errors corrupt the data set: broken roots of fully mature P_4 teeth were graded as immature in two cases (giving dental ages below 13 to the 14 and 17 year olds. The 10 year old, a boy of normal size and dental development, was assigned a dental age of 8.4 even though the Clegg and Aiello data actually average to 9.9; their assignment of a skeletal age of 6 years to him is equally questionable. Thus, comparisons of dental versus skeletal age by Clegg and Aiello (1999) are founded on faulty or irrelevant data.

Once again, there is no point in claiming that extremes cannot happen; the point is, simply, that Nariokotome's fit into human growth and development standards is uneasy at best, overlapping less than 1% of the well fed boys in Western growth studies; cemetery samples or the living malnourished should drift even further from Nariokotome because disease and malnutrition delay skeletal maturation disproportionately. As Smith (1993) concluded, while Nariokotome's size and maturation might be matched in some aspects at some human percentile, *he cannot be made ordinary*.

A Comparative Hypothesis

Comparing Nariokotome only to humans and only to relative scales, while instructive, involves us in an endless series of "if thens" (e.g., Smith, 1993; S.L. Smith, 2004). Only by adding a comparison to our closest relatives (Table 10.1) do we begin to generate a choice between alternatives: if Nariokotome is an odd fit to human growth standards, is he a better fit to something else? Briefly (see Smith, 1993), if assessed by chimpanzee growth and development standards, Nariokotome has a dental age of ~7, a skeletal age of 7.5, and a stature age (in terms of percent growth attained) like a 7 year old. Thus, his overall dental and skeletal development appears to be in synchrony by chimpanzee standards.[3]

The question that cuts to the heart of the matter is: what was his true chronological age at death? After a thorough review of possible explanations for Nariokotome's maturation, Shelley Smith (2004) states: "If he was younger than 10 years old, his skeletal development is indeed anomalous for a human child" (p. 108) and "if 8–9 years old when he died, perhaps he is better assessed by a nonhuman primate model" (p. 117).

Thus we have straightforward expectations: If Nariokotome was drawn from a population with growth and development resembling a chimpanzee, we would expect his true age to be near 7.5; if drawn from a population with fully human growth and development, we would expect true age to be much higher: an age of 12.5–13 years would minimize multiple discrepancies, although not make them disappear.

Pattern of Tooth Formation

A number of studies have found that *Homo erectus*, including KNM-WT 15000, does not share the primitive patterns of tooth formation observed in *Australopithecus* (Dean, 1987a; Beynon and Dean, 1988; Smith, 1986, 1993, 2004; Bermudez de Castro et al., 1999). *Homo erectus* shows the earliest clearly recognizable step towards a more human-like pattern of tooth maturation, particularly in early canine crown completion relative to other teeth. Even *Homo habilis*, judging from limited material, appears to echo *Australopithecus* in the pattern of synchrony of tooth maturation (Smith, 1993). This does not mean, however, that the pattern or sequence of tooth development is fully modern, indeed, the few juveniles we know share a tendency towards a more primitive pattern of dental development in which posterior teeth are slightly advanced compared to anterior teeth – i.e., when aged by human standards one consistently gets older ages for molars than incisors (e.g., see Table 10.2).

Although a shift in pattern of maturation may make us suspect that an underlying basic growth rate has also shifted (Smith, 1992), better evidence can be found in dental microanatomy.

Age of Death from Microanatomy

Microanatomical studies of enamel and dentine are beginning to yield real estimates of the age of attainment of marker events in the life history of *Homo erectus* and other hominin species (Bromage and Dean, 1985; Dean et al., 1993b, 2001; Smith et al., 2007a). In the best cases, thin sections of teeth can be analyzed to determine age of death, as well as the timing of certain stressful life history events, with an astonishing accuracy. Schwartz et al. (2006; Schwartz and Dean, 2008) were

[3] See Smith (1993) for a discussion of the internal pattern of tooth development in KNM-WT 15000, which shows a step towards a human condition.

able to date some life events to within a day by counting growth increments preserved in dentine and enamel of immature teeth of a juvenile captive gorilla. Accuracy on such a scale demands that material is well preserved, teeth can be thin sectioned, and juveniles are young (older juveniles may require cross-matching across two or more teeth). To date, studies with careful methods have produced estimates within 2–5% of the known age at death in modern human samples (Antoine, 2001; Antoine et al., 2000, 2009). Thus, when well preserved teeth can be sectioned, we may hope to achieve comparable accuracy for fossils. Nariokotome, however, remains too complete and too precious to section. Thus, although maximum precision is out of our reach at present, microanatomy can still be used to provide a good estimate of his age of death.

The basis for estimating the age at death of an individual from histological sections of teeth is the record of daily incremental markings that exists in enamel. Other coarser increments also exist several days growth apart called striae of Retzius. A notable feature of stria of Retzius, however, is that each emerges at the tooth surface in the form of a perikyma (plural perikymata; Gk. = waves around a tooth) creating an alternating pattern of troughs and grooves over the surface of the crown (Hillson, 1996; Hillson and Bond, 1997). Perikymata, often visible to the naked eye, can be counted on the tooth surface. Within all the teeth of an individual, the striae of Retzius and perikymata are set the same number of days apart (Smith, 2008). As first shown by Bromage and Dean (1985) and then subsequently by others (Beynon and Wood, 1987; Dean et al., 1993a; Moggi-Cecchi et al., 1998), perikymata visible on tooth surfaces of fossil hominins can be used to count time elapsed in tooth development without the need to make a histological section.

The catch is that the number of daily increments between adjacent perikymata has a range *across* individuals known as the periodicity. In modern humans this is between 6 and 12 days but more usually 7, 8, or 9 days with a mean and mode of 8 days (Dean, 1987b; FitzGerald, 1998; Smith et al., 2007b). While sectioning is required to determine the periodicity of each individual's striae of Retzius or perikymata, the range of expected values is relatively small in closely related groups. In many hominins, striae of Retzius are formed each week: Lacruz et al. (2008) reported that 59% of 29 australopiths examined showed a mean and mode periodicity of 7 days. To date, only seven specimens of early *Homo* fossils have been examined microscopically; two of these had a periodicity of 7 days, four of 8 days, and one of 9 days (Lacruz et al., 2008).

On the face of it, a modal value of 8 days seems the most likely choice for KNM-WT 15000, but a higher periodicity is probable for several reasons. First, the total perikymata counts on the surfaces of each of the Nariokotome anterior teeth is comparatively low (Dean and Reid, 2001a, b). Reid and colleagues have shown that teeth with widely spaced perikymata that are few in number tend to have high periodicities (Reid and Ferrell, 2006; Reid et al., 2008), whereas those with many more tightly spaced perikymata tend to have low periodicities. Secondly, an 8-day periodicity leads to unrealistic estimates of root growth, exceeding rates seen in *Gorilla* and *Pongo* (see Dean and Vesey, 2008 and below). Indeed, a periodicity of 10 days for KNM-WT 15000 gives the most parsimonious estimates for both molar and anterior crown formation times (see below). In the following analysis, therefore, periodicities of both 8 and 10 days have been used to provide what are more likely to be reasonable upper and lower estimates of the age at death of Nariokotome.

Because Nariokotome is an older juvenile, no single tooth crown records time over his entire life. It is therefore necessary to begin counts on an early forming tooth, then cross match to a tooth which overlapped, but continued forming later. We must also extend these counts beyond crown completion of the later forming tooth and make an estimate for the period of subsequent root growth up until the time of death (see below and Table 10.4).

The upper right canine tooth, unerupted in life and now isolated from the skull, provides an ideal starting point. In all primates, this tooth begins to mineralize within a few months of birth (we here used the estimate for initiation of 274 days given in Dean et al., 2001; and in Dean and Reid, 2001a, b). Deep within the enamel under the cusp of the tooth, a small thickness of enamel forms prior to that visible on the tooth surface (we here used the estimate of 266 days for this as calculated in Dean et al., 2001). We can count 100 perikymata between the cusp tip and the cervix on this canine tooth,

Table 10.4 Estimating age of death of KNM-WT 15000 as the sum of five segments of time represented in microanatomy of the upper canine and second molar

		Days	
Time segment	Perikymata count	8-day periodicity	10-day periodicity
1. Birth to initiation of UC[a]		274	274
2. Formation of hidden UC cusp enamel[a]		266	266
3. Formation of visible UC crown to second hypoplastic line	100	800	1,000
4. M^2 formed from second hypoplastic line to crown completion	25	200	250
5. Time to form 9.3 mm of M^2 root[a,b]		1,241–1,424	1,241–1,424
Total days		2,781–2,964	3,031–3,214
Total years		**7.6–8.1 years**	**8.3–8.8 years**

[a]Estimated from comparative studies of humans and apes (Dean and Reid, 2001a).

[b]Range based on time to form the first 10 mm of M2 roots form in great apes (6.5 μm/day) and in humans (7.6 μm/day) (Dean and Vesey, 2008).

which with periodicities of either 8 or 10 days equals 800 or 1,000 days. The sum of these intervals equals the age at canine crown completion and Table 10.4 shows this would have been 1,340 days (3.67 years) with an 8 day periodicity or 1,540 days (4.22 years) with a 10 day periodicity.

Several of the teeth of KNM-WT 15000 show evidence of two linear hypoplastic bands, or accentuated lines, across their crowns or roots (Fig. 10.4). These represent periods of slowed enamel or dentine formation that correspond with an illness or physiological upset that interrupted tooth growth (Hillson and Bond, 1997; Ritzman et al., 2008). In the upper right canine, the last of these bands coincides exactly with the end of enamel formation at the cervix (both the buccal and lingual), and the first occurred a mere 15 perikymata earlier. Both linear hypoplastic bands extend from enamel onto root dentine on the mesial (interproximal) surface (Fig. 10.4). In addition, they are preserved on the roots of most of the lower incisors and on the lateral enamel of some of the premolar and second permanent molar crowns that were growing at the same time (Figs. 10.5 and 10.6). The position of these bands on each of these teeth clinches the fact that they were caused by two consecutive interruptions to tooth formation (Dean et al., 1993a).

Using SEM, 15 perikymata can be counted between the start of the first band and the end of the second on the right upper canine crown as well as the left P^4 and right M^2 of KNM WT-15000 (Figs. 10.4 and 10.6). The severity of this kind of disturbance is expressed differently on each tooth type and may be influenced by the rate of enamel secretion at the time (Hillson and Bond, 1997). The fact that at least three crowns (the upper right canine, left P^4 and right M^2) all show these bands to be 15 perikymata apart is again a clear indication that they record the same two growth disturbances. Thus, by cross-matching the twin disturbances, we can continue the perikymata counts from the canine onto the right M^2 crown. Here we see an additional 25 perikymata beyond the second linear hypoplastic band on both the P^4 and M^2 to the end of enamel formation, which puts M^2 and P^4 crown completion at 4.2 years (8 day periodicity) or as much as 4.9 years (10 day periodicity) (see Table 10.4).

To estimate an age at death for KNM WT-15000 it becomes necessary to calculate the root formation time for one of the developing teeth. Several teeth in KNM-WT 15000 have roots that were still growing at the time of death. Of these, the isolated upper right canine root and the exposed distal aspect of the right M^2 root are most accessible, but the M^2 root is shorter (9–10 mm depending on where it is measured) and also represents a smaller interval of time. For these reasons alone it makes sense to use the M^2 root to estimate the time beyond enamel completion up to death in this specimen.

There are now a number of approaches to estimating rates of root growth and the time taken to grow roots in humans and great apes (Dean, 1995; Smith et al., 2007a, c). As with many fossil hominin teeth, periradicular ("around the root") bands are clearly visible on some tooth roots of KNM WT-15000. More than 80 can be counted on the distal aspect of the palatal root of the right M^2 for example. In places, as on other roots of KNM-WT 15000, the bands are spaced roughly 5–6 per millimetre. However, there are regions where finer bands (as many as 15 per millimetre) can be counted (Fig. 10.7). This raises problems about defining exactly which periradicular root bands are equivalent to perikymata on the enamel. Some

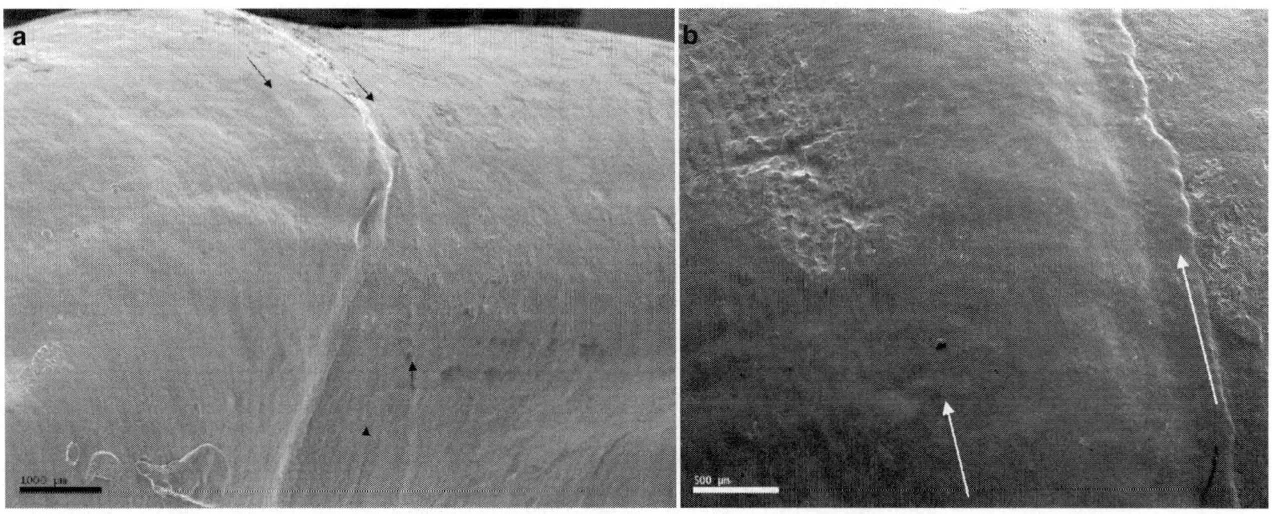

Fig. 10.4 **a**) A low power SEM image shows two hypo-plastic bands (arrows) on the isolated and unerupted upper right permanent canine of KNM-WT 15000. The first band formed runs over the cervical enamel and interproximal root dentine; the second also runs over the interproximal root dentine, coinciding with the last formed increment of enamel at the cervix. **b**) A higher power SEM image of the cervical enamel of the same tooth in the mid buccal cervical region. Two white arrows indicate the position of the hypoplastic bands, which have been traced across from the low power image. Perikymata are just visible across the tooth surface.

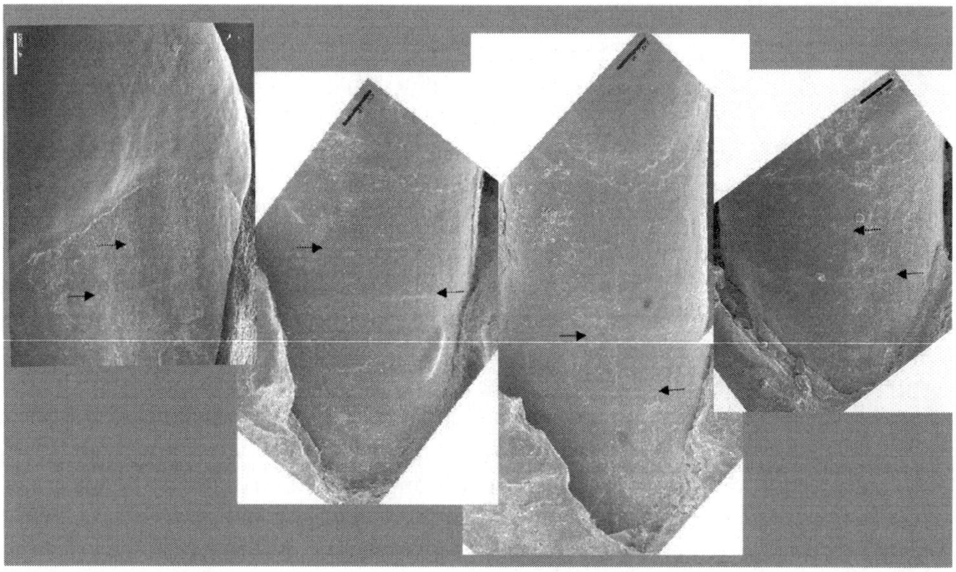

Fig. 10.5 Low power SEM images showing the twin hypoplastic bands identifiable on four teeth of KNM-WT 15000 (arrows). From left to right: interproximal root dentine of the permanent upper right canine, the lower right lateral incisor, lower right central incisor and lower left lateral incisor. These bands were not visible on the lower left central incisor, which is poorly preserved. Expoy resin casts, sputter coated with gold, were prepared for SEM from Coltene moulds made from the original specimen.

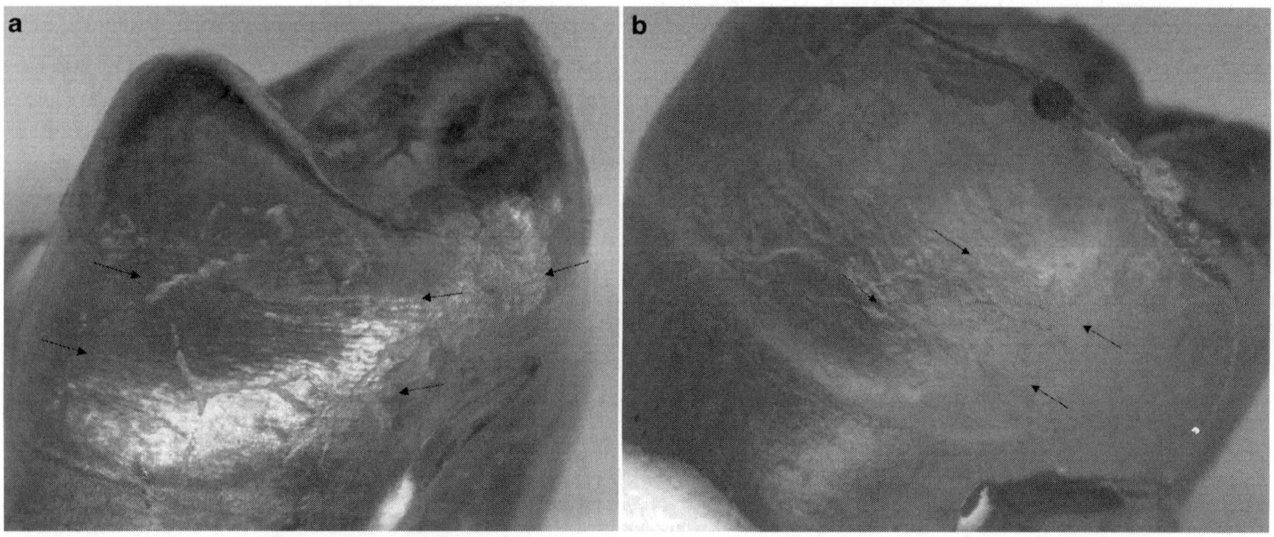

Fig. 10.6 Recurrence of twin hypoplastic bands (arrows) on **a**) the distopalatal aspect of the upper left P^4 and **b**) the upper right M^2. Both teeth were oriented with occlusal surfaces towards the upper right of the images. Epoxy resin casts of both teeth were made from Coltene moulds of the original teeth and sputter coated with gold before being photographed under a binocular microscope in oblique incident light.

of the best-preserved periradicular bands in KNM WT-15000 occur at the cervix of the left I_2 where ten clear widely spaced bands (each ~100 µm wide) exist between the enamel cervix and the first linear hypoplastic band 1,100 µm beyond this (Fig. 10.8).

Immediately beneath this, between the two hypoplastic bands on the left I_2 root, one may also count ten similarly-spaced periradicular bands (Fig. 10.8). But the expectation here is 15, based on 15 perikymata counted between the equivalent hypoplastic bands on the enamel surfaces of the canine, P^4 and M^2. One can then make two estimates for the rate of root growth in the I_2 root, one based on a periodicity of 10 days between periradicular bands (10.0 µm/day) and a second based on 15 perikymata being equivalent to the time between the twin hypoplastic root bands (7.5 µm/day). Thus, while it is tempting to use periradicular bands in the same way as perikymata to estimate M^2 root formation time in KNM-WT 15000, there are clearly problems in defining which bands to count as well as in confirming the periodicity of the bands.

Another approach to estimating duration of root formation is to make use of root extension rates determined from histology. We know from Dean and Vesey (2008) that the first 10 mm of M^2 root forms at a rate of 7.5 µm/day in *Pan* and at 6.5 µm/day in *H. sapiens*. These very similar rates in close

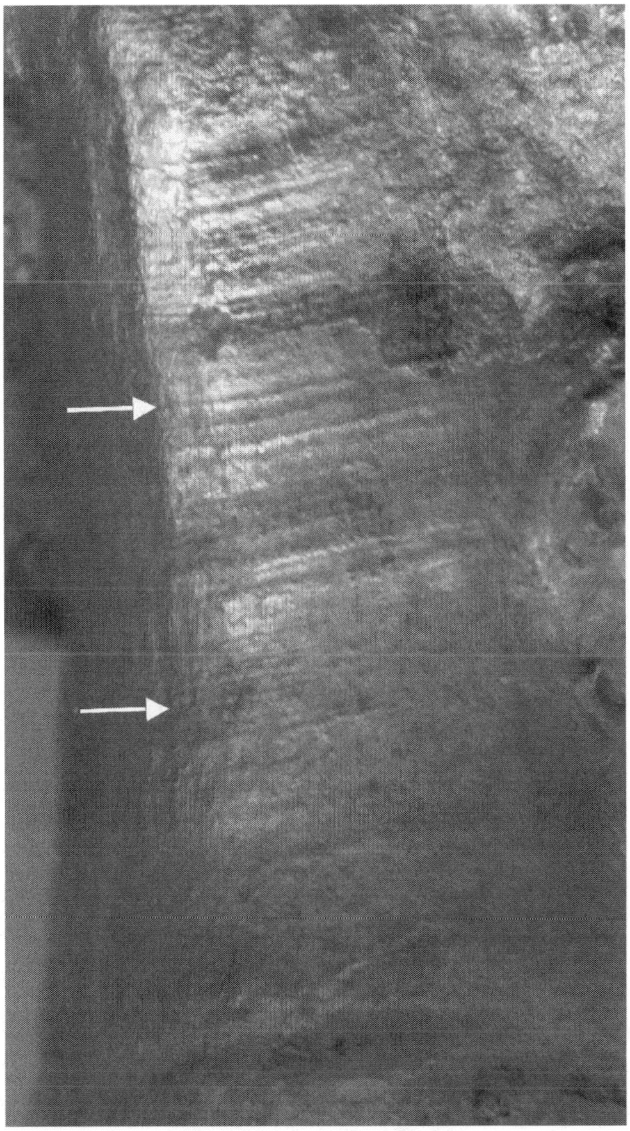

Fig. 10.7 Periradicular bands on the distopalatal root of the upper right M^2 visible on an epoxy resin cast sputter coated with gold. The enamel cervix is visible in the lower part of the image and 9 mm of root extends upward (an apical 1 mm not shown). White arrows indicate two of the regions where periradicular bands are either very widely spaced (upper arrow) or very narrowly spaced (lower arrow). In excess of 80 bands (perhaps 100) can be estimated over most of the root length but these are indistinct and uncountable at both the root cervix and root apex.

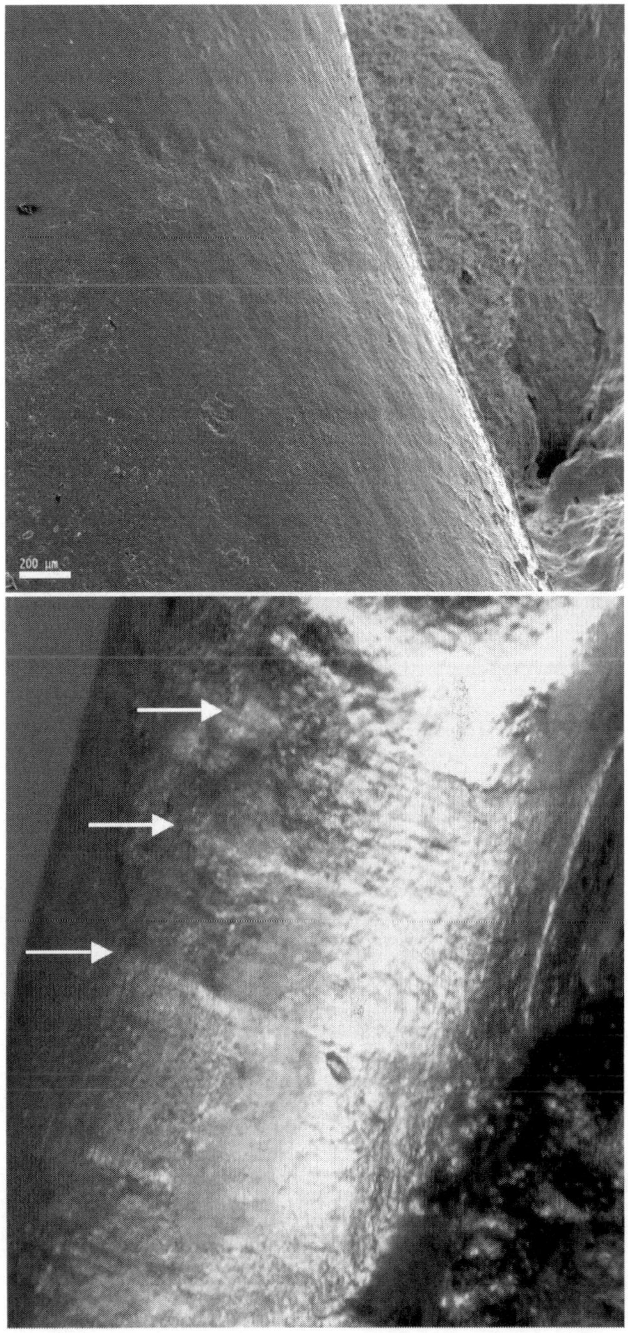

Fig. 10.8 Above, an SEM micrograph and below, an epoxy cast of the enamel cervix and cervical portion of the root of the lower left lateral incisor. In the SEM micrograph the root is tilted to the left of the image to cast a shadow of the widely spaced periradicular bands. The expoxy resin cast, sputter coated with gold, is tilted to the right and illuminated with oblique incident light. The upper white arrow indicates the enamel cervix. The middle and lower white arrows indicate the twin hypoplastic root bands approximately 1 mm apart. Ten periradicular bands can be counted between the cervix and the first formed band and a further ten bands between the twin hypoplastic bands.

relatives of *H. erectus* are reasonable to use to estimate the duration of formation of the M^2 root in KNM-WT 15000. Here we can measure 9.3 mm of root between the last formed enamel and the incomplete apex on the distobuccal aspect of the M^2. At 6.5 µm/day, 9,300 µm of root would form in 1,430 days (3.9 years) and at 7.6 µm/day, would form in 1,240 days (3.4 years). Adding together these estimates for root formation time with those made to the end of enamel completion on the M^2 provides the widest range of estimates for the age at death of KNM-WT 15000. As Table 10.4 shows, these extend from 7.6–8.8 years. In other words, using an 8 day periodicity (for the enamel portion of the estimate) gives estimates that center around 7.9 years of age, whereas using a 10 day periodicity gives estimates that center around 8.5 years.

Building an Ontogenetic Chronology for *Homo erectus*

Each fossil of a *H. erectus* child is a snapshot of stages of growth and development. Some, like KNM-ER 820, a mandible of a young *H. erectus*, show a series of simultaneous stages of tooth formation (Dean, 1987a). In other cases, linear hypoplastic bands mark growth interruptions, defining concurrent stages of tooth growth across the dentition at a particular chronological age. KNM-WT 15000 and another *H. erectus* individual, S7–37 from Sangiran, Java, are examples of where such evidence has been useful in reconstructing dental development in *H. erectus*. Histological analysis of S7–37 allowed Dean et al. (2001) to cross-match the internal microstructure of the developing M^1 with the developing Pm^4, even though the specimen was adult. Combining information from all these specimens is gradually allowing us to build a chronology of tooth formation and eruption for the species. A preliminary chart, based on many underlying comparisons, appears in Fig. 10.9. Molar and premolar crown formation times in early hominins were less than those in modern humans (Beynon and Wood, 1987; Beynon and Dean, 1987, 1988; Ramirez Rozzi, 1993, 1995; Reid and Dean, 2006; Lacruz and Bromage, 2006; Lacruz et al., 2006; Lacruz, 2007; Mahoney, 2008). However, they are unlikely to have been less than 2.5 years in *H. erectus* and there is little evidence of any substantial overlap in the crown formation periods of the M1 and M2 in *H. erectus* (Dean et al., 2001). These facts suggest the end of M2 crown completion would have been closer to 5 years of age than to 4 years, which lends support for seriously considering a 10 day periodicity in this analysis. As this chart is refined, it can be used to age other juveniles, and as the basis for more technical studies of the evolution of tooth formation in Hominidae, and to more fully describe the early life of *Homo erectus*. In the Nariokotome youth, we can bracket the age of the two growth disturbances that occurred just as the upper canine crown completed: For an 8-day periodicity these occurred at an estimated 1,220 and 1,390 days (3.3 and 3.8 years), and for a 10-day periodicity, at 1,340 and 1,540 days (3.7 and 4.2 years). Growth disturbances at this age may well indicate adjustment to a post-weaning diet, or at least, periods of risk in infant health (see Katzenberg et al., 1996; Guatelli-Steinberg et al., 2004). Learning more about the infancy of *Homo erectus* will eventually add knowledge of the kind that may allow us to reconstruct key demographic variables for these early hominins.

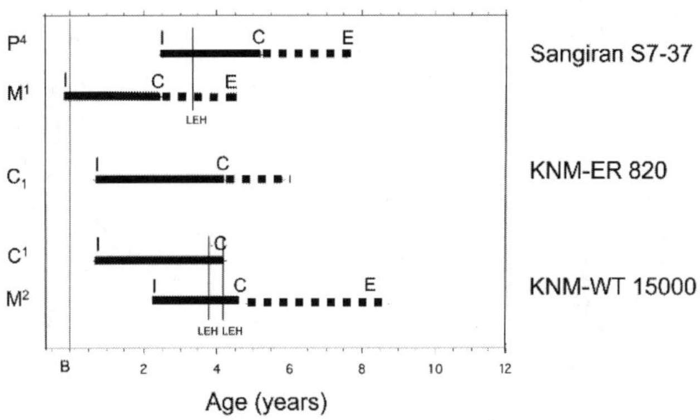

Fig. 10.9 Estimates for the age of some events in the dental developmental of three *Homo erectus* specimens made from microanatomy/histology. Sangiran S7–37 lived to become an adult, whereas KNM-ER 820 and KNM-WT 15000 died as juveniles. The period of crown formation is indicated by the thick solid horizontal lines and of root formation by the thick dotted horizontal lines. B = birth, I = initiation of mineralization of the crown, C = crown completion, E = estimate of age of gingival eruption for some teeth. LEH = linear hypoplastic lines that cross-match the time of events in different tooth types of the same individual (shown as thin vertical lines). Note that the estimated age of gingival emergence for M^1 in Sangiran S7–37 is close to the age estimated for canine crown completion in KNM-WT 15000. Canine crown completion in KNM-WT 15000 may then also have occurred close to the time of M1 eruption in this specimen. M^2 crown completion in KNM-WT 15000 was calculated from perikymata counts on the upper canine and M^2 to be 4.9 years using a 10 day periodicity. However, M^2 initiation and crown formation time in KNM-WT 15000 are estimated here to be the same as in S7–37. KNM-WT 15000 died before completing M^2 development at ca. 8.3–8.8 years (see Table 10.4); KNM-ER 820 died at closer to 6 years (assuming some detailed similarities with KNM-WT 15000). Each new fossil examined may extend the chart to new teeth or ages until a chart can be built of the entire dentition and growth period.

Summary of the Evidence from Enamel and Dentine Microanatomy

At present the balance of the histological data, therefore, favors a younger dental development age than an older one for Nariokotome, one that pulls away from his skeletal age rather than bringing the two closer together. Further, study of Nariokotome makes sense of other observations in early *Homo* that suggested an overall fast developing dentition, such as low values for root cone angles seen on radiographs of KNM-ER 820 and the acute orientation of accentuated lines in a fragment of root dentine in OH 16 (Dean, 2000).

The broadest estimates, made using both 8 and 10 day periodicities and a range of root extension rates estimated for *Pan* and *H. sapiens*, point towards an age of death between 7.6–8.8 years for the Nariokotome specimen. When tooth wear is taken into consideration, gingival emergence of the M2s seems more likely to have been closer to 8 than to 12 years of age. This places growth and development in KNM-WT 15000 at a fairly primitive level, certainly closer to the expectation for an ape of comparable dental and skeletal maturity (ca. 7.5) than for a human (ca. 10–15). Given better comparative material (Table 10.3) and evidence of rapid maturation in other *Homo erectus* individuals, it is hard to argue that it is appropriate to assess Nariokotome by any modern human standard. As Smith (2004: 240) stated, if Nariokotome was 8 years old "...he simply does not belong on human growth standards" either to assess his actual maturity or ultimate predicted growth. The more relevant question is can we distinguish *Homo erectus* growth from the great apes (see Zihlman et al., 2004) or even from earlier australopiths (see Moggi-Cecchi et al., 1998).

How Does Life History Theory Contribute to This Debate?

Life history can be defined as the allocation of an organism's energy for growth, maintenance, and reproduction (Smith, 1992); it is fundamentally a life strategy adopted by an organism to maximize fitness in a world of limited energy (Stearns, 1992; Charnov, 1991, 1993). What, then, is the human strategy? Developmental anatomists began to tie together elements of human life in a scientific context years ago (e.g. Schultz, 1956; Keith, 1949; Washburn and Avis, 1968). It is an exercise greatly enhanced by field work by evolutionary biologists who observed traditional societies with an eye to the components of fitness (e.g. Hill and Hurtado, 1996; Hawkes et al., 1998), a growing data base on primates and other mammals (Harvey and Clutton-Brock, 1985; Promislow and Harvey, 1990; Harvey and Nee, 1991) and coalescence of life history theory (Stearns, 1992). Important elements are still being newly recognized (Gurven and Walker, 2006).

Thus, by combining many sources of information, we can describe the more distinctive features of human life history strategy (Table 10.5). While many of our most extreme modern human characteristics can be grouped under the heading "live slow, die old," a few characters trend in an opposing direction. These, grouped as "live fast, die young" outline a reproductive system that works in double time compared to our closest relatives, the great apes. Human females tend to wean early, shorten birth spacing, and stack multiple dependent offspring. Yet a third set of characteristics is not easily

Table 10. 5 Key features of the human life history strategy fall into at least three categories

Live slow, die old	Live fast	Novelties
Long gestation	Early weaning	Concealed ovulation
Large brains	Short birth spacing	Helpless young
Slow maturation	Multiple dependents	Paternal care
Extended dependence		Linear adolescent growth spurt
Long life		Menopause before senescence
		Grandmothering

swept into a category of fast or slow, but instead might be called "novelties" (see Stearns, 1992); these include aspects of growth, maturation, and reproduction.

What shapes life history strategy in humans or other mammals? Two major contributors are extrinsic mortality and available energy. The human strategy of high quality (large brained) offspring, slow growth and development, and long life depends, in theory, on low mortality and rich resources. When mortality rates are low, reproductive success may be maximised by postponing age at first reproduction for as long as possible and by spreading reproductive output into adulthood over as long a period as possible (Charnov, 1991, 1993; Harvey and Nee, 1991; Stearns, 1992). Yet quality offspring, a prolonged life history profile and a long period of dependency add up to a costly and risky strategy. In parallel with our prolonged life history, however, humans appear to have evolved mechanisms for reducing costs or constraints imposed by extremely slow maturation (Lovejoy, 1981; Smith and Tompkins, 1995; Allman and Hasenstaub, 1999; Hawkes, 2006; Robson et al., 2006). Examples are adaptations in Table 10.5 columns "live fast" and "novelties" – adaptations that increase reproductive success in the face of a basic overall "live slow, die old" strategy. Human females start to reproduce late compared to great apes and have children who mature even more slowly. Despite these constraints, human females cut off reproduction in the fourth decade just as do chimpanzees. Human females compensate for this cost to lifetime fertility by weaning infants very early, before they can feed themselves. While this reduces the interval to next birth and ramps up reproduction, it also leaves mothers with multiple dependants and concomitant increased energetic demands as well as the task of keeping infants and toddlers safe simultaneously. It is increasingly recognized that mothers cannot do this alone: adaptations like paternal care, grandmothering, and provisioning reduce the energetic demands on mothers, enabling them to stack multiple dependents (Kaplan, 1997; Kaplan et al., 2000; Gurven and Walker, 2006; Robson et al., 2006; Hawkes, 2006). Bogin (1999) has also pointed out that older siblings and adolescents often assume a 'caretaker' role, which takes additional pressure off nursing mothers. A particular slowdown in human growth between weaning and puberty (later made up in the adolescent growth spurt) may be another novelty that reduced energetic demands of multiple dependents (Bogin, 1999).

Menopause takes on particular importance in this scenario, since in many ways it seems to be so counterproductive: in chimpanzees, diminishing fertility parallels that of other physical systems in general, whereas fertility fails in human females while they are otherwise still vigorous (Hill and Hurtado, 1991). Whether menopause is an adaptation or epiphenomenon is not yet resolvable (Hill and Hurtado, 1991; Peccei, 2001). But, whatever the cause of menopause - whether genes that shut down reproduction were favored

(Hawkes et al., 1998; O'Connell et al., 1999), or if human life span simply crossed a natural limit to mammalian reproduction of age 50 – grandmothers who cannot reproduce themselves can increase their reproductive fitness by investing in their children's offspring, thereby improving both quality and survival of descendants.

It is important to realize that the characteristics in Table 10.3 evolved at sometime, but not necessarily at the same time – much current research leaves open the possibility that our life history was still taking shape through much of the Pleistocene (Smith and Tompkins, 1995; Dean et al., 2001; Thompson et al., 2003; Dean, 2006; Smith et al., 2007a). The fossil record has, and will continue to provide evidence on when humans adopted each of the elements that make up our life history strategy (see Skinner and Wood, 2006). The most obvious lines of evidence are brain size (Smith, 1989, 1991), ages of attainment of marker events (Bromage and Dean, 1985) and records of weaning (Humphrey et al., 2007) – or even other reproductive events – that may be preserved in teeth. The Nariokotome juvenile, so complete and caught at such distinctive skeletal and dental stages, provides a window on adolescence, growth, and even the attainment of independence in *Homo erectus*.

An Alternative Model for Growth and Development in *Homo erectus*

Once we stop trying to force Nariokotome into a human mold, we can explore a more evolutionary approach. Comparative study of living mammals has long suggested that important aspects of maturation evolve in parallel with brain size (Sacher, 1959; Smith, 1989 and references therein). The strongest evidence for such a correspondence is found in primates, where the age of emergence of the first permanent molar (M1) correlates with cranial capacity at r = 0.98 or higher (Smith, 1989; Smith et al., 1995). Even if both are ultimately responses to some third factor (e.g., mortality), cranial capacity may be a window on growth and development of extinct hominids. If we use regression to predict age of tooth emergence from cranial capacity, early *Homo erectus* (with a cranial capacity of 810 cc) is expected to erupt M1 at 4.5 years, with M2 following at approximately 9 years of age (Smith, 1993; Smith and Tompkins, 1995; Smith et al., 1994). Such timing is not typical for any living primate, but would describe a maturation rate intermediate between living apes and humans. These predictions are remarkably close to those derived from enamel and dentine histology in *Homo erectus*.

Smith (1993) and Smith and Tompkins (1995) have also discussed the issue of the adolescent growth spurt in *Homo erectus* and questioned the likelihood of it being present 1.6 million years ago, although this has generated considerable debate. The human adolescent growth spurt – or rather the very slow period of growth between weaning and puberty that precedes the spurt, is unique. The spurt has variously been seen as a kind of catch-up growth following a period of prolonged and intensive calorific investment in brain growth, or alternatively as the end of a period when it is socially advantageous to be small during the long human learning process (Bogin, 1990, 1999). Gurven and Walker (2006), however, argue that the very slow period of growth between weaning and puberty in modern humans enables mothers to support more infants at any one time because their combined body mass is small and their energy requirements are thus lower for as long as possible before sexual maturity. For a slowed late childhood growth to be favored, we suspect that stacking multiple dependents in a family unit must have increased survival for both toddler and adolescent, while moderating energy demands. Either way, the adolescent growth spurt is arguably intricately involved with prolonged childhood dependency. If *Homo erectus* offspring were energetically independent of their mothers to a greater degree at an earlier age than in modern humans, then the advantage of the slow growth period between weaning and puberty becomes less obvious.

Aiello and Key (2002) have previously argued that the energetic costs of being a *Homo erectus* mother with a body size ~50% greater than a female australopith mother during lactation, gestation and non-reproductive periods would have been considerably higher and would have required "a revolution" in the way in which females obtained and utilised energy. Lieberman et al. (2008, 2009) have argued that the transition from *Australopithecus* to *Homo* was indeed characterized by a new strategy for acquiring and using energy in open habitats, and that this transition was almost certainly related to a profound behavioral shift characterized by an increase in meat acquisition through scavenging and/or hunting and the regular manufacture of stone tools designed for food extraction and food processing. These lines of evidence, together with those set out in this review, suggest a strategy of co-operative provisioning and food sharing where *Homo erectus* offspring were able to contribute from an early age to their own energetic requirements.

Conclusion

The young dental age, the older skeletal age and the apparent large body mass and stature of the Nariokotome youth show that a greater proportion of adult body mass and stature had been attained at an earlier age than would be expected for a modern human. While it can be argued that KNM-WT 15000 falls within the total range for modern humans if we include growth disorders, this is increasingly awkward if the

Nariokotome boy is to be considered in the normal range, and verging on the inexplicable if his true age of death is 7.6–8.8. To us, the most parsimonious explanation for this combination of facts is that the growth curve of early *Homo erectus* was *more* like that of modern chimpanzees. More specifically, whereas the precise growth curve of *Homo erectus* was likely unique, it apparently differed from ours in the direction of chimpanzees. If anything, this review, combined with data emerging from dental microanatomy, provides increasing support for the model proposed by Smith (Figure 9.7 in Smith, 1993), in which early *Homo erectus* is predicted on average to erupt M1 at around 4.5 years, M2 around 9 years and M3 around 14.5 years, with a late childhood lacking a slowdown in growth, and a lifespan potentially some 15 years longer than a modern chimpanzee. One of the more interesting recent ideas is that the human childhood slowdown/adolescent spurt in growth is connected with the human predisposition to stack multiple dependents (Bogin, 1999; Gurven and Walker, 2006). Relatively rapid attainment of adult size, then, logically indicates a relatively early attainment of independence. The Nariokotome youth, then, an early adolescent at 8 or 9, may have been much more independent than a comparable human adolescent of 12 or 13 in a traditional society. Thus the Nariokotome youth may throw light on the family structures of early *Homo erectus*.

Fossils of individuals who died as juveniles provide snapshots of growth and maturation of extinct species; their analysis has so far proved to be our best chance of working out the appearance of the elements of human life history; this knowledge, in turn, provides us with clues to explore larger questions that lie behind the appearance and evolution of the human species.

Acknowledgments We thank the organisers of the Stony Brook Human Evolution Symposium and Workshop for inviting us to contribute a review article to this volume. We thank the Government of Kenya and the Kenya National Museums for granting us permission to work on valuable fossil material in their care. We thank Emma Mbua, Meave Leakey, Dan Lieberman, Fred Spoor, and Louise Humphrey for their help with this project and we thank Fred Grine for patiently awaiting the manuscript. In particular we are grateful to Helen Liversidge for allowing us to cite unpublished data in this review and for many helpful discussions. We thank Alan Walker and Richard Leakey and Harvard University Press for permission to reproduce Fig. 10.2 in this review from the Nariokotome monograph (1993).

References

Aiello, L.C., Key, C., 2002. The energetic costs of being a *Homo erectus* female. American Journal of Human Biology 14, 551–565.

Agossou-Voyeme, A.K., Fachehoun, C.R., Boco, V., Hounnou, G.M., Biaou, O., 2005. Osseous age of the black children of Benin. A population study of 600 children aged from 9 years to 18 years and living in Cotonou. Morphologie 89(285), 64–70.

Allbrook, D., 1961. The estimation of stature in British and East African males. Journal of Forensic Medicine 8, 15–28.

Allman, J., Hasenstaub, A., 1999. Brains, maturation time and parenting. Neurobiology of Aging 20, 447–454.

Anderson, D.L., Thompson, G.W., Popovich, F., 1976. Age attainment of mineralization stages of the dentition. Journal of Forensic Science 21, 191–200.

Antoine, D., 2001. Evaluating the periodicity of incremental structures in dental enamel as a means to studying growth in children from past human populations. Ph.D. thesis, Institute of Archaeology, University College London, University of London.

Antoine, D., Dean, M.C., Hillson, S., 2000. The periodicity of incremental structures in dental enamel based on the developing dentition of post-medieval known-age children. In: Mayhall, J.T., Heikkinen T. (Eds), Dental Morphology 1998. Proceedings of the 11th International Symposium on Dental Morphology. Oulu University Press, Oulu, pp. 48–55.

Antoine, D., Hillson, S., Dean, M.C., 2009. The developmental clock of dental enamel: a test for the periodicity of prism cross-striations and an evaluation of the likely sources of error in histological studies of this kind. Journal of Anatomy 214, 45–55.

Antón, S.C., Leigh, S.R., 2003. Growth and life history in *Homo erectus*. In: Thompson, J.L., Krovitz, G.E., Nelson, A.J. (Eds), Patterns of Growth and Development in the Genus *Homo*. Cambridge University Press, Cambridge, pp. 219–245.

Bermudez de Castro, J.M., Rosas, A., Carbonell, E., Nicolas, M.E., Rodriguez, J., Arsuaga, J.-L., 1999. A modern human pattern of dental development in Lower Pleistocene Hominids from Atapuerca-TD6 (Spain). Proceedings of the National Academy of Science USA 96, 4210–4213.

Beynon A.D., Wood, B.A., 1987. Patterns and rates of enamel growth on the molar teeth of early hominids. Nature 326, 493–496.

Beynon, A.D., Dean, M.C., 1987. Crown formation time of a fossil hominid premolar tooth. Archives of Oral Biology 32, 773–780.

Beynon, A.D., Dean, M.C., 1988. Distinct dental development patterns in early fossil hominids. Nature, 335, 509–514.

Black, S., Scheuer, L., 1996. Age changes in the clavicle from the early neonatal period to skeletal maturity. International Journal of Osteoarchaeology 6, 425–434.

Bogin, B., 1990. The evolution of human childhood. BioScience 40, 16–25.

Bogin, B., 1999. Patterns of Human Growth. Second edition. Cambridge Studies in Biological Anthropology. Cambridge University Press, Cambridge.

Bogin, B., Smith, B.H., 2000. Evolution of the human life cycle. In: Stinson, S., Bogin, B., Huss-Ashmore, R., O'Rourke, D. (Eds), Human Biology: An Evolutionary and Biocultural Perspective. Wiley-Liss, New York, pp. 377–424.

Bromage, T.G., Dean, M.C., 1985. Re-evaluation of the age at death of immature fossil hominids. Nature 317, 525–528.

Brown, B., Walker, A., 1993. The dentition. In: Leakey, R., Walker, A. (Eds) The Nariokotome *Homo erectus* Skeleton. Harvard University Press, Cambridge, MA, pp. 161–192.

Brown, F.H., Harris, J., Leakey, R., Walker, A., 1985. Early *Homo erectus* from West Lake Turkana. Nature 316, 788–792.

Cameron, N., 2002. Human Growth and Development. Academic, New York.

Charnov, E.L.R., 1991. Evolution of life history variation among female mammals. Proceedings of the National Academy of Sciences USA 88, 1134–1137.

Charnov, E.L.R., 1993. Life-History Invariants. Oxford University Press, Oxford.

Clegg, E.J, Pawson, I.G., Ashton, E.H., Flinn, R.M., 1972. The growth of children at different altitudes in Ethiopia. Philosophical Transactions of the Royal Society London B 264, 403–437.

Clegg, M., Aiello, L.C., 1999. A comparison of the Nariokotome *Homo erectus* with juveniles from a modern human population. American Journal of Physical Anthropology 110, 81–93.

Dean, M.C., 1987a. The dental developmental status of six juvenile fossil hominids from Koobi Fora and Olduvai Gorge. Journal of Human Evolution 16, 197–213.

Dean, M.C., 1987b. Growth layers and incremental markings in hard tissues: a review of the literature and some preliminary observations about enamel structure in *Paranthropus boisei*. Journal of Human Evolution 16, 157–172.

Dean, M.C., 1995. The nature and periodicity of incremental lines in primate dentine and their relationship to periradicular bands in OH 16 (*Homo habilis*). In: Moggi-Cecchi, J. (Ed.), Ninth International Symposium on Dental Morphology. Aspects of Dental Biology: Palaeontology, Anthropology and Evolution. International Institute for the Study of Man, Florence, pp. 239–265.

Dean, M.C., 2000. Progress in understanding hominoid dental development. Journal of Anatomy 197, part 1, 77–101.

Dean, M.C., 2006. Tooth microstructure tracks the pace of human life history evolution. Proceeding of the Royal Society Series B, 273, 2799–2802.

Dean, M.C., Reid, D.J., 2001a. Perikymata spacing and distribution on hominoid anterior teeth. American Journal of Physical Anthropology 116, 209–215.

Dean, M.C., Reid, D.J., 2001b. Anterior tooth formation in *Australopithecus* and *Paranthropus*. In: Brook, A. (Ed.), Dental Morphology 2001. Sheffield Academic Press, Sheffield, pp.135–149.

Dean, M.C., Vesey, P., 2008. Preliminary observations on increasing root length during the eruptive phase of tooth development in modern humans and great apes. Journal of Human Evolution 54, 258–271.

Dean, M.C., Beynon, A.D., Reid, D.J., Whittaker, D.K., 1993a. Longitudinal study of tooth growth in a single individual based on long and short period incremental markings in dentine and enamel. International Journal of Osteoarchaeology 3, 249–264.

Dean, M.C., Beynon, A.D., Thackeray, J.F., Macho, G., 1993b. Histological reconstruction of dental development and age at death of a juvenile *Paranthropus robustus* specimen, SK 63, from Swartkrans, South Africa. American Journal of Physical Anthropology 91, 401–419.

Dean, M.C., Leakey, M.G., Reid, D.J., Schrenk, F., Schwartz, G.T., Stringer, C., Walker, A., 2001. Growth processes in teeth distinguish modern humans from *Homo erectus* and earlier hominins. Nature 414, 628–631.

Demirjian, A., Goldstein, H., Tanner, J.M., 1973. A new system of dental age assessment. Human Biology 45, 211–227.

FitzGerald, C.M., 1998. Do enamel microstructures have regular time dependencies? Conclusions from the literature and a large scale study. Journal of Human Evolution 35, 371–386.

Flecker, H., 1932/1933. Roentgenographic observations of the times of the appearance of the epiphyses and their fusion with the diaphyses. Journal of Anatomy 67, 118–164.

Garn S.M., Lewis, A.B., Blizzard, R.M, 1965. Endocrine factors in dental development. Journal of Dental Research 44, 43–258.

Garn, S.M., Rohmann, C.G., Silverman, F.N., 1967. Radiographic standards for postnatal ossification and tooth calcification. Medical Radiography and Photography 43, 45–66.

Goodall, J., 1986. The Chimpanzees of Gombe: Patterns of Behavior. Belknap, Cambridge MA.

Gray, S.J., Wiebusch, B., Akol, H.A., 2004. Cross-sectional growth of pastoralist Karimojong and Turkana children. American Journal of Physical Anthropology 125, 193–202.

Grøn, A.-M., 1962. Prediction of tooth emergence. Journal of Dental Research 41, 573–585.

Guatelli-Steinberg, D., Larsen, C.S., Hutchinson, D.L., 2004. Prevalence and the duration of linear enamel hypoplasia: a comparative study of Neandertals and Inuit foragers. Journal of Human Evolution 47, 65–84.

Gurven, M., Walker, R., 2006. Energetic demand of multiple dependents and the evolution of slow human growth. Proceeding of the Royal Society, Series B 273, 835–841.

Haas, J.D., Campirano, F., 2006. Interpopulation variation in height among children 7 to 18 years of age. Food and Nutrition Bulletin 27 supplement, S212–S223.

Hamada, Y., Udono, T., 2002. Longitudinal analysis of length growth in the chimpanzee (*Pan troglodytes*). American Journal of Physical Anthropology, 118, 268–284.

Harvey, P.H., Nee, S., 1991. How to live like a mammal. Nature 350, 23–24.

Harvey, P.H., Clutton-Brock, T.H., 1985. Life history variation in primates. Evolution 39, 559–581.

Hawkes, K., 2006. Life history theory and human evolution: a chronicle of ideas and findings. In: Hawkes, K., Paine, R.R. (Eds), The Evolution of Human Life History. School of American Research Press, Santa Fe, NM, pp. 45–44.

Hawkes, K., O'Connell, J.F., Blurton Jones, N.G., Charnov, E.L., Alvarez, H., 1998. Grandmothering, menopause and the evolution of human life histories. Proceedings of the National Academy of Sciences USA 95, 1336–1339.

Hill, K., Hurtado, A.M., 1991. The evolution of premature reproductive senescence and menopause in human females: an evaluation of the "grandmother hypothesis." Human Nature 2, 313–350.

Hill, K., Hurtado, A.M., 1996. Ache Life History: The Ecology and Demography of a Foraging People. Aldine de Gruyter, New York.

Hillson, S., 1996. Dental Anthropology. Cambridge University Press, Cambridge.

Hillson, S., Bond, S., 1997. Relationship of enamel hypoplasia to the pattern of tooth crown growth: a discussion. American Journal of Physical Anthropology 104, 89–103.

Holderbaum, R.M., Veeck, E.B., Olivera, H.W., da Silva, C.M., Fernandes, A., 2005. Comparison among dental, skeletal and chronological development in HIV-positive children: a radiographic study. Brazilian Oral Research 19(3), 209–215.

Humphrey, L.T., 2003. Linear growth variation in the archaeological record. In: Thompson, J.L., Krovitz, G.E., Nelson, A.J., (Eds) Patterns of Growth and Development in the Genus *Homo*. Cambridge University Press, Cambridge, pp. 191–218.

Humphrey, L.T., Dean, M.C, Jeffries, T.E., 2007. An evaluation of changes in strontium/calcium ratios across the neonatal line in human deciduous teeth. In: Bailey, S.E., Hublin, J.J. (Eds), Dental Perspectives on Human Evolution. Springer, New York, pp. 301–317.

Kaplan, H.S., 1997. The evolution of the human life course. In: Wacter, K., Finch, C.E. (Eds), Between Zeus and Salmon: The Biodemography of Aging. National Academy of Sciences, Washington, DC, pp. 175–211.

Kaplan, H.S., Hill, K.R., Hurtado, A.M., Lancaster, J.B., 2000. A theory of human life history evolution: diet, intelligence, and longevity. Evolutionary Anthropology 9, 156–185.

Katzenberg, M.A., Herring, D.A., Saunders, S.R., 1996. Weaning and infant mortality: evaluating the skeletal evidence. Yearbook of Physical Anthropology 39, 177–199.

Keith, A., 1949. A New Theory of Human Evolution. Watts, London.

Kerley, E.R., 1966. Skeletal age changes in the chimpanzee. Tulane Studies in Zoology 13, 71–80.

Krogman, W.M., 1962. The Human Skeleton in Forensic Medicine. Charles C. Thomas, Springfield, IL.

Kuykendall, K.L., 2003. Reconstructing australopithecine growth and development: what do we think we know? In: Thompson, J.L., Krovitz, G.E., Nelson, A.J. (Eds), Patterns of Growth and Development in the Genus *Homo*. Cambridge University Press, Cambridge, pp. 191–218.

Lacruz, R.S., 2007. Enamel microstructure of the hominid KB 5223 from Kromdraai, South Africa. American Journal of Physical Anthropology 132, 175–182.

Lacruz R.S., Bromage, T.G., 2006. Appositional enamel growth in molars of South African fossil hominids. Journal of Anatomy 209, 13–20.

Lacruz R.S., Ramirez Rozzi F., Bromage T.G., 2006. Variation in enamel development of South African fossil hominids. Journal of Human Evolution 51, 580–590.

Lacruz, R.S., Dean M.C., Ramirez Rozzi, F., Bromage, T.G., 2008. Patterns of enamel secretion in fossil hominins. Journal of Anatomy 213, 148–158.

Lewis, A.B., 1991. Comparisons between dental and skeletal ages. Angle Orthodontist 61, 87–92.

Lieberman, D.E., Pilbeam D.R., Wrangham, R.W., 2008. The transition from *Australopithecus* to *Homo*. In: Shea, J.J., Lieberman, D.E. (Eds), Transitions in Prehistory: Essays in Honor of Ofer Bar-Yosef. Oxbow Press, Oxford, pp. 1–24.

Lieberman, D.E., Bramble, D.M., Raichlen, D.A., Shea, J.J., 2009. Brains, brawn, and the evolution of endurance running capabilities. In: Grine, F.E., Fleagle, J.G., Leakey, R.E. (Eds), The First Humans: Origin and Early Evolution of the Genus *Homo*. Springer, Dordrecht, pp. 77–92.

Liversidge, H.M., 2003. Worldwide variation in human dental development. In: Thompson, J.L., Nelson, A., Krovitz, G. (Eds), Growth and Development in the Genus *Homo*. Cambridge University Press, Cambridge, pp. 73–113.

Liversidge, H.M., 2008a. Dental age revisited. In: Irish, J.D., Nelson, G.C. (Eds), Technique and Application in Dental Anthropology. Cambridge University Press, Cambridge, pp. 234–252.

Liversidge, H.M., 2008b. Timing of human mandibular third molar formation. Annals of Human Biology 35, 294–321.

Liversidge, H.M., Townsend, G.C., 2005. Tooth formation in Australian Aborigines. In: Żądzińska, E. (Ed), Current Trends in Dental Morphology Research. 13th International Symposium on Dental Morphology, Wydawnictwo Uniwersytetu Lodzkiego Lodz, Poland, pp. 405–410.

Liversidge, H.M., Chaillet, N., Mörnstad, H., Nyström, M., Rowlings, K., Taylor, J., Willems, G., 2006. Timing of Demirjian's tooth formation stages. Annals of Human Biology 33, 454–470.

Lovejoy, C.O., 1981. The origin of man. Science 21, 341–350.

Lundy J.K., Feldesman, M.R., 1987. Revised equations for estimating living stature from the long bones of South African Negroes. South African Journal of Science 78, 54–55.

Mackay, D.H., Martin, W.J., 1952. Dentition and physique of Bantu children. Journal of Tropical Medicine and Hygiene 55, 265–275.

Mahoney, P., 2008. Intraspecific variation in M_1 enamel development in modern humans: implications for human evolution. Journal of Human Evolution 55, 130–146.

Modi, J.P., 1957. Medical Jurisprudence and Toxicology, 12th edition. Tripathy Private Ltd., Bombay.

Moggi-Cecchi, J., Tobias, P.V., Beynon, A.D., 1998. The mixed dentition and associated fragments of a juvenile fossil hominid from Sterkfontein, South Africa. American Journal of Physical Anthropology 106, 425–465.

Molinari, L., Gasser, T., 2004. The human growth curve: distance, velocity and acceleration. In: Haudpie, R.E., Cameron, N., Molinari, L. (Eds), Methods in Human Growth Research, Cambridge University Press, Cambridge, pp. 27–54.

Molleson, T., Cox, M., 1993. The Spitalfields Project. The Society for Medieval Archaeology Monograph Series 4, London.

Moorrees, C.F.A., Fanning, E.A., Hunt, E.E., 1963. Age variation of formation stages for ten permanent teeth. Journal of Dental Research 42, 1490–1502.

O'Connell, J.F., Hawkes, K., Blurton Jones, N.G., 1999. Grandmothering and the evolution of *Homo erectus*. Journal of Human Evolution 36, 461–485.

Peccei, J.S., 2001. Menopause: adaptation or epiphenomenon? Evolutionary Anthropology 10, 43–57.

Promislow, D.E.L., Harvey, P.H., 1990. Living fast and dying young: a comparative analysis of life-history variation among mammals. Proceedings of the National Academy of Sciences USA 88, 1134–1137.

Ramirez Rozzi, F., 1993. Tooth development in East African *Paranthropus*. Journal of Human Evolution 24, 429–454.

Ramirez Rozzi, F., 1995. Time of crown formation in Plio-Pleistocene hominid teeth. In: Moggi-Cecchi, J. (Ed), Ninth International Symposium on Dental Morphology. Aspects of Dental Biology: Palaeontology, Anthropology and Evolution. International Institute for the Study of Man, Florence, pp. 217–238.

Reid, D.J., Dean, M.C., 2006. Variation in modern human enamel formation times. Journal of Human Evolution 50, 329–346.

Reid, D.J., Ferrell, R.J., 2006. The relationship between the number of striae of Retzius and their periodicity in imbricational enamel formation. Journal of Human Evolution 50, 195–202.

Reid, D.J., Guatelli-Steinberg, D., Walton, P., 2008. Variation in modern human premolar enamel formation times: implications for Neandertals. Journal of Human Evolution 54; 225–235.

Ritzman, T.B., Baker, B.J., Schwartz, G.T., 2008. A fine line: a comparison of methods for estimating ages of linear enamel hypoplasia formation. American Journal of Physical Anthropology 135, 348–361.

Roberts, D.F., 1960. Effects of race and climate on human growth as exemplified by studies on African children. In: Tanner, J.M. (Ed.), Human Growth. Pergamon, Oxford, pp. 59–72.

Roberts, D.F., Bainbridge, D.R., 1963. Nilotic physique. American Journal of Physical Anthropology 21, 341–370.

Robson, S.L., van Schaik, C.P., Hawkes, K., 2006. The derived features of human life history. In: Hawkes, K., Paine, R.R. (Eds), The Evolution of Human Life History. School of American Research Press, Santa Fe, NM, pp. 17–44.

Ruff, C., 2007. Body size prediction from juvenile skeletal remains. American Journal of Physical Anthropology 133, 698–716.

Ruff, C.B., Walker, A., 1993. Body size and body shape. In: Leakey, R., Walker, A. (Eds), The Nariokotome *Homo erectus* Skeleton. Harvard University Press, Cambridge, MA, pp. 234–265.

Sacher, G.A., 1959. Relation of lifespan to brain weight and body weight in mammals. In: Wolstenholme, G.E.W., O'Connor, M. (Eds), Ciba Foundation Colloquia on Ageing, Volume 5, The Lifespan of Animals. Churchill, London, pp. 115–133.

Scheuer, L., Black, S., 2000. Developmental Juvenile Osteology. Academic, London.

Schultz, A.H., 1956. Postembryonic age changes. In: Hofer, H., Schultz, A.H., Starck, D. (Eds), Primatologia, Volume 1. Karger, Basel, pp. 887–964.

Schwartz, G.T., Dean M.C., 2008. Charting the chronology of developing teeth. In: Irish, J.D., Nelson, G.C. (Eds), Technique and Application in Dental Anthropology, Studies in Biological Anthropology. Cambridge University Press, Cambridge, pp. 219–233.

Schwartz, G.T., Dean, M.C., Reid, D.J., Zihlman, A.L., 2006. A faithful record of stressful life events preserved in the dental developmental record of a juvenile gorilla. International Journal of Primatology 27(3), 1–19.

Skinner, M.M., Wood, B.A., 2006. The evolution of modern human life history. A paleontological perspective. In: Hawkes, K., Paine, R.R. (Eds), The Evolution of Human Life History. School of American Research Advanced Seminar Series. School of American Research Press, Sanata Fe, NM, pp. 331–364.

Smith, B.H., 1986. Dental development in *Australopithecus* and early *Homo*. Nature 323, 327–330.

Smith, B.H., 1989. Dental development as a measure of life history in primates. Evolution 43, 683–688.

Smith, B.H., 1991. Dental development and the evolution of life history in Hominidae. American Journal of Physical Anthropology 86, 157–174.

Smith, B.H., 1992. Life history and the evolution of human maturation. Evolutionary Anthropology 1, 134–142.

Smith, B.H., 1993. The physiological age of KNM-WT 15000. In: Leakey, R., Walker, A. (Eds), The Nariokotome *Homo erectus* Skeleton. Harvard University Press, Cambridge, MA, pp. 195–220.

Smith, B.H., 1994. The sequence of eruption of teeth in *Macaca, Pan, Australopithecus* and *Homo*. Its evolutionary significance. American Journal of Human Biology 6, 61–76.

Smith, B.H., 2004. The paleontology of growth and development. Evolutionary Anthropology 13, 239–241.

Smith, B.H., Tompkins, R.L., 1995. Toward a life history of the Hominidae. Annual Review of Anthropology 24, 257–279.

Smith, R.J., Gannon, P.J., Smith, B.H., 1995. Ontogeny of australopthecines and early *Homo*: evidence from cranial capacity and dental eruption. Journal of Human Evolution 29, 155–168.

Smith, S.L., 2004. Skeletal age, dental age, and the maturation of KNM-WT 15000. American Journal of Physical Anthropology 125, 105–120

Smith, T.M., 2008. Incremental dental development: methods and applications in hominoid evolutionary studies. Journal of Human Evolution 54, 205–224.

Smith, T.M., Tafforeau, P., Reid, D.J., Grun, R., Eggins, S. Boutakiout, M. and Hublin, J.-J., 2007a. Earliest evidence of modern human life history in North African early *Homo sapiens*. Proceedings of the National Academy of Sciences USA 104, 6128–6123.

Smith, T.M., Reid, D.J., Dean, M.C., Olejniczak, A.J., Ferrell, R.J., Martin, L.B., 2007b. New perspectives on chimpanzee and human molar crown development. In: Bailey, S.E., Hublin, J.-J., (Eds), Dental Perspectives on Human Evolution. Springer, New York, pp.177–192.

Smith, T.M., Reid, D.J., Dean, M.C., Olejniczak, A.J., Martin, L.B., 2007c. Molar development in common chimpanzees (*Pan troglodytes*). Journal of Human Evolution 52, 2001–2216.

Stearns, S., 1992. The Evolution of Life Histories. Oxford University Press, Oxford.

Tanner, J.M., 1962. Growth at Adolescence. 2nd Edition. Blackwell Scientific, Oxford.

Tanner, J.M., Whitehouse, R.H., 1982. Atlas of Children's Growth: Normal Variation and Growth Disorders. Academic, London.

Tardieu, C., 1998. Short adolescence in early hominids: infantile and adolescent growth of the human femur. American Journal of Physical Anthropology 107, 163–178.

Terry, R.J., 1932. The clavicle of the American Negro. American Journal of Physical Anthropology 16, 351–379.

Thompson, J.L., Krovitz, G.E., Nelson, A.J., 2003. Patterns of Growth and Development in the Genus *Homo*. Cambridge University Press, Cambridge.

Vallejo-Bolaños, E., España-López, A.J., Muñoz-Hoyos, A., Fernandez-Garcia, J.M., 1999. The relationship between bone age, chronological age and dental age in children with isolated growth hormone deficiency. International Journal of Paediatric Dentistry 9, 201–206.

Walker, A., Leakey, R., 1993. The postcranial bones. In: Leakey, R., Walker, A. (Eds), The Nariokotome *Homo erectus* Skeleton. Harvard University Press, Cambridge, MA, pp. 95–160.

Washburn, S.L., Avis, V., 1958. Evolution of human behavior. In: Roe, A., Simpson, G.G. (Eds), Behavior and Evolution. Yale University Press, New Haven, CT, pp. 421–436.

Welbourn, H.F., 1956. The teeth of children attending Kampala child welfare clinics and schools. East African Medical Journal 33, 181–187.

Zihlman, A., Bolter, D., Boesch, C., 2004. Wild chimpanzee dentition and its implications for assessing life history in immature hominin fossils. Proceedings of the National Academy of Sciences USA 101, 10541–10543.

Chapter 11
Dental Evidence for Diets of Early *Homo*

Peter S. Ungar and Robert S. Scott

Keywords Dental topographic analysis • molars • microwear texture analysis • diet • *Homo habilis* • *Homo erectus*

Introduction

The evolution of diet in the earliest members of our genus, *Homo rudolfensis*, *H. habilis* and *H. erectus* has received increased attention over the past few years (see Ungar et al., 2006a for review). Many models have been constructed, based largely on nutritional studies combined with direct analogy (with living peoples or non-human primates) or on contextual evidence, such as archeological and paleoenvironmental indicators. These models suggest hypotheses, some of which can be tested with the fossil evidence for the hominins themselves.

In this paper we review and evaluate some recent models for the dietary adaptations of early *Homo*. While there are real intractable limits to what we can learn, the dental remains of these hominins offer some clues to the diets of these species. Results of a recent study on molar occlusal functional morphology (Ungar, 2004) will be reviewed, along with results from a recent study of dental microwear of early *Homo* (Ungar et al., 2006b). In addition, new data on dental microwear textures for early *Homo* cheek teeth will be presented for comparison with results for extant primates and other fossil hominins.

A synthetic view of this evidence, in the context of paleoenvironmental reconstructions and archeological remains, suggests that the origin and early evolution of *Homo* were most likely associated with biological and cultural adaptations for a more flexible, versatile subsistence strategy. This strategy would have put the earliest members of our genus at an advantage given climatic fluctuation and a mosaic of different microhabitats in Africa during the Plio-Pleistocene.

P.S. Ungar (✉)
Department of Anthropology, University of Arkansas, Old Main 330, Fayetteville, AR 72701, USA
e-mail: pungar@uark.edu

R.S. Scott
Department of Anthropology, Rutgers, The State University of New Jersey, 131 George Street, New Brunswick, NJ 08901-1414, USA
e-mail: robertsc@rci.rutgers.edu

Environmental Dynamics and Diets of Early *Homo*

Environmental dynamics likely played an important role in dietary changes related to the origin and evolution of early *Homo*. Cerling (1992) has noted, for example, that after 2.5 Ma, C_4 grasslands spread across East Africa, concomitant with periodic fluctuations in climate. If any of the early hominins had critical keystone foods (those essential for survival and reproduction) found only in more closed habitats, then migration or extinction may well have followed. In contrast, hominins able to face environmental change with a more flexible diet would more likely have survived in place (Teaford et al., 2002; Wood and Strait, 2004; Ungar et al., 2006a). This would have been advantageous given variable, unpredictable environments, or a mosaic environment comprised of many different microhabitats (Behrensmeyer et al., 1997; Potts, 1998; Wood and Strait, 2004).

The Archeological Record

The earliest evidences for tool manufacture and use are stone artifacts from Gona, Ethiopia and faunal remains with cut marks from Bouri, Ethiopia, both dated to about 2.5 Ma (Semaw et al., 1997; de Heinzelin et al., 1999 Roche et al., 2009). Oldowan tools were almost certainly used to process a wide variety of foods, both animal and plant (Keeley and Toth, 1981). Further, perishable implements were likely also used to prepare foods (Panger et al., 2002). In fact, it is likely that hominins made and used such tools long before 2.5 Ma, as chimpanzees and orangutans do today. Thus, early stone tools are best seen not as evidence of the origins of tool use, but as indicators of an expanded toolkit that included durable implements, reflecting increased dietary versatility and flexibility (e.g., Mann, 1972; Schick and Toth, 1993). It has also been suggested that increasing distances of stone tool transport, and the recovery of artifacts from a broader range of environments through the Late Pliocene may further reflect increasing adaptive versatility for early hominins (Rogers et al., 1994; Potts, 1998; Roche et al., 2009).

While the apparent near synchrony of appearances of *Homo* and the first stone tools and cut-marked bones have been connected for some time (Leakey et al., 1964), we must remember that there were other early hominins in Africa around 2.4–2.5 Ma, and we do not know for certain which one(s) were responsible for these artifacts.

What about associations between major archaeological advances and the appearance of *Homo erectus*? While the earliest Acheulean tools postdate the appearance of *H. erectus* (e.g., Asfaw et al., 1992), early large concentrations of stone tools and modified bones at sites such as DK and FLK 22 at Olduvai and FxJj 1 at Koobi Fora are approximately coincident with the appearance of this hominin (Blumenschine and Masao, 1991). This implies that animal tissues had become an important part of early hominin diets by the beginning of the Pleistocene (Potts, 1983; Shipman, 1983; Blumenschine, 1995; Bunn, 2001). While there were several early hominin species moving about the landscape at this time it is reasonable to assume that *H. erectus* was at least partially responsible for these larger accumulations given similar sites outside of Africa, where no australopiths have been found (e.g., Dennell et al., 1988; Gabunia and Vekua, 1995).

Diet Models and Early *Homo*

The apparent contemporaneity of the earliest *Homo*, Oldowan technology and the spread of C_4 grasslands across East Africa makes for a compelling, if not well-supported, model. The argument is that environmental change during the late Pliocene (whether directional or an increase in variability) would have led to changes in resources available to hominins. If early *Homo* used a greater range of habitats, or more variable habitats, an expanded toolkit would have allowed these hominins to process and consume foods that would have otherwise been unavailable. Environmental change provides the motive, and technological innovation offers the opportunity for new dietary adaptations. Both grassland adapted ungulates, and open savanna plants (especially underground storage organs) have been suggested as possible keystone foods.

Increased consumption of meat has been thought for more than half a century to have played an important role in human evolution (e.g., Dart, 1953). As forest resources became scarce, hominins presumably consumed the increasingly abundant grassland-adapted ungulates. A feedback loop of increasing meat protein and fat consumption, improving intelligence and hunting abilities, and an expanding toolkit was said to follow; along with a division of labor, more complex social systems and selection for yet greater intelligence (Washburn, 1963; Lee and Devore, 1968; Isaac, 1971). Researchers to this day continue to develop and refine these models by incorporating nutritional studies and energetics theory (e.g., Hayden, 1981; Milton, 1987; Speth, 1989; Leonard and Robertson, 1992, 1994; Aiello and Wheeler, 1995; Stanford, 2001; Eaton et al., 2002).

Others have proposed an alterative to this "Man the Hunter" scenario, focusing on xeric plants rather than animals as key to the evolution of diet in early *Homo* (Linton, 1971; Coursey, 1973; Wolpoff, 1973). As Zihlman and Tanner (1978) noted, plants often account for 60–70% of the human forager diet. O'Connell et al. (1999) and Wrangham et al. (1999) both proposed that underground storage organs (USOs) might have been keystone resources for these hominins, as did Hatley and Kappleman (1980) before them. These authors all suggest a transition from ape-like *H. habilis* and *H. rudolfensis* subsistence strategies to a more human-like strategy for *H. erectus*.

A Review of Dental Evidence for Diet in Early *Homo*

While many have proposed that fundamental changes in diet accompanied the origin and early evolution of the genus *Homo*, fewer have attempted to test such hypotheses using the fossil remains of the hominins themselves. Most efforts have focused on teeth, which are not only part of the digestive system, but also the most commonly preserved elements in hominin fossil assemblages. Work has focused on (1) species-level (genetic) adaptations such as tooth size, shape and structure, and (2) epigenetic lines of evidence such as dental microwear and mineralized tissue chemistry, which relate to effects of foods on individuals during their lifetimes. Historical perspectives on many of these studies are presented in detail elsewhere (Ungar et al., 2006a; Ungar, 2007c). Here we focus on one adaptive line of evidence (tooth shape), and one non-adaptive line (tooth wear). Recent dental topographic analysis and dental microwear results will be reviewed and new microwear texture analysis data will be presented for available, useable early *Homo* specimens. These will be interpreted in light of models for the origins and evolution of diet in early *Homo*.

The first thing to note is that there are not very many specimens of *Homo habilis, H. rudolfensis* or early African *H. erectus* to study. The available sample of Plio-Pleistocene *Homo* permanent cheek teeth from Ethiopia, Kenya, Tanzania, Malawi, and South Africa includes about 83 separately numbered specimens (see Ungar et al., 2006b). While this might seem at first to be a reasonable sample, most of these teeth exhibit postmortem chipping, cracking or surface erosion. This damage can make dental topographic analysis difficult and microwear texture analysis impossible. Further, a lack of consensus on taxonomic attribution of some of these specimens (e.g., see Grine, 2005) makes it even more challenging to assemble reasonable samples for individual species. Nevertheless, congeners can be pooled[1] and occlusal topography and microwear texture of the combined early *Homo* sample can be compared and contrasted with those of other early hominins and extant primates to gain some insights into the diets of the earliest members of our genus.

Dental Topographic Analysis

Primate molar teeth are adapted to the fracture properties of foods consumed (Lucas, 2004). As Spears and Crompton (1996) have noted, dental morphology affects the nature, magnitude and distribution of stresses on food particles. Extant species known to consume tough foods, for example, have more occlusal relief than those adapted to eating hard, brittle objects (e.g., Kay, 1984; Meldrum and Kay, 1997; Lucas, 2004). Most studies to date have focused on unworn molar teeth (usually M_2s) to allow comparisons without the confounding affects of wear on morphology. While differences in occlusal morphology between early hominin species are also likely to relate to diet, there are few if any unworn, undamaged M_2s of early *Homo* available for us to study!

Dental topographic analysis, a technique designed to allow inclusion of worn teeth in functional analyses, was developed with just this in mind (Ungar and Williamson, 2000; M'Kirera and Ungar, 2003; Ungar and M'Kirera, 2003; Dennis et al., 2004; Ungar, 2004, 2007a, b; King et al., 2005). The idea behind dental topographic analysis is that teeth should evolve to wear such that they preserve aspects of shape that keep them efficient for fracturing whatever foods they are adapted to comminute. More specifically, because dental tissues differ in their resistances to wear, the distribution of enamel in the crown and dentin underneath can be integrated to sculpt the occlusal surface in a specific manner as a tooth wears. We should therefore be able to compare at least some functional aspects of shape in similarly worn teeth across species.

Dental topographic analysis characterizes crown morphology without the need to measure between specific landmarks that change or are lost with wear. The usual procedure begins with a cloud of points representing the occlusal surface of a tooth. First, elevation maps of occlusal surfaces of teeth (original or high-resolution replicas) are created using a laser scanner. We have used a Surveyor 500 Scanner with an RPS 150 laser probe (LaserDesign, Inc) and an XSM multi-sensor scanning machine (Xystum Corp) with an integrated OTM3 laser head (Dr. Wolf & Beck GmbH). Other devices, such as an electromagnetic digitizer, a piezo touch-probe scanner, a reflex microscope and a confocal microscope have also been used with some success. All of the data reported here were collected using the Surveyor 500 with RPS 150 system, producing point clouds with lateral sampling intervals and vertical resolutions of 25.4 μm.

Individual point clouds are opened in GIS software and a digital elevation model of the occlusal surface is interpolated by inverse distance weighting. We use Arcview 3.2 (ESRI Corp) because of its ease of use and broad availability, though other GIS packages would work equally well. Each surface model is then cropped to the lowest point on the central basin to maximize comparability between specimens, and various descriptive statistics are calculated to characterize functional aspects of the surface. Average surface slope, angularity, and relief are recorded, along with a gross wear score for the tooth (modified from Scott, 1979; see Ungar, 2004). Slope is defined as the average change in elevation between adjacent points representing the crown surface. Angularity is the average change in slope across the surface. Finally, relief is the ratio of occlusal table surface area divided by the planimetric area, or two-dimensional area as measured in top view. Data for taxa are then compared using conventional statistical analyses.

Results for longitudinal series of primates in the wild have shown that individuals within a species wear their teeth down in consistent ways (Dennis et al., 2004; King et al., 2005). This suggests that different individuals with variably worn teeth can be used to construct a "species-specific" dental wear sequence. This is a necessary prerequisite to the study of dental functional morphology of variably worn hominin teeth. Likewise, studies of individuals at a given stage of wear show that species differ in predictable ways consistent with dietary differences (M'Kirera and Ungar, 2003; Ungar and M'Kirera, 2003; Ungar, 2007a, b). For example, at a given stage of wear, gorilla molars evince steeper slopes and occlusal relief than chimpanzees. Orangutans have intermediate slopes. This is consistent with differences in ratios of leaf and stem to fruit consumption reported in the literature, and with our understanding of dental biomechanics – more occlusal relief means more efficient fracture of tough foods.

[1] Small sample sizes for each species of early *Homo* makes statistical comparisons of congeners impractical. The justification for combining species is their evident closer evolutionary relationship with one another than with any known australopith species (Strait and Grine, 2001).

Dental Topography in Early *Homo*

A dental topographic analysis for early *Homo* specimens was recently published (Ungar, 2004). High resolution replicas of all available undamaged M_2s of *Homo erectus* (KNM-ER 806, KNM-ER 992, KNM-WT 15000, OH 22), *H. habilis* (OH 16), *H. rudolfensis* (KNM-ER 1506, KNM-ER 1802), and early *Homo* sp. indet. (KNM-ER 3734) were included in this study.

Data for these specimens were pooled into a single sample given the number of individuals of each species and compared with results for *Australopithecus afarensis* (n = 15 individuals from Hadar, Ethiopia and Laetoli, Tanzania) and the extant African apes *Gorilla gorilla gorilla* (n = 47) and *Pan troglodytes troglodytes* (n = 54). *Australopithecus afarensis* was selected as a reasonable model for a generalized australopith occlusal morphology from which early *Homo* might have been derived (see Ungar, 2004 for discussion). The western lowland gorilla and central African chimpanzee were chosen as the extant baseline for comparison because of moderate differences in their diets (see M'Kirera and Ungar, 2003; Ungar and M'Kirera, 2003 for references and details), and because their teeth follow the same general "Bauplan" as those of early hominins. More information on these samples can be found in Ungar (2004).

Results for comparisons of occlusal slope are shown in Figs. 11.1 and 11.2 and Table 11.1. A two-way analysis of variance was run on ranked slope data with taxon and wear stage as the factors. There is significant variation in both taxon and wear stage, but no significant interaction between the factors. This means that the species differ in average slope, and that slope varies by wear stage. On the other hand, differences between species remain consistent across wear stages, suggesting that slopes can be compared between species at given stages of dental wear.

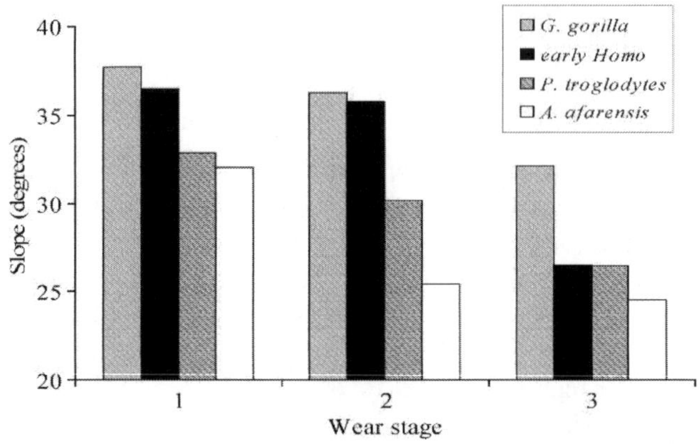

Fig. 11.2 Comparisons of mean surface slopes for taxa at given wear stages (Modified from Ungar, 2004).

Multiple comparisons tests indicate significant differences between most pairs of species examined. Early *Homo* occlusal slope values are intermediate between those of gorillas and chimpanzees, suggesting an adaptation for foods with fracture properties intermediate between those of gorillas and chimpanzees. Both *Pan troglodytes troglodytes* and *Gorilla gorilla gorilla* prefer soft, ripe fruits when available, though the steep slopes of the gorilla molars allow them to efficiently shear tough, less preferred leaves and stems at times of resource scarcity (see Ungar and M'Kirera, 2003 and references therein). This suggests that early *Homo* molars could have fractured tough foods more efficiently than can chimpanzees but less efficiently than can gorillas.

Fig. 11.1 Digital elevation models of representative specimens of (**a**) *Homo erectus* (KNM-WT 15000), (**b**) *Homo rudolfensis* (KNM-ER 1506), (**c**) *Gorilla gorilla* (CMNH B1781), and (**d**) *Pan troglodytes* (CMNH B3437).

Table 11.1 Results for the dental topography study (Data from Ungar, 2004)

A. Summary statistics for occlusal slope (in degrees). Dashes indicate no data at that wear stage

	Australopithecus			*Gorilla*			*Homo*			*Pan*		
Wear	Mean	SD	*N*	Mean	SD	*N*	Mean	SD	*N*	Mean	SD	*N*
1	37.89	–	1	–	–	–	37.92	–	1	–	–	–
2	32.01	6.154	4	37.75	5.036	7	36.5	4.677	2	32.88	5.859	5
3	25.46	3.81	7	36.29	2.665	10	35.83	9.519	3	30.15	5.771	28
4	25.21	2.45	2	32.13	5.069	14	26.52	3.296	2	26.48	4.68	18
5	–	–	–	27.53	4.29	13	–	–	–	25.69	8.607	2
6	32.05	–	1	32.14	6.445	3	–	–	–	29.31	–	1

B. Analyses of rank-transformed slope data

Two factor ANOVA results

Effect	*df*	*F*	*p*
Taxon	3.90	8.607	0.00
Wear class	2.90	8.409	0.00
Interaction	6.90	0.683	0.66

Pairwise mean differences (Bonferroni test results)

	Australopithecus	*Gorilla*	*Homo*
Gorilla	52.295*		
Homo	34.493**	−17.802	
Pan	19.172	−33.123*	−15.321

*$p \leq 0.05$, **$p \leq 0.1$.

Of particular interest is the comparison with *Australopithecus afarensis* molars. The shallow sloped australopith molars were not well-suited to fracturing tough foods, but would have been better at crushing hard, brittle items. Indeed, the average slope difference between *A. afarensis* and the combined sample of early *Homo* is on the same order of magnitude as that between the extant chimpanzee and gorilla samples examined. Perhaps then, differences between these hominins relate to adaptations for fallback resource differences, as has been suggested for chimpanzees and gorillas.

Ungar (2004) suggested that both *Australopithecus afarensis* and early *Homo* probably preferred nutrient rich foods with less challenging fracture properties, but that early *Homo* might have relied more on tough fallback foods, whereas their australopith predecessors would have consumed more hard, brittle items at "crunch times." Tougher foods eaten by early *Homo* might have included pliant plant parts and/or meat (see Lucas and Peters, 2000).

Dental Microwear

The patterns of microscopic use-wear on primate molar teeth have also been related to food preferences. Diets dominated by hard, brittle foods tend to leave relatively large pits in teeth, whereas those dominated by tougher foods result in more striations and perhaps smaller pits (Teaford, 1988; Teaford and Runestad, 1992). The basic idea is that striations are formed as tough foods are fractured between opposing molar crests and abrasives are dragged across the wear facets, much as they would be if sliced between opposing scissor blades. In contrast, large pits are formed as molars come into direct opposition while hard, brittle foods are crushed between opposing facets, as if between a hammer and anvil. Small pits are also said to form with the consumption of tough foods, presumably as prisms are "plucked" from their surrounding matrix due to friction as opposing surfaces slide past one another (Walker, 1984; Teaford and Runestad, 1992).

Only one comprehensive study of dental microwear in early *Homo* has been published to date. Ungar et al. (2006b) examined all available Plio-Pleistocene *Homo* cheek-teeth, and found 18 that preserved antemortem microwear. Results of that study indicate that early *Homo* (*H. habilis*, *H. erectus*, and specimens from Sterkfontein Member 5 and Swartkrans Member 1 of uncertain taxonomic affinity) have moderate pit percentages, intermediate between extant primates that eat more hard objects such as nuts and seeds (*Cebus apella* and *Lophocebus albigena*) and those that consume more tough foods, including leaves and stems (*Gorilla gorilla*). Early *Homo* pit widths also fall near the middle of the extant primate range, suggesting that the individuals examined did not specialize on extremely hard, stiff, or tough foods, at least in the days or weeks prior to death.

While sample sizes for each group are quite small, results also hint at possible variation in diet between early *Homo* congeners. *H. erectus* and specimens from Swartkrans Member 1 have modestly higher pit percentages on average than do *H. habilis* and specimens from Sterkfontein Member

5C. This led Ungar and coauthors (2006b) to opine that while early *Homo* as a group probably preferred less fracture resistant foods, *H. erectus* and Swartkrans Member 1 individuals may have eaten more tough or hard, brittle foods prior to death than did the *H. habilis* and Sterkfontein Member 5C individuals studied.

Microwear Texture Analysis

One of the principal advantages of microwear analysis is that because it reflects diet over the course of days or months, a sufficient sample should allow us to gauge within species variation in diet and to evaluate hypotheses concerning diet overlap and fallback resource exploitation. Unfortunately, conventional microwear analyses may involve high observer error rates (Grine et al., 2002) that make within species variation difficult to assess. With this in mind, our research group developed a new, objective and repeatable approach to dental microwear analysis that allows us to document and analyze within group variation. This new technique, dental microwear texture analysis, combines white-light confocal microscopy with scale-sensitive fractal analyses to characterize microwear surface textures (Ungar et al., 2003, 2007; Scott et al., 2005, 2006).

Microwear texture analyses confirm conventional microwear results that distinguish primates by broad diet category (see Scott et al., 2005, 2006). Texture analyses make it clear, however, that central tendencies are merely the beginning of the microwear story; as the dispersions around those central tendencies are often more interesting and more informative. This is especially so for primate generalists, for whom niche differentiation manifests itself largely on those occasions when preferred resources are scarce (Lambert et al., 2004; Ungar, 2005; Wrangham, 2005). It should come as no surprise then that *Alouatta palliata*, *Trachypithecus cristatus*, *Cebus apella* and *Lophocebus albigena* all have overlapping values for microwear surface texture fractal complexity (Table 11.2 and Fig. 11.3a). That said, capuchins and mangabeys, two species known to fall back on hard, brittle foods, have much greater dispersions, with clusters of less complex surfaces, but a few specimens showing very rough, "beaten up" microwear facets. This is exactly the pattern expected of species that occasionally fall back on hard, brittle objects such as nuts and seeds. A similar pattern has been observed for *Paranthropus robustus* when compared with *Australopithecus africanus* (Scott et al., 2005), suggesting that diet-related adaptive differences between "gracile" and "robust" australopiths may well reflect fallback resource adaptations rather than fundamental differences in food preferences.

Microwear Texture Analysis of Early *Homo*

Here we present the first dental microwear texture analysis of early *Homo* to put it context with conventional microwear studies and other lines of evidence for the diets of these hominins.

Materials and Methods

Available replicas produced from molds used by Ungar et al. (2006b) were examined using white-light confocal microscopy. Eighteen specimens in total were studied (Table 11.3). This sample included most of the same individuals analyzed by Ungar et al. (2006b). Three specimens from that study (OH 7, Stw 82, and SK 2635) were unavailable for analysis, but three others were added (KNM-BK 8518, KNM-ER 992, and KNM-ER 1808) with the detection of previously unidentified antemortem microwear using the confocal microscope. Any study with limited sample sizes can be affected by the exclusion of a few specimens and inclusion of others, so comparisons between results presented here and those in Ungar et al. (2006b) should be approached with this in mind. Nevertheless, the consistencies of results presented below with those from Ungar et al. (2006b) give us confidence that the difference in sample composition has little effect on our interpretations.

As with the feature-based microwear analysis, the specimens used in this study come from Plio-Pleistocene deposits at Olduvai Gorge, Koobi Fora, Baringo, West Turkana, Sterkfontein and Swartkrans. These can be divided into four groups: *Homo habilis* ($n = 5$), *Homo erectus* ($n = 8$), *Homo* sp. indet. from Sterkfontein Member 5C ($n = 2$), and *Homo* sp. indet. from Swartkrans Member 1 ($n = 3$). The taxonomic attributions of the individual specimens are presented in Table 11.3. Explanations of and rationale for these assignments are presented by Ungar et al. (2006b) for specimens used in that study, and by Wood and Van Noten (1986) and Wood (1991) for the others.

Methods of specimen preparation followed conventional microwear procedures, except that replicas were not coated or mounted on stubs (as would be necessary for typical SEM study). Original fossils were cleaned with cotton swabs soaked in alcohol or acetone. Molds of occlusal crowns were prepared using President's Jet regular body polyvinylsiloxane dental impression material (Coltène-Whaledent Corp) and casts were poured using Epotek 301 (Epoxy Technologies, Inc.) epoxy resin and hardener. Resulting epoxy replicas were mounted directly on the confocal microscope stage with plasticine.

Table 11.2 Descriptive microwear texture statistics

Taxon	Statistic	*n*	Complexity	Scale of maximum complexity	Anisotropy	Heterogeneity (=HAsfc$_{9\ cells}$)[a]	Textural fill volume[b]
Alouatta palliata	Mean	11	0.3603	53.4269	0.0058	0.6924	3871
	Median		0.3149	0.2668	0.0057	0.5211	482
	Standard deviation		0.1834	175.2870	0.0021	0.3827	4,583
	Skewness		0.4020	3.3165	−0.3064	1.2619	0.7520
Cebus apella[c]	Mean	13	5.4658	31.9023	0.0037	0.7863	9,683
	Median		2.8818	0.2666	0.0029	0.7603	9,707
	Standard deviation		6.3043	91.1688	0.0019	0.3342	4,924
	Skewness		1.6238	3.2418	0.7579	0.5457	0.7458
Lophocebus albigena	Mean	15	1.7687	28.4170	0.0038	0.5350	12,369
	Median		1.0181	7.6932	0.0035	0.5049	12,669
	Standard deviation		1.7398	52.5858	0.0020	0.2764	2,374
	Skewness		2.3044	2.7791	1.2128	1.6457	−0.4201
Trachypithecus cristatus	Mean	12	0.7337	1.2233	0.0048	0.6124	8,342
	Median		0.5141	0.2386	0.0036	0.5659	8,492
	Standard deviation		0.6603	2.5286	0.0026	0.2855	4,365
	Skewness		2.6182	3.0584	0.8262	0.8144	−0.4098
Early *Homo*	Mean	18	1.4173	1.0906	0.0034	0.4587	8,552
	Median		1.0855	0.3451	0.0029	0.4228	10,096
	Standard deviation		0.9365	2.0868	0.0019	0.1663	5,500
	Skewness		1.3978	3.6264	0.5210	1.1522	−0.0363
Homo erectus	Mean	8	1.7339	0.4782	0.0037	0.4327	7,078
	Median		1.4553	0.1808	0.0032	0.4144	6,220
	Standard deviation		1.1968	0.7798	0.0015	0.1142	5,422
	Skewness		0.8158	2.7827	0.3960	0.3386	0.8405
Homo habilis	Mean	5	0.9605	2.7615	0.0038	0.3717	13,006
	Median		0.7469	1.3690	0.0042	0.3695	12,639
	Standard deviation		0.4392	3.6889	0.0028	0.0696	2,994
	Skewness		1.4656	1.9876	0.0476	−0.1320	0.7865

[a] Heterogeneity reported in Scott et al. (2006) was Hasfc$_{9\ cells}$ some parts of the text erroneously referred to Hasfc$_{81\ cells}$.

[b] Descriptive statistics for *Tfv* of extants species differ slightly from those reported in Scott et al. (2006) and were calculated using the most recent and improved (non-directional) volume filling algorithm in Sfrax.

[c] The *Cebus apella* sample used here and in Scott et al. (2006) includes only specimens form Bahia, Brazil. These specimens have also been attributed to *Cebus xanthosternos* and *Cebus nigritus robustus*.

The microwear texture analysis followed procedures described by Scott et al. (2006). Three dimensional point clouds representing each specimen were generated using a Sensofar Plμ confocal microscope (Solarius, Inc.) with an integrated white light vertical scanning interferometer (see Fig. 11.4). Data were collected for four adjacent fields on a "phase II" facet (usually facet 9) using a 100x long working distance objective. This generated about 1,738,000 elevations for each surface analyzed, sampled at intervals of 0.18 μm along both the x- and y-axes with vertical resolution specified to be five nanometers (0.005 μm). The combined work envelope of the fields examined was 276 × 204 μm. This resolution is better and work envelope is larger than those reported for the recent SEM-based microwear study of these hominins (e.g., Ungar et al., 2006b).

Resulting point clouds were analyzed using scale-sensitive fractal analysis (*SSFA*) software (ToothFrax and SFrax, Surfract Corp). The premise of *SSFA* is that a given surface can look different at different scales – an asphalt road may look smooth at coarse scales (to a motorist driving along it), but rough and bumpy at finer scales (to an ant trying to cross it). Several *SSFA* texture attributes identified as useful for microwear analysis were considered for early *Homo*. These include area-scale fractal complexity (*Asfc*), anisotropy (*epLsar*$_{1.8\mu m}$), scale of maximal complexity (*Smc*), textural fill volume (*Tfv*), and heterogeneity of complexity (*HAsfc*). Each is described in detail elsewhere (Ungar et al., 2003, 2007; Scott et al., 2006).

Area-scale fractal complexity is change in roughness of a surface across scales of observation. The faster a measured surface area increases with scale, the higher the *Asfc*. Anisotropy can be measured as variation in lengths of transect lines sampled at given scales across surfaces in different orientations. A highly anisotropic scratched surface, for example, will have shorter transects when sampled parallel to

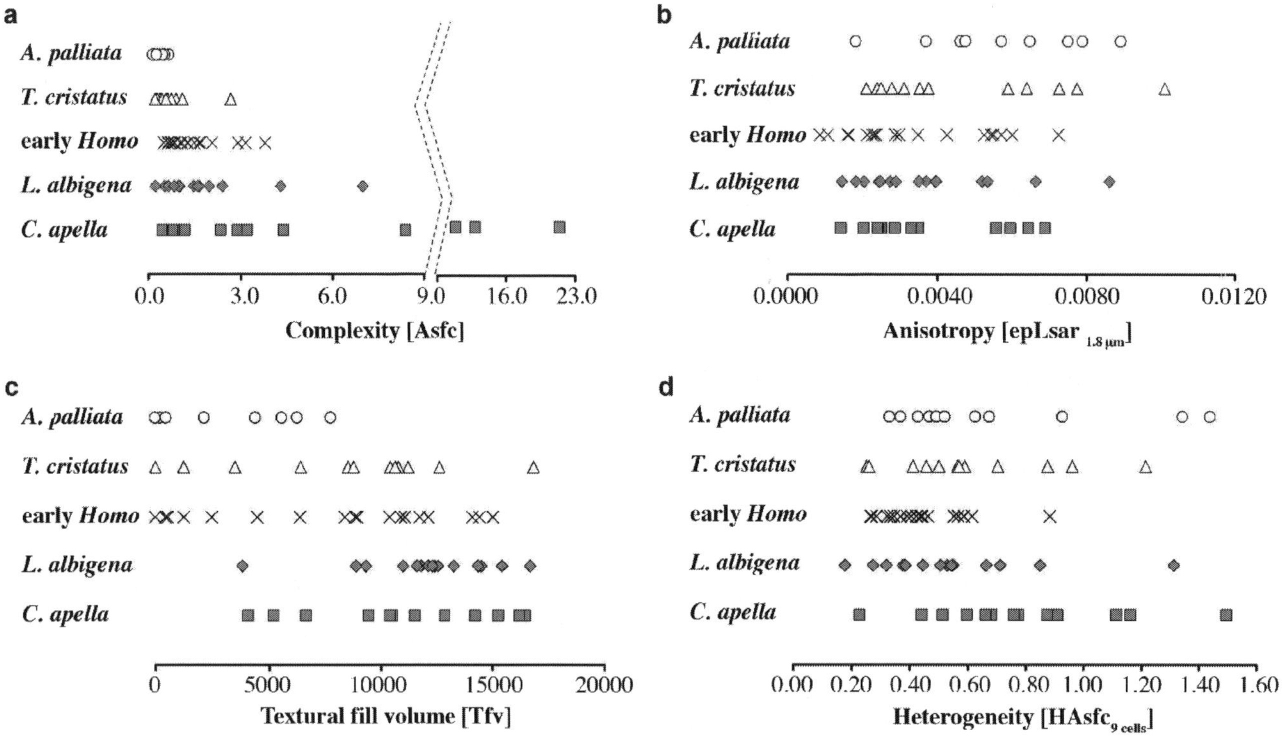

Fig. 11.3 Distributions of the microwear texture values of (**a**) complexity, (**b**) anisotropy, (**c**) textural fill volume, and (**d**) heterogeneity for extant the comparative samples and the combined early *Homo* sample.

Table 11.3 Microwear texture data for individual early *Homo* specimens

Taxon	Specimen	*Asfc*	$epLsar_{1.8\mu m}$	*Smc*	*Tfv*	$HAsfc_{9\,cells}$
H. erectus	KNM-BK 8518	0.51	0.0035	0.35	1,552.14	0.57
H. erectus	KNM-ER 807	1.65	0.0023	0.21	10,049.87	0.46
H. erectus	KNM-ER 820	3.77	0.0055	0.15	17,395.20	0.33
H. erectus	KNM-WT 15000	1.26	0.0024	0.15	4,314.10	0.39
H. erectus	OH 60	2.06	0.0021	0.15	10,141.65	0.27
H. erectus	SK 15	0.62	0.0057	2.40	5,454.38	0.61
H. erectus	KNM-ER 1808	3.16	0.0052	0.15	6,985.56	0.42
H. erectus	KNM-ER 992	0.85	0.0030	0.27	729.81	0.41
H. habilis	OH 4	1.68	0.0042	2.34	12,639.34	0.44
H. habilis	OH 15	0.65	0.0054	0.27	17,438.06	0.37
H. habilis	OH 16	0.75	0.0011	0.62	10,307.38	0.29
H. habilis	OH 41	1.09	0.0008	1.37	14,302.75	0.32
H. habilis	Stw 19	0.65	0.0072	9.21	10,340.10	0.44
early *Homo*, Sterkfontein Mb. 5C	SE 1508	0.72	0.0060	0.43	11,922.00	0.35
early *Homo*, Sterkfontein Mb. 5C	SE 1579	1.04	0.0023	0.42	0.00	0.27
early *Homo*, Swartkrans Mb. 1	SK 27	0.84	0.0016	1.21	1,878.74	0.58
early *Homo*, Swartkrans Mb. 1	SK 45	1.48	0.0016	0.21	5,640.57	0.55
early *Homo*, Swartkrans Mb. 1	SK 847	1.28	0.0029	0.67	12,849.83	0.88

the preferred orientations of scratches than sampled across those scratches. Transects perpendicular to striations are longer because they must move in and out of individual features as they cut across them. Thus, heavily pitted surfaces tend toward high complexity and low anisotropy, whereas surfaces dominated by shallow, parallel striations are the opposite, with lower complexity and high anisotropy.

Other attributes allow us to fine-tune descriptions of surface texture. Scale of maximal complexity identifies the scale range over which *Asfc* is calculated, such that larger values for *Smc* should correspond to more, deeper features at coarser scales. Textural fill volume increases as more filling elements can be packed into features at a given scale. Textural fill volume tends to be larger as features become larger

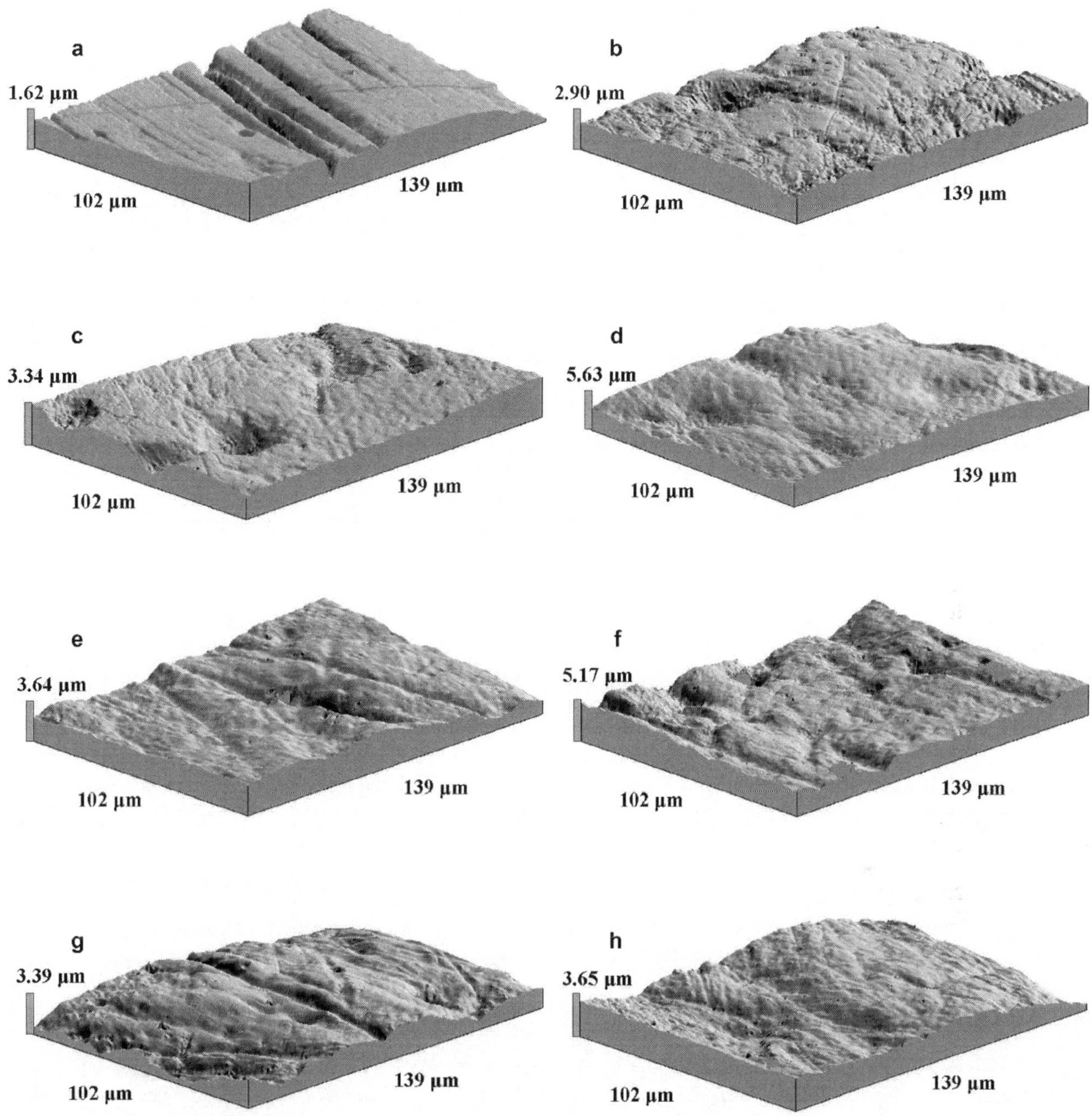

Fig. 11.4 Three-dimensional photosimulations derived from microwear surface point clouds for (**a**) *Alouatta palliata*, (**b**) *Cebus apella*, (**c**) *Trachypithecus cristatus*, (**d**) *Lophocebus albigena*, (**e**) *Homo erectus*, (**f**) *Homo habilis*, (**g**) early *Homo* Swartkrans Mb. 1, and (**h**) early *Homo* Sterkfontein Mb. 5C.

and/or more square or circular as opposed to linear. Heterogeneity of complexity is simply a measure of how much *Asfc* tends to vary across a given surface at a given scale. Thus, surfaces with wear that is similar in degree and type (homogeneous) will have lower values for *HAsfc*.

Early *Homo* specimens were compared first to a baseline series of extant taxa published recently by Scott and coauthors (2006), and then to one another. The extant baseline series included two pair of contrasting species, *Cebus apella* (n = 13) and *Alouatta palliata* (n = 11) and *Lophocebus albigena* (n = 15) and *Trachypithecus cristatus* (n = 12). The techniques used for data collection were the same as those described here for early *Homo*. Details about these specimens can be found in Scott et al. (2006). Suffice it to say that these are all specimens are all wild-shot individuals, and that each pair contained a species considered to be frugivorous with hard-object components (*C. apella, L. albigena*) and a species reported to be more foliviorous (*A. palliata,*

T. cristatus). While a primate species that specializes on underground storage organs would be a valuable addition to the extant baseline, such microwear data are not currently available. Extant species examined here nevertheless demonstrate how primates with diets dominated by tough foods differ in microwear textures from those that consume more hard, brittle ones. And it is these fracture properties, rather than food types themselves, that microwear differences should reflect. It would be useful to examine dental microwear in terrestrial species too, as microwear patterns may also be affected by differences in exogenous grit related to substrate preferences (Daegling and Grine, 1999).

First, a MANOVA was performed on ranked data for all variables (*Asfc*, $epLsar_{1.8\mu m}$, *Smc*, $Hasfc_{9\ cells}$ and *Tfv*) to compare a combined sample of early *Homo* to the extant baseline taxa. Individual ANOVAs and multiple comparisons tests were used to determine the sources of significant variation. Both Tukey's Honestly Significant Difference (HSD) and Fisher's Least Significant Difference (LSD) tests were used to balance risks of Type I and Type II errors (Cook and Farewell, 1996). Values of $p \leq 0.05$ for Tukey's HSD tests may be assigned significance with some confidence, whereas values of $p \leq 0.05$ on Fisher's LSD tests but not Tukey's HSD tests are here considered to suggest possible but unimpressive differences between pairs.

Early *Homo* specimens were then divided into four groups following Ungar et al. (2006b): (1) *Homo habilis*; (2) *H. erectus*; (3) *Homo* from Sterkfontein Mb. 5C; and (4) *Homo* from. Swartkrans Mb. 1. These groups were compared using the same scheme described above – a MANOVA on ranked data for all variables, and individual ANOVAs and multiple comparisons tests to determine sources of significant variation. While expectations of significant variation between the groups are admittedly optimistic given the small samples and resulting low power, significant variation in pit percentage was reported for the conventional microwear study (Ungar et al., 2006b), thus differences in microwear textures could indicate directions for future research.

Results

The area-scale fractal complexity (*Asfc*), anisotropy ($epLsar_{1.8\mu m}$), scale of maximal complexity (*Smc*), textural fill volume (*Tfv*), and heterogeneity of complexity (*HAsfc*) values obtained for the individual early *Homo* specimens are recorded in Table 11.3. Descriptive statistics for both central tendencies and distributions of data for each variable are listed in Table 11.2, and illustrated in Fig. 11.3. Results for the statistical comparisons of early *Homo* with the extant taxa are presented in Table 11.4. The MANOVA showed significant variation in the overall model, and individual ANOVAs indicated that differences among taxa exist for all variables examined except for *Smc*.

Early *Homo* specimens are on average intermediate in microwear texture complexity, both in terms of central tendency and dispersion. Early hominin *Asfc* values are significantly higher than those of *Alouatta* and *Trachypithecus* according to Tukey's HSD tests and lower than those of *Cebus* according to Fisher's LSD test results. Early *Homo* specimens have fairly low anisotropy values on average, comparable to those of *L. albigena* and *C. apella* and significantly lower than those of *Alouatta*. Early *Homo* has variable but intermediate textural fill volume values, suggesting features of moderate size on average. The early hominin *Tfv* values are significantly higher than those of *Alouatta*, and significantly lower than those of *Lophocebus*. Early *Homo* has somewhat lower values for heterogeneity (at least at the scale considered here) than do the other taxa, suggesting relatively homogenous microwear surfaces. The early hominin $Hasfc_{9\ cells}$ values are significantly lower than those for *Cebus* and are marginally lower than those for *Alouatta*.

Results for the statistical comparisons of early *Homo* groups with each other are presented in Table 11.5. Descriptive statistics for *Homo erectus* and *Homo habilis* are shown in Table 11.2. The MANOVA again showed significant variation in the overall model, but this time, only *Smc* and $Hasfc_{9\ cells}$ ANOVAs showed significant variation. Significance levels for these two variables are hardly impressive, as might be expected given the small samples available for each group. The only significant difference found for *Smc* was that *H. habilis* had higher values than did *H. erectus*. This suggests somewhat larger, deeper features on average

Table 11.4 Comparisons of early *Homo* with extant baseline taxa

A. MANOVA results

	Statistic	*F*	*df*	*P*
Wilks' Lambda	0.221	5.754	20.199	0.00
Pillai Trace	1.075	4.63	20.252	0.00
Hotelling-Lawley	2.329	6.14	20.234	0.00

B. Univariate ANOVA results

Effect	*df*	*F*	*p*
Asfc	4.64	14.946	0.00
Lsar	4.64	2.57	0.05
Smc	4.64	1.832	0.13
Tfv	4.64	9.502	0.00
$HAsfc_9$	4.64	3.562	0.01

C. Pairwise mean differences between early *Homo* and extant taxa

	Asfc	Lsar	Tfv	$HAsfc_9$
Alouatta	29.035**	−21.157**	20.376**	−15.449*
Cebus	−12.440*	−2.38	−11.498	−24.107**
Lophocebus	1.589	3.544	17.206**	6.522
Trachypithecus	−17.278**	11.611	0.847	13.222

*Fisher's LSD $p \leq 0.05$, ** Tukey's HSD $p \leq 0.05$

Table 11.5 Comparisons of early *Homo* groups to each other

A. MANOVA results

	Statistic	*F*	*Df*	*p*
Wilks' Lambda	0.044	3.938	15.28	0.00
Pillai Trace	1.693	3.108	15.36	0.00
Hotelling-Lawley	4.173	4.173	15.26	0.00

B. Univariate ANOVA results

Effect	*Df*	*F*	*P*
Asfc	3.14	0.534	0.67
Lsar	3.14	0.807	0.51
Smc	3.14	3.258	0.05
Tfv	3.14	2.19	0.14
HAsfc$_9$	3.14	3.564	0.04

C. Pairwise mean differences between early *Homo* taxa

1. Smc

	H. erectus	*H. habilis*	Sterkfontein 5C
H. habilis	7.800**		
Sterkfontein M.5C	4.500	−3.300	
Swarktrans M.1	5.000	−2.800	0.500

2. HAsfc$_9$

	H. erectus	*H. habilis*	Sterkfontein 5C
H. habilis		−2.35	
Sterkfontein M.5C	−5.75	−3.4	
SwarktransM.1	6.25	8.6	12.000**

** Tukey's HSD $p \leq 0.05$

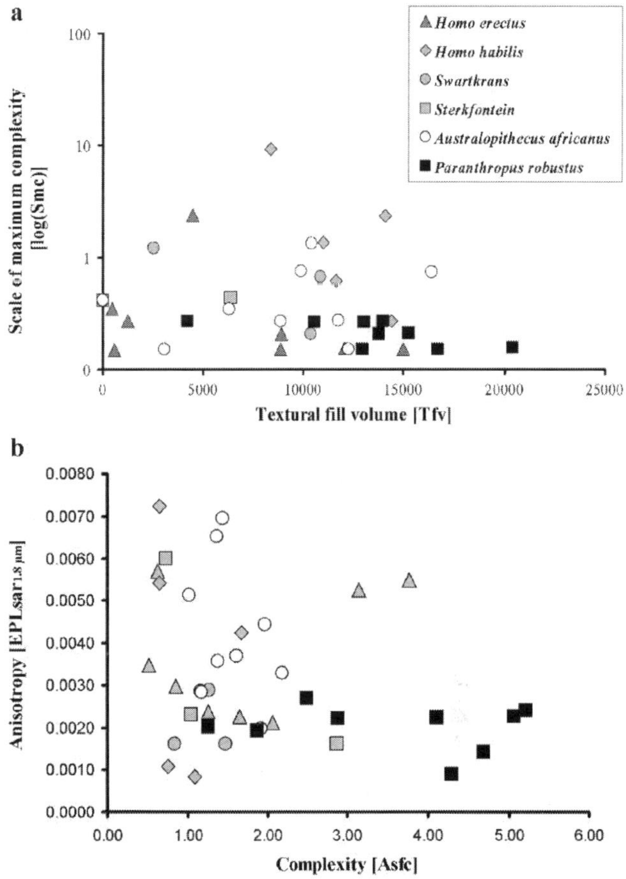

Fig. 11.5 Bivariate plots of (**a**) *Smc* and *Tfv* and (**b**) *Asfc* and *epLsar*$_{1.8\mu m}$ including all four early *Homo* groups, *Paranthropus robustus*, and *Australopithecus africanus*.

for the former. The other difference was that Sterkfontein *Homo* specimens both had much lower *Hasfc*$_9$ values than any of the Swarktrans individuals. Given the sample sizes of two and three individuals respectively, we will not pretend that this actually means anything, but we note it as something to keep in mind for the future as more specimens are recovered and their attributions become clearer.

It may also be that other multivariate analyses will allow us to better separate the early *Homo* groups by their dental microwear textures. For example, *H. habilis* and *H. erectus* separate completely in bivariate space when *Smc* and *Tfv* are considered together (Fig. 11.5a). Further, characteristics of the distributions of microwear texture values for a sample may ultimately be of equal or greater value for reconstructing dietary behaviors than are measures of central tendency. Some samples seem to separate best at the tails of their distributions. The descriptive statistics reported in Table 11.2 are presented with this in mind.

Discussion

Compared with the extant baseline series, the combined early *Homo* sample tends to have moderate levels of microwear surface complexity and feature sizes (judging from *Tfv* values). Early *Homo* specimens also tend to have low levels of microwear anisotropy or feature orientation alignment, and low variability in feature complexity across the surface when compared with the extant taxa. Further, while parsing the early *Homo* specimens into groups makes for abysmally small sample sizes, the texture data do hint at some differences between these early hominins. Both the *Smc* results and the plot of *Smc* and *Tfv* suggest textures of larger, deeper features for *H. habilis* than for *H. erectus*. The lack of a significant *Asfc* difference between the two suggests no substantive differences in complexity measured as scale-related change in surface roughness.

These findings are quite consistent with results from conventional feature-based microwear analyses (Ungar et al., 2006b). Thus, both studies indicate that early *Homo* as a group had fairly unremarkable microwear patterns. The hominins in these samples evidently did not process mechanically challenging foods on a regular basis shortly before death. Results of both feature-based and texture microwear analyses suggest that they did not prefer to chew extremely tough, or hard and brittle foods.

Moreover, both microwear studies show some variation between the groups of early *Homo*. *Homo habilis* had lower pit percentages but higher scales of maximal complexity than did *H. erectus*. This implies that *H. erectus* had relatively more pitting, but overall smaller features than did *H. habilis*. This is consistent with observations that early *Homo* microwear pits tend not to be very large (Ungar et al., 2006b). Both results suggest that while early *Homo* as a group probably preferred less fracture resistant foods, *H. erectus* individuals studied may have eaten a somewhat higher proportion of tough or hard and brittle foods prior to death than did the *H. habilis* sample.

We can also compare these results to those published by Scott et al. (2005) for *Australopithecus africanus* and *Paranthropus robustus* (Fig. 11.5b). The most notable differences between the two South African australopiths were in *Asfc* and $epLsar_{1.8\mu m}$. *Paranthropus robustus* had a greater range of values and a higher average *Asfc* than did *A. africanus*. The "gracile" australopiths, in contrast, had a higher average and greater range of values for *epLsar*. This is consistent with more *P. robustus* surfaces dominated by pitting, and more *A. africanus* specimens showing shallow, uniformly oriented striations (Grine, 1986).

Microwear texture variable distributions for early *Homo* are fairly scattered in $Asfc\text{-}epLsar_{1.8\mu m}$ bivariate space compared with the australopiths. The $epLsar_{1.8\mu m}$ values for early *Homo* are greater on average than those for *P. robustus*, with a range closer too, but even exceeding that of *A. africanus*. Early *Homo* specimens tend to have lower *Asfc* values on average than *Paranthropus*, but their range is greater than that of *A. africanus*. Early *Homo* specimens are also scattered in *Tfv-Smc* space, though *H. habilis* and *H. erectus* specimens separate nicely from one another (Fig. 11.5a). Still, the combined early *Homo* sample and the australopiths overlap greatly in *Tfv-Smc* space.

These comparisons suggest to us that early *Homo* had a fairly variable diet in at least some material properties. Early *Homo* has greater ranges of complexity than *A. africanus* and greater ranges of scales of maximum complexity than *P. robustus*. This should come as no surprise, especially given that the *Homo* sample represents at least two and perhaps more species from sites separated by thousands of kilometers. On the other hand, the early *Homo* ranges for *Asfc* and *Tfv* do not extend to the *P. robustus* extreme. This may suggest that early *Homo* consumed foods with extremely challenging fracture properties less often.

Conclusions

With all this work in progress, what, if anything, can we say at this point? First, both the occlusal topography and microwear texture data vitiate notions that the origin and early evolution of the genus *Homo* were marked by major shifts towards specialization for mechanically challenging preferred foods. It is reasonable to assume that the spread of C_4 grasslands across parts of Africa during the late Pliocene made savanna resources, such as underground storage organs or large mammal prey, more prevalent. We can also assume that an expanded toolkit including Oldowan technology facilitated inclusion of these savanna resources in the diets of early *Homo* species.

On the other hand, both the microwear and occlusal morphology evidence are more consistent with dietary versatility than with specialization. It is most likely that early *Homo* continued to prefer high energy yield foods with less challenging fracture properties, such as soft, sugary fruits when these were available – as has been observed for living gorillas (see Wrangham, 2005 for review). It may be though, that early *Homo* dental-dietary adaptations reflect fallback on less preferred, difficult to obtain, or difficult to fracture, savanna resources. Perhaps this explains why early *Homo* specimens show more sloping occlusal surfaces than do their australopith predecessors.

We expect that more work on microwear and occlusal morphology integrated with a better understanding of food fracture mechanics, will provide additional insights. A larger baseline series including terrestrial primate species and an understanding of the fracture properties of and microwear patterns left by specific food items suggested to have been eaten by early *Homo* will undoubtedly be of value. Other lines of evidence, such as enamel thickness and mandibular corpus shape may allow us to better distinguish the feeding adaptations of these taxa (see Ungar et al., 2006a for discussion). Studies of stable isotopes and trace elements also hold the potential to help us to better reconstruct the diets of these hominins. Consilience of all of these lines of evidence promise even more robust models and an even better understanding of the role diet played in the origin and early evolution of our genus (Ungar, 2007c).

Acknowledgments We thank Richard Leakey, John Fleagle and Fred Grine for their invitation to participate in this volume and the workshop that led to it. We are especially grateful to Fred Grine, Mark Teaford, and Alejandro Pérez-Pérez for their help collecting impressions of specimens used in this study. We also thank Francis M'Kirera and Sarah Taylor for their assistance collecting dental topography data for the extant primate baseline series. We acknowledge the curators at the National Museums of Ethiopia, Kenya and Tanzania, the Transvaal Museum and the University of the Witwatersrand for allowing us to study the specimens in their care, and the anonymous reviewers for their helpful comments. PSU also thanks Fred Grine, Richard Kay, Mark Teaford and Alan Walker for their collegiality and for the keen insights on the evolution of human diets they have shared over the years. The background for this paper is expanded from a recent review by Ungar and coauthors (2006a) with some text reprinted with permission from Annual Review of Anthropology, Volume 35 © 2006 by Annual Reviews, www.annualreviews.org. This work was funded in part by grants from the US National Science Foundation and the LSB Leakey Foundation.

References

Aiello, L.C., Wheeler, P., 1995. The expensive-tissue hypothesis: the brain and the digestive system in human and primate evolution. Current Anthropology 36, 199–221.

Asfaw, B., Beyene, Y., Suwa, G., Walter, R.C., White, T.D., WoldeGabriel, G., Yemane, T., 1992. The earliest Acheulean from Konso-Gardula. Nature 360, 732–735.

Behrensmeyer, A.K., Todd, N.E., Potts, R., 1997. Paleoecological implications of faunas associated with hominids in the Turkana Basin, Kenya. Journal of Human Evolution 32, A2–A3.

Blumenschine, R.J., 1995. Percussion marks, tooth marks, and experimental determinations of the timing of hominid and carnivore access to long nones at FLK *Zinjanthropus*, Olduvai Gorge, Tanzania. Journal of Human Evolution 29, 21–51.

Blumenschine, R.J., Masao, F.T., 1991. Living sites at Olduvai Gorge, Tanzania: preliminary landscape archaeology results in the basal Bed I lake margin zone. Journal of Human Evolution 21, 451–462.

Bunn, H. T., 2001. Hunting, power scavenging, and butchering by Hadza foragers and by Plio- Pleistocene *Homo*. In: Stanford, C. B., Bunn, H. T. (Eds), Meat-Eating and Human Evolution. Oxford University Press, Oxford, pp. 199–218.

Cerling, T.E., 1992. Development of grasslands and savannas in East Africa during the Neogene. Global Planetary Change 97, 241–247.

Cook, R.J., Farewell, V.T., 1996. Multiplicity considerations in the design and analysis of clinical trials. Journal of the Royal Statistical Society, Series A, 159, 93–110.

Coursey, D.G., 1973. Hominid evolution and hypogeous plant foods. Man 8, 634–635.

Daegling, D.J., Grine, F.E., 1999. Terrestrial foraging and dental microwear in *Papio ursinus*. Primates 40, 559–572.

Dart, R.A., 1953. The predatory transition from ape to man. International Anthropology and Linguistics Review 1, 201–217.

de Heinzelin, J., Clark, J.D., White, T., Hart, W., Renne, P., WoldeGabriel, G., Beyene, Y., Vrba, E., 1999. Environment and behavior of 2.5-million-year-old Bouri hominids. Science 284, 625–629.

Dennell, R.W., Rendell, H., Hailwood, E., 1988. Early tool making in Asia: two million year old artifacts in Pakistan. Antiquity 62, 98–106.

Dennis, J.C., Ungar, P.S., Teaford, M.F., Glander, K.E., 2004. Dental topography and molar wear in *Alouatta palliata* from Costa Rica. American Journal of Physical Anthropology 125, 152–161.

Eaton, S. B., Eaton S.B., Cordain, L., 2002. Evolution, diet and health. In: Ungar, P.S., Teaford, M.F. (Eds), Human Diet: its Origin and Evolution. Bergen and Garvey, Westport, CT, pp. 7–18.

Gabunia, L., Vekua, A., 1995. A Plio-Pleistocene hominid from Dmanisi, East Georgia, Caucasus. Nature 373, 509–512.

Grine, F.E., 1986. Dental evidence for dietary differences in *Australopithecus* and *Paranthropus:* a quantitative analysis of permanent molar microwear. Journal of Human Evolution 15, 783–822.

Grine, F.E., 2005. Early *Homo* at Swartkrans, South Africa: a review of the evidence and an evaluation of recently proposed morphs. South African Journal of Science 101, 43–52.

Grine, F.E., Ungar, P.S., Teaford, M.F., 2002. Error rates in dental microwear quantification using scanning electron microscopy. Scanning 24, 144–153.

Hatley, T., Kappelman, J., 1980. Bears, pigs, and Plio-Pleistocene hominids: a case for the exploitation of belowground food resources. Human Ecology 8, 371–387.

Hayden, B., 1981. Subsistence and ecological adaptations of modern hunter/gathers. In: Harding, R.S.O., Teleki, G. (Eds), Omnivorous Primates: Gathering and Hunting in Human Evolution. Columbia University Press, New York, pp. 344–421.

Isaac, G., 1971. Diet of early man: aspects of archaeological evidence from lower and middle Pleistocene sites in Africa. World Archaeology 2, 278–299.

Kay, R. F., 1984. On the use of anatomical features to infer foraging behavior in extinct primates. In: Rodman, P.S., Cant, J.G.H. (Eds), Adaptations for Foraging in Nonhuman Primates: Contributions to an Organismal Biology of Prosimians, Monkeys and Apes. Columbia University Press, New York, pp. 21–53.

Keeley, L.H., Toth, N., 1981. Microwear polishes on early stone tools from Koobi Fora, Kenya. Nature 293, 464–465.

King, S.J., Arrigo-Nelson, S.J., Pochron, S.T., Semprebon, G.M., Godfrey, L.R., Wright, P.C., Jernvall, J., 2005. Dental senescence in a long-lived primate links infant survival to rainfall. Proceedings of the National Academy of Sciences USA 102, 16579–16583.

Lambert, J.E., Chapman, C.A., Wrangham, R.W., Conklin-Brittain, N.L., 2004. The hardness of cercopithecine foods: implications for the critical function of enamel thickness in exploiting fallback foods. American Journal of Physical Anthropology 125, 363–368.

Leakey, L.S.B., Tobias, P.V., Napier, J.R., 1964. New species of genus *Homo* from Olduvai Gorge. Nature 202, 7–9.

Lee, R.B., DeVore, I., 1968. Man the Hunter. Aldine, Chicago, IL.

Leonard, W.R., Robertson, M.L., 1992. Nutritional requirements and human evolution - a bioenergetics model. American Journal of Human Biology 4, 179–195.

Leonard, W.R., Robertson, M.L., 1994. Evolutionary perspectives on human nutrition: the influence of brain and body size on diet and metabolism. American Journal of Human Biology 6, 77–88.

Linton, S., 1971. Woman the gatherer: male bias in anthropology. In: Jacobs, S. E. (Ed), Women in Cross-Cultural Perspective: a Preliminary Sourcebook. University of Illinois Press, Urbana, IL, pp. 9–21.

Lucas, P.W., 2004. Dental Functional Morphology: How Teeth Work. Cambridge University Press, New York.

Lucas, P.W., Peters, C.R, 2000. Function of postcanine tooth shape in mammals. In: Teaford, M.F., Smith, M.M., Ferguson, M.W.J. (Eds), Development, Function, and Evolution of Teeth. Cambridge University Press, Cambridge, pp. 282–289.

Mann, A.E., 1972. Hominid and cultural origins. Man 7, 379–386.

Meldrum, D.J., Kay, R.F., 1997. *Nuciruptor rubricae*, a new pitheciin seed predator from the Miocene of Colombia. American Journal of Physical Anthropology 102, 407–427.

Milton, K., 1987. Primate diets and gut morphology: implications for hominid evolution. In: Harris, M., Ross, E.B. (Eds), Food and Evolution: Toward a Theory of Human Food Habits. Temple University Press, Philadelphia, PA, pp. 93–115.

M'Kirera, F., Ungar, P.S., 2003. Occlusal relief changes with molar wear in *Pan troglodytes troglodytes* and *Gorilla gorilla gorilla*. American Journal of Primatology 60, 31–41.

O'Connell, J.F., Hawkes, K., Jones, N.G.B., 1999. Grandmothering and the evolution of *Homo erectus*. Journal of Human Evolution 36, 461–485.

Panger, M.A., Brooks, A.S., Richmond, B.G., Wood, B., 2002. Older than the Oldowan? Rethinking the emergence of hominin tool use. Evolutionary Anthropology 11, 235–245.

Potts, R., 1983. Foraging for faunal resources by early hominids. In: Clutton-Brock, J., Grigson, C. (Eds), Animals and Archaeology, Volume 1. British Archaeological Reports, London, pp. 51–62.

Potts, R., 1998. Variability selection in hominid evolution. Evolutionary Anthropology 7, 81–96.

Roche, H., Blumenschine, R.J., Shea, J., 2009. Origins and adaptations of early *Homo*: what archeology tells us. In: Grine, F.E., Fleagle, J.G., Leakey, R.E. (Eds), The First Humans: Origin and Early Evolution of the genus *Homo*. Springer, Dordrecht, pp. 135–147.

Rogers, M.J., Harris, J.W.K., Feibel, C.S., 1994. Changing patterns of land use by Plio-Pleistocene hominids in the Lake Turkana Basin. Journal of Human Evolution 27, 139–158.

Schick, K. D., Toth, N., 1993. Making Silent Stones Speak. Orion Books, Phoenix.

Scott, E.C., 1979. Dental wear scoring technique. American Journal of Physical Anthropology 51, 213–217.

Scott, R.S., Ungar, P.S., Bergstrom, T.S., Brown, C.A., Grine, F.E., Teaford, M.F., Walker, A., 2005. Dental microwear texture analysis reflects diets of living primates and fossil hominins. Nature 436, 693–695.

Scott, R.S., Ungar, P.S., Bergstrom, T.S., Brown, C.A., Childs, B.E., Teaford, M.F., Walker, A., 2006. Dental microwear texture analysis: technical considerations. Journal of Human Evolution 51, 339–349.

Semaw, S., Renne, P., Harris, J.W.K., Feibel, C.S., Bernor, R.L., Fesseha, N., Mowbray, K., 1–2 3–1997. 2.5-million-year-old stone tools from Gona, Ethiopia. Nature 385, 333–336.

Shipman, P., 1983. Early hominid lifestyle: Hunting and gathering or foraging and scavenging? In: Clutton-Brock, J., Grigson, C. (Eds), Animals and Archeology, Volume 1. British Archeological Reports, London, pp. 31–50.

Spears, I.R., Crompton, R.H., 1996. The mechanical significance of the occlusal geometry of great ape molars in food breakdown. Journal of Human Evolution 31, 517–535.

Speth, J.D., 1989. Early hominid hunting and scavenging: the role of meat as an energy source. Journal of Human Evolution 18, 329–343.

Stanford, C. B., 2001. A comparison of social meat-foraging by chimpanzees and human foragers. In: Stanford, C. B., Bunn, H. T. (Eds), Meat-Eating and Human Evolution. Oxford University Press, Oxford, pp. 122–140.

Strait, D.S., Grine, F.E., 2001. The systematics of *Australopithecus garhi*. Ludus Vitalis 9, 109– 135.

Teaford, M.F., 1988. A review of dental microwear and diet in modern mammals. Scanning Microscopy 2, 1149–1166.

Teaford, M.F., Runestad, J.A., 1992. Dental microwear and diet in Venezuelan primates. American Journal of Physical Anthropology 88, 347–364.

Teaford, M.F., Ungar, P.S., Grine, F.E., 2002. Paleontological evidence for the diets of African Plio-Pleistocene hominins with special reference to early *Homo*. In: Ungar, P.S., Teaford, M.F. (Eds), Human Diet: its Origin and Evolution. Bergin and Garvey, Westport, CT, pp. 143–166.

Ungar, P.S., 2004. Dental topography and diets of *Australopithecus afarensis* and early *Homo*. Journal of Human Evolution 46, 605–622.

Ungar, P.S., 2005. Dental evidence for the diets of fossil primates from Rudabánya, northeastern Hungary with comments on extant primate analogs and "noncompetitive" sympatry. Paleontographica Italica 90, 97–111.

Ungar, P.S., 2007a. Dental functional morphology: the known, the unknown and the unknowable. In: Ungar, P.S. (Ed), Early Hominin Diets: The Known, the Unknown and the Unknowable. Oxford University Press, Oxford, pp. 39–55.

Ungar, P.S., 2007b. Dental topography and human evolution: with comments on the diets of *Australopithecus africanus* and *Paranthropus robustus*. In: Bailey, S., Hublin, J.J. (Eds), Dental Perspectives on Human Evolution: State of the Art Research in Dental Anthropology. Springer, New York, pp. 321–344.

Ungar, P.S., 2007c. Limits to knowledge on the evolution of hominin diet. In: Ungar, P.S. (Ed), Early Hominin Diets: the Known, the Unknown and the Unknowable. Oxford University Press, Oxford, pp. 395–408.

Ungar, P.S., M'Kirera, F., 2003. A solution to the worn tooth conundrum in primate functional anatomy. Proceedings of the National Academy of Sciences USA 100, 3874–3877.

Ungar, P.S., Williamson, M., 2000. Exploring the effects of tooth wear on functional morphology: a preliminary study using dental topographic analysis. Paleontologia Electronica 3, 18 pp.

Ungar, P.S., Brown, C.A., Bergstrom, T.S., Walkers, A., 2003. Quantification of dental microwear by tandem scanning confocal microscopy and scale-sensitive fractal analyses. Scanning 25, 185–193.

Ungar, P.S., Grine, F.E., Teaford, M.F., 2006a. Diet in early *Homo*: a review of the evidence and a new model of adaptive versatility. Annual Reviews of Anthropology 35, 209–228.

Ungar, P.S., Grine, F.E., Teaford, M.F., El Zaatari, S., 2006b. Dental microwear and diets of African early *Homo*. Journal of Human Evolution 50, 78–95.

Ungar, P.S., Scott, R.S., Scott, J.R., Teaford, M.F., 2007. Dental microwear analysis: historical perspectives and new approaches. In: Irish, J.D., Nelson, G.C. (Eds), Dental Anthropology. Cambridge University Press, Cambridge, pp. 389–425.

Walker, A., 1984. Mechanisms of honing in the male baboon canine. American Journal of Physical Anthropology 65, 47–60.

Washburn, S. L., 1963. Behavior and human evolution. In: Washburn, S.L. (Ed), Classification and Human Evolution. Aldine, Chicago, IL, pp. 190–203.

Wolpoff, M.H., 1973. Posterior tooth size, body size, and diet in South African gracile australopithecines. American Journal of Physical Anthropology 39, 375–393.

Wood, B.A., 1991. Hominid Cranial Remains. Koobi Fora Research Project, Volume 4. Clarendon Press, Oxford.

Wood, B.A., Strait, D.S., 2004. Patterns of resource use in early *Homo* and *Paranthropus*. Journal of Human Evolution 46, 119–162.

Wood, B.A., Van Noten, F.L., 1986. Preliminary observations on the BK 8518 mandible from Baringo, Kenya. American Journal of Physical Anthropology 69, 117–127.

Wrangham, R.W., 2005. The Delta Hypothesis: hominoid ecology and hominin origins. In: Lieberman, D.E, Smith, R.J., Kelley, J. (Eds), Interpreting the Past: Essays on Human, Primate and Mammal Evolution in Honor of David Pilbeam. Brill Academic Publishers, Boston, MA, pp. 231–242.

Wrangham, R.W., Jones, J.H., Laden, G., Pilbeam, D., Conklin-Brittain, N., 1999. The raw and the stolen: cooking and the ecology of human origins. Current Anthropology 40, 567–594.

Zihlman, A. and Tanner, N. M., 1978. Gathering and hominid adaptation. In: Tiger, L., Fowler, H.T. (Eds), Female Hierarchies. Beresford Food Service, Chicago, IL, pp. 163–194.

Chapter 12
Origins and Adaptations of Early *Homo*: What Archeology Tells Us

Hélène Roche, Robert J. Blumenschine and John J. Shea

Keywords Oldowan • Acheulean • lithic technology • flake production • hammerstones • carnivory • scavenging • butchery • preservation bias

Introduction

Brain enlargement, reduction in molar tooth size, increased stature and other features of early *Homo* did not evolve in a vacuum. These evolutionary changes reflect shifts in a complex web of relationships among their populations, between early *Homo* and other hominin species, and between their biotic community and abiotic forces (i.e., climate change). Archeological evidence complements and balances inferences from hominin fossil remains, non-hominin vertebrate paleontology, geology, and other component fields of paleoanthropology. This paper represents an attempt to pull together the various strands of its authors' expertise to shed light on the origins and adaptations of early *Homo*. It is not intended to be a comprehensive review of Oldowan sites, their chronology, lithic typology, paleontological associations, and interpretive issues. For recent overviews of these subjects, see Plummer (2004), Schick and Toth (2006) as well as papers in Toth and Schick (2006), Ungar (2007) and Hovers and Braun (2009). The coincidence of knapped stone tools, butchery-marked bones and fossil remains of early *Homo* is usually linked to increased hominin carnivory. This paper reviews evidence for this hypothesis, and considers alternative hypotheses as well.

H. Roche (✉)
Directeur de recherches au CNRS, UMR 7055 du CNRS, Université Paris X – Nanterre, Maison de l'Archéologie et de l'Ethnologie, 21 allée de l'Université, 92023 – Nanterre, France
e-mail: helene.roche@mae.u-paris10.fr

R.J. Blumenschine
Center for Human Evolutionary Studies, Department of Anthropology, Rutgers University, The State University of New Jersey, 131 George Street, New Brunswick, NJ 08901-1414, USA
e-mail: rjb@rci.rutgers.edu

J.J. Shea
Department of Anthropology and Turkana Basin Institute, Stony Brook University, Stony Brook, NY 11794–4364, USA
e-mail: John.Shea@sunysb.edu

The Nature of the Earliest Oldowan

The scope of this paper is the Oldowan in Africa during the Late Pliocene and Early Pleistocene, roughly 2.7–1.6 Ma. For convenience and clarity, we distinguish this as the "Earliest Oldowan." Since it was first proposed by Leakey (1936), the term Oldowan has gradually eclipsed "Chellean," "Pre-Acheulean" and similar local terms for Early and Middle Pleistocene Eurasian stone tool industries that lack evidence for Acheulean bifacial technology. Although the adoption of this name implies a comparable antiquity to the earliest Oldowan contexts in East Africa, many of these Eurasian "Oldowan" industries have turned out to be much younger. The relationship, if any, between these younger Eurasian "Oldowan" assemblages and their Plio-Pleistocene counterparts in Africa lies beyond the scope of this paper.

The defining technological feature of the Oldowan Industry is the production of flakes from pebbles, cobbles, and angular rock fragments by hard hammer percussion, or knapping. The earliest evidence of stone knapping appears as early as 2.6 Ma in East Africa (Semaw et al., 1997, 2003). The end of the Oldowan is less clear. For the purposes of this paper, we shall fix it at ca. 1.6–1.7 Ma, prior to the widespread appearance of Acheulean assemblages throughout Africa. Lasting approximately 1 million years, the Oldowan is not the static entity as which it is sometimes portrayed (Semaw, 2000).

The Earliest Oldowan Occurrences

The known spatial and temporal distribution of earliest Oldowan sites suggests that the initial stages of stone knapping were isolated and sporadic. Whether the gaps among these early sites are real, or artifacts of preservation and discovery remains unclear. Insufficient field research in appropriate deposits (which are rare) may play a major role in the sporadic aspect of the data. So far, the 2.6–2.5 Ma evidence is limited to four sites located at Gona in the Hadar Formation (Upper Awash, Ethiopia) (Roche and Tiercelin, 1980; Harris, 1983; Semaw et al., 1997, 2003). There is a great potential in the Gona area where more sites have been identified, and it is likely that still more sites – including older ones – may be found in the future. At 2.5 Ma, within the Hata Member of the Bouri Formation, bone fragments reported to bear cut marks and percussion marks have been found in the same general area as fossil remains of *Australopithecus garhi* (Asfaw et al., 1999; de Heinzelin et al., 1999). No stone tools were found with these bones, and they are not in close association with hominin remains. Thus, their bearing on stone-tool production by *A. garhi* remains unresolved.

More sites are found between 2.4 and 2.3 Ma. In Hadar, there are two excavated sites, A.L. 666 and A.L. 184, (Kimbel et al., 1996; Hovers et al., 2002). De la Torre (2004) has reassessed the lithic assemblages from the Lower Omo Valley Shungura Formation Members E and F that were excavated in the 1970s (Howell et al., 1987). Only three sites qualified with good context and intentionally knapped lithic material (Omo 57, Omo 123 and FtJi 2), as well as two channel sites with unabraded stone tools recovered in situ (FtJi1 and FtJi5). All of these sites are from Member F, and are thus bracketed between 2.34 and 2.32 Ma. In the Nachukui Formation, in the western part of the Turkana Basin, the two Late Pliocene sites of Lokalalei 1 and Lokalalei 2C are dated at 2.3 Ma (Kibunjia et al., 1992; Kibunjia, 1994; Roche et al., 1999). Less than ten excavated archeological sites are thus recorded for the 2.4–2.3 Ma time period. Unlike the earlier sites, these are often characterized by association or at least spatial proximity between hominin remains and stone tools concentrations: *Homo* aff. *H. habilis* at site A.L. 666 in Hadar (Kimbel et al., 1996), and early *Homo* at Lokalalei 1 in West Turkana (Prat et al., 2005). There is presence but no proximity of *Paranthropus boisei/aethiopicus* for the Omo sites in the Shungura Formation (Member G, 2.32–1.9 Ma) and for Lokalalei sites in West Turkana (KNM-WT 17000, Lokalalei Member, 2.5 Ma) (Walker et al., 1986).

What is the nature of and basis for variation among the earliest Oldowan assemblages? Variation among Oldowan knapping sequences is thought to reflect cognitive and motor abilities, raw materials, and other situational variables. Oldowan knapping skills appear somewhat limited by lithic raw material properties in the earlier assemblages. Later Oldowan assemblages exhibit greater freedom from raw material constraints. This is illustrated, for instance, when knappers become able to rectify a knapping accident and thus to go on with the knapping process, showing a real improvement in decision making. To be able to create new striking platforms and to be less constrained by the shape of the original raw material is another example. This suggests the development of an ability to create morphology that is totally independent of the shape of the clast; an ability seen later, and somewhat more clearly, for shaped Acheulean handaxes.

The Leakeys (Leakey, 1936, 1971) originally treated the Oldowan as an undifferentiated entity, whose stability contrasted with both the Developed Oldowan and Acheulean. Following discoveries of very early occurrences of stone knapping (Roche and Tiercelin, 1977; Harris, 1983; Kibunjia et al., 1992), several researchers distinguished a break (ca. 2.0 Ma) between a "Pre-Oldowan" and an Oldowan based on technological patterns (Roche, 1989; Kibunjia, 1994; Roche and Kibunjia, 1996). Discoveries made in the mid 1990s, mainly in Gona (Semaw et al., 1997, 2003), Hadar (Kimbel et al., 1996) and West Turkana (Roche et al., 1999) challenged this distinction, and few researchers today retain it. At the very most, an "Early Oldowan" (prior to 2 Ma) could be distinguished. However, the Oldowan lithic assemblages are not homogenous, whether one considers the raw material procurement strategies, the operational sequences (*chaînes opératoires*), or the by-products that are obtained. The Oldowan is not only marked by the emergence of a behavior which put into action skills never used by other animal species, it is also characterized by a wide range of techno-economic patterns, some of them being more productively efficient than others. There is no single explanation for this variability. Functional adaptation for food acquisition, response to environmental constraints (including raw material availability), and variations in inter- or intra-specific cognitive and motor abilities are all viable and non-exclusive hypotheses for Oldowan technological variability.

The appearance of the Oldowan also marks a drastic niche shift within the lineage of its makers. This shift involves not only expansion of the diet to include foods from larger mammals (>5 kg body mass), but also an expanded set of relationships with members of the larger carnivore guild they joined. The presence of stone tool cut marks (e.g., Bunn, 1981; Potts and Shipman, 1981) and percussion marks (e.g., Blumenschine and Selvaggio, 1988; Blumenschine, 1995) on bones of animals ranging in size up to elephants demonstrates that Oldowan hominins had become the first primate to consume meat, marrow and perhaps other edible tissues from animals greater than or equal to their own body weight. Cut-marked bones associated with stone artifacts from Gona, Ethiopia (Domínguez-Rodrigo et al., 2005) suggest that this dietary shift was coincident with the earliest stone artifacts at 2.6 Ma (Semaw et al., 1997). This dietary niche expansion placed Oldowan hominins as the only predominantly plant-eating members of the large carnivore guild of the East African Late Pliocene. This large carnivore guild included a richer range of medium to large-bodied felids, hyaenids, canids and crocodilians than are extant today (Werdelin and Lewis, 2005). While the Oldowan hominins likely remained prey for all of these carnivores, they also became direct or indirect food competitors for some of the carnivores. The nature of these intra-guild interactions is central to understanding the behavioral, technological and social capabilities of Oldowan hominins. Historically, these interactions have been conceptualized as the hunting vs. scavenging debate, and investigated in terms of co-occurrences of butchery marks and tooth marks in bone assemblages (and sometimes on the same bone). There is a growing recognition that framing such ecological relationships in terms of a simple dichotomy is an oversimplification that does justice neither to the evidence, nor to the behavioral variability of the species under consideration (cf. Washburn and Lancaster, 1968; Isaac, 1978; Shipman and Walker, 1989; Aiello and Wheeler, 1995; Foley, 2001; Blumenschine and Pobiner, 2007; Bunn, 2007).

Nature of the Oldowan-Acheulean Transition

Approximately 1.7 Ma can be considered the transition interval between the Oldowan and the Acheulean phases of techno-cultural evolution. This period is marked by the appearance of a new hominin species (*Homo ergaster/erectus*). In stone knapping, this period witnesses the beginnings of purposeful bifacial shaping (*façonnage*) of large core tools. A third major aspect of this transitional period is the apparent increase in hominin population size, as shown by the extension of the geographic range of *Homo ergaster/erectus* beyond Africa, and by the increase in the number of identifiable Acheulean archaeological sites throughout Africa. This population growth seems to start *circa* 1.8 Ma in East Africa, at least in certain favorable environments. The colonization of new lands and territories is another noteworthy fact of the post-Oldowan period. Because hominins fossils are rare or absent from these contexts, this expansion of the geographic range of *Homo ergaster/erectus* within Africa and beyond is traced primarily through the archaeological record.

After 2.0–1.9 Ma, the whole of the East African Rift Valley area is populated by greater numbers of archeological sites than was seen earlier. Similar increases in archeological "visibility" are apparent in South Africa (Kuman, 1994, 2003) and North Africa (Sahnouni, 2005, 2006). Beyond Africa, the two most secure occurrences of early hominins are Dmanisi in Georgia (Gabunia et al., 2001; Lordkipanidze et al., 2006) and Ubeidiya in Israel (Bar-Yosef and Goren-Inbar, 1993; Guérin et al., 2003). For recent overviews of other claimed Plio-Pleistocene sites in Eurasia, see Antón and Swisher (2004), Langbroek (2004), Dennell and Roebroeks (2005), and Wang et al. (2007).

Lithic Technology

Technical behavior is documented through analysis of the lithic component of the hominin's technical system. Technological analysis takes us back to the cognitive abilities and motor skills of early hominins, from the individual performance of a single stone-knapper to the level of competence of a group at a given time. Studies of raw material sources and procurement/distribution also inform us about interactions with environment and land use (Blumenschine et al., 2008). Stone knapping is the earliest known technical behavior that strongly contrasts with technical behaviors of nonhuman primates observed in the wild (Roche, 2005). The ability to intentionally modify a natural homogenous, isotropic, hard and rigid mineral material (e.g., hard rock) is a technical behavior only shared by humans and their congeneric ancestors. It can be contrasted to the faculty of using unmodified mineral or organic natural elements, or to the faculty of altering natural soft or plastic organic material, both of which are shared by a number of animal species. Hundreds of well-known examples of stone-using by chimpanzees in the wild have been documented, including archeological evidence (Boesch and Boesch, 1984; McGrew, 1992; Mercader et al., 2002). A case of hunting with tools by *Pan troglodytes verus* confirms the ability of chimpanzees to transform soft vegetal material (Pruetz and Bertolani, 2007). Stone knapping by any animal species other than humans or their ancestrally-related forms has never been reported. Moreover, there is a fundamental shift in technical behavior between using natural stone tools and processing soft material, on the one hand, and knapping hard rocks on the other. Beyond the remarkable cognitive and motor abilities that stone knapping implies, it also results in drastic changes in the hominin's relation with their external milieu. From handling natural elements (stones, leaves, branches, etc.) to having direct access to nutrients or for defense, as is the case with the rest of tool using animal species, we shift to a technical system that implies an increasing number of steps (each step consisting in a chain of actions, underpinned by decision-making), the second step being a consequence of the first and allowing the third, and so on, until the anticipated goal is achieved.

From the very beginning, the aim of stone knapping was to create, and to infinitely reproduce a physical function which is absent, or extremely rare, in the natural world: the cutting function. This is obtained by collecting raw material (first step), extracting flakes bearing sharp cutting edges from a block of raw material (the second step), which in itself constitutes most of the innovation of this new behavior. Flake production is made with a natural intermediary tool, a stone hammer. These objects will, in turn (third step), act on the external milieu by cutting organic soft materials. Used on animal tissues, they give direct access to nutrients (the case of breaking bones is treated below). Used to process vegetal components, they can also give direct access to nutrients (edible plants, fruits, etc.), but they can also be used to build shelter with branches and foliage. Lastly, they can be used as intermediary tools to handcraft wooden weaponry (fourth step) or traps, which expand access to food, improve either self or group protection during encounters with predators (which can also involve natural stones), and many other individual or social behaviors (see below).

Generally speaking, Oldowan assemblages are comprised of pebbles and cobbles or blocks of raw material reduced by percussion-controlled conchoidal fracture. The knapping sequences result in flakes (the intended products) and cores (usually the waste), and sometimes chopper-cores, i.e. objects which could either be flake-producing cores and/or heavy-duty cutting tools. Unworked cobbles are very often present at sites, sometimes in large amounts, and are considered as part of the lithic assemblage and as hammerstones. Hammerstones and any tools that could have been used for percussive activities, such as breaking bones (see below) indisputably play an important role in hominin activity (Mora and de la Torre, 2005). The status of the unmodified material is more subject to debate (de la Torre and Mora, 2005; Delagnes and Roche, 2005). Among the two unique knapping modes (percussion and pressure) used during prehistoric times (Inizan et al., 1999), only percussion was used during the Oldowan. Among the three specific knapping actions (flaking, shaping and retouching), flaking is most prominent. Retouching is found in limited occurrences, and only towards the end of the Oldowan do we find polyhedric shaping (see Fig. 3.1 in Roche, 2005). Among the dozens of techniques invented for modifying stones, only three of them were put to use during the Oldowan: (1) direct percussion with a hand held stone hammer, (2) direct percussion on an anvil (no hammerstone; the block of raw material is held in hand), and (3) bipolar percussion between a hand-held stone hammer and anvil, the raw material to be fractured being held with the second hand (analogous to the technique used to crack a hard-shelled nut). The first technique is by far the one most often employed; it can be precisely controlled and is, therefore, more efficient (Inizan et al., 1999; Pelegrin, 2005; Roche, 2005). It is not yet certain that any of the knapping sequence patterns in evidence for the Oldowan can be qualified as a method (that is, an "orderly sequence of actions carried out according to one or more techniques and guided by a rational plan" (Inizan et al., 1999: 145), such as the "Kombewa method" or the "Levallois method"), but it can be applied to some shaping operational sequences.

When stone knapping becomes visible in the archeological record, this behavior is no longer in its very earliest possible form – a stage that will be extremely difficult, if not impossible, to identify. To the contrary, the earliest record of Oldowan tool-making shows an already well-mastered technical behavior shared by small, possibly isolated groups of hominins. However, the mode, action, techniques and methods show a limited range of technological possibilities compared to what stone knapping would eventually become. What then accounts for the variability in Oldowan technical behavior? Some of the avenues for research on this question include the investigation of raw material procurement strategies and of stone tool production techniques.

Raw Material Procurement Strategies

For the few Early Oldowan sites at 2.7–2.6 and 2.4–2.3 Ma, there is no evidence of long distance transport of raw material from source to the place of use, loss and/or discard. To the contrary, it has been shown that raw materials were collected from immediate-to-local sources less than a few hundreds meters distant (Harmand, 2005; Stout et al., 2005; Goldman and Hovers, 2009). In some instances, comparisons of raw materials at potential sources and the material recovered from archeological sites has shown an obvious selection for particular rock types. Selection criteria appear to have included different rock types and petrographic, structural and granular patterns (Harmand, 2004, 2005, 2009a,b; Stout et al., 2005), as well as morphology and size of the clasts. Systematic knapping experiments confirm the superior knapping properties of the selected *vs.* non-selected raw material (Harmand, 2005; Stout et al., 2005). Rock type selection varies according to the petrographic substratum of each region. All these substrates are dominated by lavas, but of variable quality. Archeological assemblages feature conspicuously high proportions of the rare, high-quality raw materials from the surrounding environment. When it is possible to conduct a diachronic study within a single area, as in West Turkana, which yields archeological sites through a sequence from 2.3 to 0.7 Ma (Roche et al., 2003), the source-to-site distance remains the same, but there is more selection in term of rock type, granularity and size of clasts during later Oldowan (1.8–1.7 Ma) (Harmand, 2005, 2009a). Within the Oldowan, from 2.0 Ma onwards, and with an increasing number of sites, there is evidence for transport of raw materials from sources to places of use over longer distances (up to 15–20 km). This applies, for instance, to Koobi Fora (Isaac et al., 1997), Olduvai (Hay, 1976; Blumenschine et al., 2003, 2008), and Kanjera (Plummer et al., 1999; Braun et al., 2009). Where studied, the same patterns of raw material selectivity are indicated.

Stone Tool Production Techniques

Oldowan stone tool production techniques show a range of different patterns. The more common of these include the following:

- Very limited production of flakes by blow-after-blow randomfl aking of a cobble

- Very limited production of flakes by one to several contiguous or alternating removals on a side of a core, creating a strong cutting edge (chopper-core or core tool)
- More abundant production of flakes with simple and non-organized *débitage* of an ordinary core (i.e., a core for which there was no morphological selection of the original clast)
- Elevated production of flakes with an organized *débitage*, with successive, multiple and orderly series of removals on a specific core (i.e., a core whose form indicates morphological selection of the original clast)

These patterns might be too similar to be differentiated from one another. This is true in terms of general tendency, and when compared with later stages of stone knapping. However, a careful examination of the Oldowan lithic assemblages shows these different knapping sequences vary according to raw material availability and, more importantly, are linked to the level of the abilities implied in the technical actions. Most of the hominin groups chose only one knapping process or, when two or (seldom) more were chosen, one predominates. This is the case in the two late Pliocene sites of Lokalalei in West Turkana, dated at 2.3 Ma.

At Lokalalei 1 (Kibunjia, 1998) flakes were produced by simple and non-organized *débitage* of globular cores, while at Lokalalei 2C (Delagnes and Roche, 2005) an organized *débitage* predominated. Moreover, the simple *débitage* at Lokalalei 1 was inadequately implemented, considering the petrographic organization of the raw material (which, by the way, is of good knapping quality and has been selected for this purpose). The result is low productivity, with a majority of accidentally broken flakes, and with cores bearing many scars of knapping accidents. In contrast, at Lokalalei 2C, controlled *débitage* has been conducted, following constant rules applied to a good quality raw material. Blocks and cobbles were also chosen with peculiar natural angular morphologies (i.e., cobbles with a flat surface as opposed to a convex one), thus providing directly serviceable striking angles (<90°). Sometimes, the flat surface was obtained by splitting a rounded cobble or was formed on the ventral face of large flakes. This flat surface was then exploited as a flaking surface, by means of successive and multidirectional series of invasive and sub-parallel flakes, a practice that maintained the flaking surface flat and allowed the production of large numbers of flakes (Roche et al., 1999; Delagnes and Roche, 2005). This very specific reduction sequence, and the ensuing high productivity have been documented by dynamic reconstruction of entire cobbles on the basis of particularly informative refitting sets (where 12% of a total of 2,614 artifacts may refit to one another). On average, 18 flakes were knapped from each core, and up to 73 flakes were removed from a single block of raw material. The production of such a large number of flakes shows foresight and anticipation while the whole operational sequence is in progress. This cognitive ability goes with a controlled motor skill, shown by the high control of percussion gestures, which can be seen on flakes and cores, and with a very circumscribed and limited area bearing percussion marks on the hammerstones (Delagnes and Roche, 2005).

It has not yet been demonstrated that a similar level of raw material management, anticipation, and manual dexterity exists elsewhere during Oldowan times. At A.L. 666 in Hadar, which is penecontemporaneous to the Lokalalei sites, the same morphotype of clast (a flat surface opposed to a convex one) was flaked, but with much stronger blows such that each flake removed was thick, and the flaking surface was not maintained flat (Hovers, 2001). This stopped the flaking process and curtailed core productivity. At Gona, flaking appears to have been more simple, at least to judge from the published descriptions (Semaw, 2000; Semaw et al., 2003). At Olduvai, production is mainly from core tools, and there are few, if any, examples of real *débitage* (i.e., systematic flake production, as opposed to the purposeful shaping [*façonnage*] involved in chopper-core or core-tool flaking) (Leakey, 1971; Stiles et al., 1974).

Although they are not always easy to decipher, the different knapping sequences allow us to characterize the competence of group or individual performance, and it is through the whole operational sequence that planning and foresight can be evaluated. Several hypotheses can be proposed to explain the diversity of Oldowan technical behaviors. At present, variation in cognitive and motor abilities as implied by the knapping action seems more likely than any other factor. This is not to say that other factors (e.g., raw material, function), lack explanatory power; rather, they are relatively difficult to investigate. In the near absence of use-wear evidence, almost nothing is known about the functionality of the lithic artifacts, apart from the indisputable cutting quality of the sharp edges of the flakes, and the pounding qualities of cobble and core forms. Nevertheless, as discussed in the next section, it is precisely these cutting and pounding aspects of the lithic artifacts that link them meaningfully to the faunal remains with which they are associated in archeological sites.

Larger Mammal Carnivory

Blumenschine and Pobiner (2007) recently reviewed the zooarchaeological evidence for large mammal carnivory in Oldowan hominins, which we summarize below. Although assemblages of larger mammal bones have been reported from almost 20 Oldowan localities (Blumenschine and Pobiner, 2007: Table 10.1), most of what is known about Oldowan hominin carnivory derives from a single site, FLK Level 22 (*Zinjanthropus* level), in Bed I of Olduvai Gorge, dated to approximately 1.8 Ma. (Bunn and Kroll, 1986; Potts, 1988; Oliver, 1994; Blumenschine, 1995; Capaldo, 1997; Selvaggio, 1998; Domínguez-Rodrigo and Barba, 2006; Blumenschine et al., 2007b; Bunn, 2007).

The animal species and tissue types consumed define a basic parameter of the Oldowan hominin carnivorous niche. Butchery marks on bone surfaces provide direct evidence for hominins extracting flesh using sharp-edged stone flakes (cut marks), and marrow using rounded hammerstones and infrequently choppers (percussion marks). To the extent that the butchery-marked bone can be identified taxonomically, traces of hominin feeding provide paleoanthropology with a more specific itemization of dietary elements than is currently possible from other lines of evidence (e.g., tooth morphology and wear, stable isotopes). Blumenschine and Pobiner (2007) compiled comprehensive lists of both published occurrences of butchery-marked bone from Oldowan assemblages, as well as the larger mammalian species (≥5 kg live body weight) in these assemblages that have been reported to be butchery marked. Two remarkable features of Oldowan hominin carnivory highlight the results of these compilations.

First, of the 16 Oldowan localities from which butchery-marking has been reported, only two, FLK Level 22 and FLK N Levels 1–2 from Bed I, Olduvai Gorge, show substantial proportions of cut-marked and/or percussion-marked bone: approximately 9% of analyzed (non-dental) larger mammal specimens from FLK Level 22 are reported to be cut-marked (Bunn and Kroll, 1986), while approximately 27% and 28% of analyzed long bone specimens from FLK Level 22 and FLK N Levels 1–2, respectively, are reported to be percussion-marked (Blumenschine, 1995; Capaldo in Blumenschine et al., 2007a). Fewer than nine butchery-marked specimens, and often only one or two, have been reported from 11 other localities, including the Olduvai assemblages from FLK N Level 6, FLK NN Level 2, HWK E Levels 1–2, and OLAPP Trench 57 (Bunn, 1982; Monahan, 1996; Blumenschine et al., 2003); the West Turkana assemblage from Lokalalei 1A (Kibunjia, 1994); the Gona localities of DAN2, EG13, OGS-6, and WG9 (Semaw et al., 2003; Domínguez-Rodrigo et al., 2005); Bouri (de Heinzelin et al., 1999); and Sterkfontein Member 5 (Pickering, 1999; Pickering et al., 2000). No butchery marks were identified on the poorly preserved bones from the three KBS Industry localities of FxJj1, 3, and 10 (Bunn, 1997; Isaac and Harris, 1997).

In some cases, the low numbers of butchery-marked specimens can be attributed to small assemblage size, poor bone-surface preservation, or incomplete analysis. Alternatively, the paucity of butchery-marking outside of the two abundantly-marked Olduvai assemblages might be a signal that large mammal carnivory was infrequent during the early stages of this adaptation, and/or that localized carcass processing leading to concentrations of butchery-marked specimens took place only in specific landscape settings such as those exposed at FLK Level 22 and FLK N Level 1–2. These landscape settings theoretically afforded hominins a long-lived grove of refuge trees adjacent to places where carcasses could be found regularly (Blumenschine and Peters, 1998). No known method can determine the frequency of large mammal carnivory by hominins. However, one relevant consideration is that the large amounts of flesh, marrow or brain from even small, gazelle-sized carcasses, and the very high rates of nutrient return from processing these tissues with Oldowan stone tools, suggest that available carcass foods should have been taken whenever encountered (for discussion, see Blumenschine and Pobiner, 2007). Nonetheless, the possibility that Oldowan hominin carnivory was uncommon appears to stand in contrast to at least some immediate post-Oldowan assemblages such as those recently described from Okote Member occurrences at East Turkana, where butchery-marking is common (Pobiner, 2007).

A second remarkable feature of Oldowan hominin carnivory is the large body size range and broad ecological range of butchery-marked ungulates that have been identified to the genus or species level (Blumenschine and Pobiner, 2007). Fourteen taxa reported to be butchery-marked encompass the full size range of larger mammals, from the small gazelle *Antidorcas recki* through *Hippopotamus gorgops* and possibly *Elephas recki*, and include seven bovids, two suids, one equid, two giraffids, one hippopotamid, and one elephantid. Among bovids for which femoral ecomorphology was measured by Kappelman et al. (1997), butchery-marked taxa are associated with open to light cover through light to heavy cover. Although they are present in the assemblages, bovids that represent open, heavy cover or forest settings have not been reported to be butchery-marked. Butchery-marked specimens derive from species identified as either grazers or browsers on the basis of carbon stable isotopes (Cerling et al., 1999, 2003; Harris and Cerling, 1999) and jaw ecomorphology (Spencer, 1997). Given that the vast majority of butchery marks occur on specimens identifiable only to the family level or above, the series of butchery-marked taxa was probably broader, including some or all of the 14 other taxa present at the Oldowan localities. Most of the butchery-marked taxa are reported only from FLK Level 22, but it is clear that here, Oldowan hominins at least occasionally fed on taxa that were prey for a full size range of predator/scavengers in the larger carnivore guild of the East African Plio-Pleistocene.

Whether Oldowan hominins acquired food from larger mammals through hunting and/or scavenging can be cast more broadly as the emerging role of hominins in the larger carnivore guild (cf. Blumenschine and Pobiner, 2007). It is during the Oldowan that members of presumably one hominin lineage expanded their range of interactions with large carnivores from sole status as prey, to include indirect and/or direct competition with at least some species for larger mammal carcasses and/or live prey, eventually leading to modern humans' status as the top predators in most ecosystems.

Passive scavenging, or the acquisition of foods from carcasses unattended by large carnivores, is an evolutionarily conservative hypothesis, requiring few derived behavioral and technological capabilities of Oldowan hominins over their direct ancestor. Many forms of passive scavenging have been hypothesized on the basis of observations of scavenging opportunities in modern settings and the inferred habits of extinct carnivores (e.g., Blumenschine, 1987; Cavallo and Blumenschine, 1989; Marean, 1989). Those hypothesized to have provided opportunities that were most frequent, moderate to high-yielding, predictably located, and of low risk from predation, disease or toxicity are scavenging from abandoned lion kills, temporarily or finally abandoned tree-stored leopard kills, and abandoned saber-tooth cat kills. Passive scavenging from abandoned felid kills can account for many aspects of Oldowan bone assemblage composition and condition, including head and limb-dominated skeletal part profiles, most of the extreme body size and ecological ranges of carcass taxa, the frequency and anatomical distribution of cut and percussion marks, and the high frequency of carnivore tooth marking on long bone midshafts from FLK Level 22 (e.g., Blumenschine, 1987, 1995; Marean et al., 1992; Capaldo, 1997; Selvaggio, 1998; Blumenschine and Pobiner, 2007). A recent claim that passive scavenging from felid kills at FLK Level 22 has been falsified (Domínguez-Rodrigo and Barba, 2006) has been invalidated on the basis of serious methodological and conceptual flaws (Blumenschine et al., 2007b).

Passive scavenging, in its most opportunistic form, involving encounters with abandoned carcasses during plant food foraging or other daily activities, requires only that Oldowan hominins carry stone for butchery-tool production and/or transport carcass parts to tool locations. Involving no direct interactions with large carnivores, it would represent conservation of the presumed predator avoidance of ancestral hominins, while allowing for the acquisition of calorie- and protein-rich foods linked to brain expansion and gut reduction (Aiello and Wheeler, 1995), among other developments.

Confrontational scavenging, or kleptoparasitism of kills from feeding large social or solitary predators, denotes more advanced behavioral and technological capabilities for Oldowan hominins. By usurping complete or nearly complete carcasses from carnivores, confrontational scavenging would yield extremely large quantities of food. For example, an adult wildebeest, the size of the animals most commonly represented at the Oldowan sites, bears approximately 70 kg of flesh, implying large hominin group sizes and/or a high proportionate contribution of meat to the diet. The presence of cut marks on "meaty" upper limb bones is often cited in support of confrontational scavenging (e.g., Domínguez-Rodrigo, 1997; Bunn, 2001). However, the relationships between the location and frequency of cut-marking, on the one hand, and the amounts of flesh removed, on the other, are currently uncertain, remaining as one of the most important issues in the early hominin hunting and scavenging debate. Unlike the predator avoidance of passive scavenging, confrontational scavenging implies that Oldowan hominins were dominant members of the larger carnivore guild, presumably achieving this status through coordinated group tactics, such as "power scavenging" (Bunn, 2001) and/or the use of effective offensive weaponry.

Like confrontational scavenging, hunting by Oldowan hominins would yield extremely large quantities of food from complete carcasses. As such, the two types of carcass acquisition would be largely indistinguishable on the basis of skeletal part profiles or patterns of cut-marking. If hunting can be shown to account for the full size and ecological range of prey species at Oldowan sites, top predator status likely involving effective projectile weaponry would be implied for Oldowan hominins, conditions that appear incongruent with their small body size and simple stone technology.

The Other Uses of Oldowan Tools

For what purposes were Oldowan tools used? The two irrefutable answers are, for stone tool production and for butchering animal carcasses (see above). But were Oldowan tools used for other cutting tasks? In trying to answer this question, one must remember that Oldowan tool use involves one of the most problematic and conjectural aspects of stone tool technology at the extreme limits of the archeological record. This record has to be augmented by middle-range theoretical principles derived from actualistic-experimental studies, from ethnoarcheology, and from studies of tool use by non-human species. Lastly, it also has to be understood that preservation bias strongly influences our perceptions of the evolutionary forces that shaped Oldowan stone tool technology.

Percussion-marked hammerstones are most clearly implicated in stone tool production. However, the kind of comminution and crushing that identifies a stone object as having been used to knap stone forms cumulatively. Stones used briefly as percussors may not preserve diagnostic traces of use. The artifact category of manuports may preserve superficially-utilized hammerstones.

We do not know which of the particular categories of sharp-edged Oldowan tools created cut marks on animal bones. Pebble cores, flakes, and retouched flake-tools all work reasonably well as butchery aids in experiments (Jones, 1980, 1994; Toth, 1987, 1997). Stone tools knapped and used for ad hoc butchery by recent mobile human groups show little attention to imposition of design beyond assuring the presence of a sharp cutting edge (e.g., Gould et al., 1971). Thus there is no compelling reason to reject the hypothesis that any or all Oldowan stone tools could have been used as butchery aids by early hominins. Whether the performance differences some experimenters have noted among different classes of replicated Oldowan tools in butchery experiments (Toth et al., 1996; Tactikos, 2005; Shea, 2007) were sufficient to influence the decisions of Oldowan tool-users remains an open question.

The use of stone tools as butchery aids is likely to have generated large numbers of flaked stone artifacts, for several reasons.

First, the energetic costs involved in gathering locally-available rocks and knapping a few flakes to be carried as "personal gear" would have been miniscule compared to the potential windfall energetic gain from meat- and fat-bearing animal carcasses encountered in daily foraging. Stone tools are durable resources that, once knapped, could have persisted on the landscape for decades or more. (Nearly every known ethnographic stone-tool-using human population treats abandoned campsites and known archaeological sites as sources of raw material for their immediate needs.) Assuming that there was some degree of local-scale continuity among Oldowan toolmakers, a strategy of returning unused flakes to central places/residential sites (Isaac, 1978) and/or one involving in-bulk stockpiling of raw materials at strategic points on the landscape (Potts, 1988) would be, in effect, a strategy with direct benefits to the knappers themselves and to their immediate descendants.

Secondly, while butchery is not necessarily a task that involves high rates of edge attrition on stone tools, it is a task that can have low thresholds for tool discard (and correspondingly high rates for tool provisioning). Prolonged and forceful contact between a stone tool and bone causes numerous minute fractures on the tool edge. These fracture scars scoop up fat, meat, and periosteum, lubricating the edge, and reducing its cutting effectiveness. Microtopographic irregularities on the surface of coarse-grained rocks function in the same way, even in the absence of microfracturing damage. The functionality of such a lubricated stone tool edge declines rapidly, requiring that the tool either be resharpened or replaced. Knapping razor-sharp stone whilst one's hands are caked in blood, dirt, hair and grease is never a good idea (JS has the scar tissue to prove it!). A far safer option would have been to simply replace the tool with a fresh one from a previously-knapped supply. It is possible that the vast quantities of knapped stone at Oldowan sites reflect the accurate perception by their makers that butchery episodes require frequent replacements of stone tools.

Lastly, in human hunter-gatherer societies, control over the distribution of meat may confer social status and significant social and reproductive advantages (Kaplan and Hill, 1985; Hill and Kaplan, 1993). The benefits of meat-sharing among chimpanzees are less clear (Mitani and Watts, 2001; Stanford and Bunn, 2001; Gilby, 2006). If, as the forgoing discussion suggests, early hominin carnivory involved regular interactions with larger carcasses than those consumed by chimpanzees, it follows that being able to knap and use stone tools as aids to butchery would have been a skill with positive fitness consequences. Knapping is a learned skill, and it further follows that there would also have been strong selective pressure for hominins to start knapping at an early age and to practice (particularly with unfamiliar materials) throughout the course of their lives. The virtually indestructible lithic byproducts of such practice knapping may form a significant portion of the stone tools at Plio-Pleistocene sites (Shea, 2006).

In textbook accounts of human origins, the inception of knapped stone technology is usually linked to increased hominin carnivory. This limited view of Oldowan tool use ignores contrary arguments in favor of a more functionally diversified early hominin stone tool technology.

First, the edges of ethnographically documented knapped stone tools are used for many other purposes than butchery, including woodworking, preparing leather from animal hides, and processing soft plant matter. Inasmuch as chimpanzees regularly shape tools out of wood and soft plant matter, it seems reasonable to suppose that early hominins would have appreciated the benefits of carving wood with stone tools, rather than with their teeth and fingernails. Experiments suggest that even simple stone tools would have dramatically increased the rate at which hominins would have been able to shape wood into useful subsistence aids, such as digging sticks, spears, clubs, or throwing sticks (Crabtree and Davis, 1968). Such increased work rates would have substantially lowered the costs involved in technologically-assisted foraging, potentially leading in turn to a broadening of early hominin subsistence strategies.

Second, the prominence of the link between stone tools and butchery is plainly a function of preservation bias. Cutmarked bones are more likely to fossilize than wooden implements. Archaeologists have not discovered preserved wooden tools in Plio-Pleistocene contexts, but this may as much reflect the absence of waterlogged contexts of this antiquity in tropical Africa. Middle Pleistocene waterlogged contexts associated with lithic traces of hominin activity, including Gesher Benot Ya'acov (Israel), Kalambo Falls (Zambia), and Schöningen (Germany) have yielded rich evidence of hominin woodworking skills (Theime, 1997; Clark, 2001; Goren-Inbar et al., 2002). This evidence includes clubs, spears, possible fragments of trays, and, especially from Kalambo Falls, objects whose purposes remain enigmatic. In view of the wide range of uses our near primate relatives make of wooden implements, stone-tool-assisted Plio-Pleistocene carpentry is difficult to dismiss out of hand. Evidence for Plio-Pleistocene woodworking comes from microwear and residue studies. Keeley and Toth (1981) report microwear polishes interpreted as woodworking traces on flake tools from Developed Oldowan (Karari Industry) contexts in East Turkana, Kenya. Domínguez-Rodrigo et al. (2001) have identified wood phytoliths preserved on the edges of early Acheulean stone tools from Peninj, Tanzania.

The only evidence we have for Plio-Pleistocene hominin use of osseous tissues as tool material are a series of bone fragments from Swartkrans (South Africa) whose ends have been abraded and polished from use as digging instruments (Brain and Shipman, 1993; D'Errico and Backwell, 2003). Working bone with stone tools is an immensely time-consuming activity. It is possible that early hominins inattention to the potential of carved bone tools may reflect either (1) that they could be rendered into useful shapes without carving, or (2) that carving them into shapes other than those in which they naturally occurred was prohibitively costly in terms of time and energy.

Of the use of stone tools to process hide or soft plant matter, the only evidence is from a small sample of tools that has been examined for lithic microwear. The scarcity of evidence for these activities, on the other hand, cannot be taken at face value. Microwear traces form slowly in most activities that involve soft materials, such as animal hide or non-lignous plant matter. If Oldowan tool use was relatively brief, a hypothesis consistent with other indications of minimal effort in tool design (i.e., retouch), there may have been insufficient time for diagnostic wear traces to form. Furthermore, many of the rocks of which Oldowan tools are made are of a friable nature, such that their edges do not preserve wear traces. Hide-working is today a uniquely human technology, and so absence of such wear traces may legitimately indicate its recent origin. Processing wood and soft plant matter, on the other hand, are activities abundantly documented among nonhuman primates. In the case of these activities, it would be foolish to equate absence of evidence with evidence of absence.

What difference does it make whether Oldowan flaked stone production was specialized, linked primarily to butchery (and thus to carnivory), or was instead a more generalized, functionally-diverse technology of which the carnivory-related dimensions are simply the best preserved remains?

The hypothesis linking stone tool production to carnivory predicts that changes in hominin predatory strategies ought to be correlated with significant variation in stone tool technology. Bramble and Lieberman (2004) have argued that the evolution of a more modern-looking postcranial skeletal form by *Homo ergaster/erectus* reflects increasing dependence on hunting strategies that involve endurance running. It is possible that the large, purposefully-shaped, bifacial core tools (handaxes, cleaver, etc.) that appear around the same time, *ca.* 1.7–1.6 Ma, are related to increasing demands for highly-portable tools that are simultaneously efficient butchery tools and (when recycled as cores) effective sources for smaller flakes (Shea, 2007).

One of us, (Shea, 2007) has proposed that we should view Oldowan stone tool production as a lithic strategy for optimizing stone tool versatility, rather than simply a response to the need for butchery tools. If this model of a functionally differentiated Oldowan is correct, then the period over which Oldowan assemblages are distributed may have witnessed not just increased carnivory among one or more hominin species, but also the emergence of a broader pattern of technologically-assisted subsistence (again, among one or more hominin species). The particular focus of Oldowan technology may have varied widely through time, across space, and among the one or more hominin species responsible for it. Some sets of Oldowan tools may reflect increasing emphasis on carnivory, others increased production of wooden tools, others both these things, and still others combinations of tool uses whose nature remains unknown. Oldowan stone tools are found over such a long period of time, and in so wide a range of contexts that any hypothesis linking their appearance to a one-time-only behavioral shift among a single hominin species is almost certainly wrong.

Conclusions

This paper has explored some of the interpretive issues surrounding early hominin stone tool production, stone tool use, and carnivory. These are the questions that most interest archeologists working on the early phases of human evolution. We have saved for last the question that we archeologists are most often asked by our physical anthropologist colleagues, "who (i.e., which hominins) made the Oldowan tools?"

If one bases an assessment of the identity of the Oldowan toolmakers strictly on chronostratigraphic associations between hominin fossils and the earliest Oldowan lithic assemblages, the putative authors include *Australopithecus garhi*, *Paranthropus aethiopicus*, *P. boisei*, *P. robustus*, *Homo habilis*, *H. rudolfensis*, and *H. ergaster/erectus*. The most repetitive pattern of association between hominin taxa and Oldowan tools is with "early *Homo*" (*H. habilis* and *H. ergaster/erectus*) and various species of *Paranthropus*. The claim that *Paranthropus* made stone tools rests upon two arguments. The first is Susman's (1991) observation that the thumb attributed to *P. robustus* exhibits morphological adaptations to a precision grasp associated with tool use. The second is the fact that *Paranthropus* fossils are stratigraphically associated with stone tools at many Plio-Pleistocene localities. The claim that *Homo* was the principal tool author rests primarily with similar repetitive stratigraphic association, and with evolutionary changes in cranial shape (brain enlargement) and dentition (molar reduction) *thought to reflect increased carnivory*, and thus greater dependence on tool use. No such trends are apparent in the *Paranthropus* lineage and, indeed, its extinction leaves not the slightest trace in the pattern of African Early Paleolithic industrial variability. Therefore, while we cannot rule out stone tool production and use by australopiths and *Paranthropus*, the principal beneficiaries of knapped stone tool technology appear to have been early representatives of the genus *Homo*.

Acknowledgments We thank Fred Grine, Richard Leakey and John Fleagle for organizing the meeting out of which this paper developed. We also thank John Fleagle, Patricia Crawford, and two anonymous reviewers for their comments on earlier versions of this paper.

References

Aiello, L.C., Wheeler, P., 1995. The expensive-tissue hypothesis: the brain and digestive system in human and primate evolution. Current Anthropology 36, 199–221.

Antón, S., Swisher, C.C., 2004. Early dispersals of *Homo* from Africa. Annual Review of Anthropology 33, 271–296.

Asfaw, B., White, T., Lovejoy, O., Latimer, B., Simpson, S., Suwa, G., 1999. *Australopithecus garhi*: a new species of early hominid from Ethiopia. Science 284, 629–635.

Bar-Yosef, O., Goren-Inbar, N., 1993. The Lithic Assemblages of 'Ubeidiya. Hebrew University Institute of Archaeology, Jerusalem.

Blumenschine, R.J., 1987. Characteristics of an early hominid scavenging niche. Current Anthropology 28, 383–407.

Blumenschine, R.J., 1995. Percussion marks, tooth marks, and experimental determinations of the timing of hominid and carnivore access to long bones at FLK *Zinjanthropus*, Olduvai Gorge, Tanzania. Journal of Human Evolution 29, 21–51.

Blumenschine, R.J., Peters, C.R., 1998. Archaeological predictions for hominid land use in the paleo-Olduvai Basin, Tanzania, during lowermost Bed II times. Journal of Human Evolution 34, 565–607.

Blumenschine, R.J., Pobiner, B.L., 2007. Zooarchaeology and the ecology of early hominin carnivory. In: Ungar, P. (Ed.), Evolution of the Human Diet: The Known, the Unknown, and the Unknowable. Oxford University Press, Oxford, pp. 167–190.

Blumenschine, R.J., Selvaggio, M.M., 1988. Percussion marks on bone surfaces as a new diagnostic of hominid behavior. Nature 333, 763–765.

Blumenschine, R.J., Peters, C.R., Masao, F.T., Clarke, R.J., Deino, A.L., Hay, R.L., Swisher, C.C., Stanistreet, I.G., Ashley, G.M., McHenry, L.J., Sikes, N.E., van der Merwe, N.J., Tactikos, J.C., Cushing, A.E., Deocampo, D.M., Njau, J.K.,Ebert, J.I., 2003. Late Pliocene *Homo* and hominid land use from western Olduvai Gorge, Tanzania. Science 299, 1217–1221.

Blumenschine, R.J., Peters, C.R., Capaldo, S.D., Andrews, P., Njau, J.K., Pobiner, B.L., 2007a. Vertebrate taphonomic perspectives on Oldowan hominin land use in the Plio-Pleistocene Olduvai Basin, Tanzania. In: Pickering, T.R., Toth, N., Schick, K, (Eds.), Breathing Life into Fossils: Taphonomic Studies in Honor of C.K. (Bob) Brain. Stone Age Institute Press, Gosport, Indiana, pp. 161–179.

Blumenschine, R.J., Prassack, K.A., Kreger, C.D., Pante, M.C., 2007b. Carnivore tooth marking, microbial bioerosion, and the invalidation of Domínguez-Rodrigo and Barba's (2006) test of Oldowan hominin scavenging behavior. Journal of Human Evolution 53, 420–426.

Blumenschine, R.J., Masao, F., Tactikos, J., Ebert, J., 2008. Effects of distance from stone source on landscape-scale variation in Oldowan artifact assemblages in the Paleo-Olduvai Basin, Tanzania. Journal of Archaeological Science, 35, 76–86.

Boesch, C., Boesch, H., 1984. Mental map in wild chimpanzees: an analysis of hammer transports for nut cracking. Primates 25, 160–170.

Brain, C.K., Shipman, P., 1993. The Swartkrans Bone Tools. Transvaal Museum Monograph 8, Transvaal Museum, Pretoria.

Bramble, D.M., Lieberman, D.E., 2004. Endurance running and the evolution of *Homo*. Nature 432, 354–352.

Braun, D.R., Plummer, T.W., Ditchfield, P.W., Bishop, L.C., Ferraro, J.V., 2009. Oldowan technology and raw material variability at Kanjera South. In: Hovers, E., Braun, D. (Eds.), Interdisciplinary Approaches to the Oldowan. Springer, Dordrecht, pp. 99–110.

Bunn, H.T., 1981. Archaeological evidence for meat-eating by Plio-Pleistocene hominids from Koobi Fora and Olduvai Gorge. Nature 291, 574–577.

Bunn, H.T., 1982. Meat-eating and human evolution: studies of the diet and subsistence patterns of Plio-Pleistocene hominids. Ph.D. dissertation, University of California, Berkeley.

Bunn, H.T., 1997. The bone assemblages from the excavated sites. In: Isaac, G., Isaac, assisted by Isaac, B. (Eds.), Koobi Fora Research Project, Volume 5. Plio-Pleistocene Archaeology. Clarendon, Oxford, pp. 402–458.

Bunn, H.T., 2001. Hunting, power scavenging, and butchery by Hadza foragers and by Plio-Pleistocene *Homo*. In: Stanford, C.B., Bunn, H.T. (Eds.), Meat Eating and Human Evolution. Oxford University Press, New York, pp. 199–218.

Bunn, H.T., 2007. Meat made us human. In: Ungar, P.S. (Ed.), Evolution of the Human Diet: The Known, the Unknown, and the Unknowable. Oxford University Press, Oxford, pp. 191–211.

Bunn, H.T., Kroll, E.M., 1986. Systematic butchery by Plio/Pleistocene hominids at Olduvai Gorge, Tanzania. Current Anthropology 27, 431–452.

Capaldo, S., 1997. Experimental determinations of carcass processing by Plio-pleistocene hominids and carnivores at FLK 22 (*Zinjanthropus*), Olduvai Gorge, Tanzania. Journal of Human Evolution 33, 555–597.

Cavallo, J.A., Blumenschine, R.J., 1989. Tree-stored leopard kills: expanding the hominid scavenging niche. Journal of Human Evolution 18, 393–399.

Cerling, T.E., Harris, J.M., Leakey, M.G., 1999. Browsing and grazing in elephants: the isotope record of modern and fossil proboscideans. Oecologia 120, 364–374.

Cerling, T.E., Harris, J.M., Passey, B.H., 2003. Diets of East African bovids based on stable isotope analysis. Journal of Mammalogy 84, 456–470.

Clark, J.D., 2001. Modified, used and other wood and plant specimens fro Bwalya Industry Acheulean horizons. In: Clark, J.D. (Ed.), Kalambo Falls Prehistoric Site, Volume III, The Earlier Cultures: Middle and Earlier Stone Age. Cambridge University Press, Cambridge, pp. 481–491.

Crabtree, D.E., Davis, E.L., 1968. Experimental manufacture of wooden implements with tools of flaked stone. Science 159, 426–428.

d'Errico, F., Backwell, L., 2003. Possible evidence of bone tool shaping by Swartkrans early hominids. Journal of Archaeological Science 30, 1559–1576.

de Heinzelin, J., Clark, D., White, T., Hart, W., Renne, P., WoldeGabriel, G., Beyene, Y.,Vrba, E., 1999. Environment and behavior of 2.5-million-year-old Bouri hominids. Science 284, 625–629.

Delagnes, A., Roche, H., 2005. Late Pliocene hominid knapping skills: the case of Lokalalei 2C, West Turkana, Kenya. Journal of Human Evolution 48, 435–472.

de la Torre, I., 2004. Omo revisited: evaluating the technological skills of Pliocene hominids. Current Anthropology 45, 439–467.

de la Torre, I., Mora, R., 2005. Unmodified lithic material at Olduvai Bed I: manuports or ecofacts? Journal of Archaeological Science 32, 273–285.

Dennell, R., Roebroeks, W., 2005. An Asian perspective on early human dispersal from Africa. Nature 438, 1099–1104.

Domínguez-Rodrigo, M., 1997. Meat-eating by early hominids at the FLK 22 *Zinjanthropus* site, Olduvai Gorge (Tanzania): an experimental approach using cut-mark data. Journal of Human Evolution 33, 669–690.

Domínguez-Rodrigo, M., Barba, R., 2006. New estimates of tooth mark and percussion mark frequencies at the FLK Zinj site: the carnivore-hominid-carnivore hypothesis falsified. Journal of Human Evolution 50, 170–194.

Domínguez-Rodrigo, M., Serralonga, J., Juan-Tresserras, J., Alcala, L., Luque, L., 2001. Woodworking activities by early humans: a plant residue analysis on Acheulian stone tools from Peninj (Tanzania). Journal of Human Evolution 40, 289–299.

Domínguez-Rodrigo, M., Pickering, T.R., Semaw, S., Rogers, M.J., 2005. Cutmarked bones from Pliocene archaeological sites at Gona, Ethiopia: implications for the function of the world's oldest stone tools. Journal of Human Evolution 48, 109–121.

Foley, R.A., 2001. The evolutionary consequences of increased carnivory in hominids. In: Stanford, C.B., Bunn, H.T. (Eds.), Meat Eating and Human Evolution. Oxford University Press, New York, pp. 305–331.

Gabunia, L., Antón, S.C., Lordkipanidze, D., Vekua, A., Justus, A., Swisher, C.C., 2001. Dmanisi and dispersal. Evolutionary Anthropology 10, 158–170.

Gilby, I.C., 2006. Meat sharing among the Gombe chimpanzees: harassment and reciprocal exchange. Animal Behaviour 71, 953–963.

Goldman, T., Hovers, E., 2009. Methodological issues in the study of Oldowan raw material selectivity: insights from A.L. 894 (Hadar, Ethiopia). In: Hovers, E., Braun, D. (Eds.), Interdisciplinary Approaches to the Oldowan. Springer, Dordrecht, pp. 71–89.

Goren-Inbar, N., Werker, E., Feibel, C.S. (Eds.), 2002. The Acheulian Site of Gesher Benot Ya'acov, Israel: The Wood Assemblage. Oxbow Books, Oxford.

Gould, R.A., Koster, D.A., Sontz, A., 1971. The lithic assemblage of the Western Desert Aborigines of Australia. American Antiquity 36, 149–168.

Guérin, C., Bar-Yosef, O., Debard, E., Faure, M., Shea, J., Tchernov, E., 2003. Oubeidiyeh, carrefour biogéographique et culturel entre l'Afrique et l'Eurasie au paléolithique ancien. In: Vandermeersch, B. (Ed.), Echanges et Diffusion dans la Préhistoire Méditerranéenne. Editions du Comité des travaux historiques et scientifiques, Paris, pp. 131–146.

Harmand, S., 2004. Raw material and economic behaviour of late Pliocene hominids: the case of Lokalalei 2C and Lokalalei 1 sites, West Turkana (Kenya). Paper presented at the Paleoanthropological Society Annual Meeting, Montreal, Canada.

Harmand, S., 2005. Matières premières lithiques et comportements techno-économiques des homininés plio-pleistocènes. Ph.D. dissertation, Université Paris X-Nanterre.

Harmand, S., 2009a. Raw materials and techno-economic at Oldowan and Acheulean sites in the West Turkana region, Kenya. In: Blades, B. (Ed.), Lithic Materials and Palaeolithic Societies. Blackwell, Oxford, pp. 3–14.

Harmand, S., 2009b. Variability in raw material selectivity at the late Pliocene sites of Lokalalei, West Turkana, Kenya. In: Hovers, E., Braun, D. (Eds.), Interdisciplinary Approaches to the Oldowan. Springer, Dordrecht, pp. 85–98.

Harris, J.M., Cerling, T.E., 1999. Dietary adaptations of extant and Neogene African suids. Journal of Zoology, London 256, 45–54.

Harris, J.W.K., 1983. Cultural beginnings: Plio-Pleistocene archaeological occurrences from the Afar, Ethiopia. African Archaeological Review 1, 3–31.

Hay, R.L., 1976. The Geology of Olduvai Gorge: a Study of Sedimentation in a Semi-arid Basin. University of California Press, Berkeley.

Hill, K., Kaplan, H., 1993. On why male foragers hunt and share food. Current Anthropology 34, 701–706.

Hovers, E., 2001. Stone knapping in the Late Pliocene of Hadar, Ethiopia. Abstract of paper presented at: Knapping Stone: A Uniquely Human Behavior? web.mae.u-paris10.fr/knappingstone/participant.htm

Hovers, E., Schollmeyer, K., Goldman, T., Eck, G., Reed, K., Johanson, D., Kimbel, W., 2002. Late Pliocene archaeological sites in Hadar, Ethiopia. Journal of Human Evolution 42, A17.

Howell, F.C., Haesaerts, P., de Heinzelin, J., 1987. Depositional environments, archeological occurrences, and hominids from Members E and F of the Shungura Formation (Omo Basin, Ethiopia). Journal of Human Evolution 16, 665–700.

Inizan, M.-L., Reduron-Ballinger, M., Roche, H., Tixier, J., Féblot-Augustins, 1999. Technology and Terminology of Knapped Stone. Cercle de Recherches et d'Etudes Préhistoriques (CNRS), Meudon.

Isaac, G., Harris, J.W.K., 1997. Sites stratified within the KBS Tuff: reports. In: Isaac, G. assisted by Isaac, B. (Eds.), Koobi Fora Research Project, Volume 5: Plio-Pleistocene Archaeology. Clarendon, Oxford, pp. 71–114.

Isaac, G., Harris, J.W.K., Kroll, E.M., 1997. The stone artefact assemblages: a comparative study. In: Isaac, G., assisted by Isaac, B. (Eds.), Koobi Fora Research Project, Volume 5: Plio-Pleistocene Archaeology. Clarendon, Oxford, pp. 262–362.

Isaac, G.L., 1978. The food sharing behavior of protohuman hominids. Scientific American 238, 90–109.

Jones, P.R., 1980. Experimental butchery with modern stone tools and its relevance for palaeolithic archaeology. World Archaeology 12, 153–165.

Jones, P.R., 1994. Results of experimental work in relation to the stone industries of Olduvai Gorge. In: Leakey, M.D., Roe, D.A. (Eds.), Olduvai Gorge, Volume 5: Excavations in Beds III, IV and the Masek Beds, 1968–1971. Cambridge University Press, Cambridge, pp. 254–298.

Kaplan, H., Hill, K., 1985. Hunting ability and reproductive success among male Ache foragers: preliminary results. Current Anthropology 26, 131–133.

Kappelman, J., Plummer, T., Bishop, L., Duncan, A., Appleton, S., 1997. Bovids as indicators of paleoenvironments in East Africa. Journal of Human Evolution 32, 229–256.

Keeley, L.H., Toth, N.P., 1981. Microwear polishes on early stone tools from Koobi Fora, Kenya. Nature 293, 464–465.

Kibunjia, M., 1994. Pliocene archaeological occurrences in the Lake Turkana basin. Journal of Human Evolution 27, 159–171.

Kibunjia, M., 1998. Archaeological investigations of Lokalalei 1 (GaJh5): a Late Pliocene site, west of Lake Turkana, Kenya. Ph.D. dissertation, Rutgers University.

Kibunjia, M., Roche, H., Brown, F.H., Leakey, R.E.F., 1992. Pliocene and Pleistocene archaeological sites west of Lake Turkana, Kenya. Journal of Human Evolution 23, 431–438.

Kimbel, W.H., Walter, R.C., Johanson, D.C., Reed, K.E., Aronson, J.L, Assefa, Z., Marean, C.W., Eck, G.G., Bobe-Quinteros, R., Hovers, E., Rak, Y., Vondra, C., Yemane, T., York, D., Chen, Y., Evenson, N.M., Smith, P.E., 1996. Late Pliocene *Homo* and Oldowan tools from the Hadar Formation (Kada Hadar Member), Ethiopia. Journal of Human Evolution 31, 549–561.

Kuman, K., 1994. The archaeology of Sterkfontein, past and present. Journal of Human Evolution 27, 471–495.

Kuman, K., 2003. Site formation in the early South African stone age sites and its influence on the archaeological record. South African Journal of Science 99, 251–254.

Langbroek, M., 2004. Out of Africa: An Investigation into the Earliest Occupation of the Old World. Archaeopress (British Archaeological Reports International Series 1244), Oxford.

Leakey, L.S.B., 1936. Stone Age Africa. Oxford University Press, London.
Leakey, M.D., 1971. Olduvai Gorge: Excavations in Beds I and II, 1960–1963. Cambridge University Press, Cambridge.
Lordkipanidze, D., Vekua, A., Ferring, R., Rightmire, P.G., Zollikofer, C.P.E., Ponce de Leon, M.S., Agusti, J., Kiladze, G., Mouskhelishvili, A., Nioradze, M., Tappen, M., 2006. A fourth hominin skull from Dmanisi, Georgia. Anatomical Record 288A, 1146–1157.
Marean, C.W., 1989. Sabertooth cats and their relevance for early hominid diet and evolution. Journal of Human Evolution 18, 559–582.
Marean, C.W., Spencer, L.M., Blumenschine, R.J., Capaldo, S.D., 1992. Captive hyaena bone choice and destruction, the schlepp effect and Olduvai archaeofaunas. Journal of Archaeological Science 19, 101–121.
McGrew, W.C., 1992. Chimpanzee Material Culture: Implications for Human Evolution. Cambridge University Press, Cambridge.
Mercader, J., Panger, M., Boesch, C., 2002. Excavation of a chimpanzee stone tool site in the African rain forest. Science 296, 1452–1455.
Mitani, J.C., Watts, D.P., 2001. Why do chimpanzees hunt and share meat? Animal Behaviour 61, 915–924.
Monahan, C.M., 1996. New zooarchaeological data from Bed II, Olduvai Gorge, Tanzania: implications for hominid behavior in the Early Pleistocene. Journal of Human Evolution 31, 93–128.
Mora, R., de la Torre, I., 2005. Percussion tools in Olduvai Beds I and II (Tanzania): implications for early human activities. Journal of Anthropological Archaeology 24, 179–192.
O'Brien, E.M., 1981. The projectile capabilities of an Acheulean handaxe from Olorgesailie. Current Anthropology 22, 76–79.
Oliver, J.S., 1994. Estimates of hominid and carnivore involvement in the FLK *Zinjanthropus* fossil assemblage: some socioecological implications. Journal of Human Evolution 27, 267–294.
Pelegrin, J., 2005. Remarks about archaeological technics and methods of knapping: elements of cognitive approach to stone knapping. In: Brill, B., Roux, V. (Eds.), Stone Knapping: the Necessary Conditions for a Uniquely Hominid Behaviour. Cambridge University Press, Cambridge, pp. 23–34.
Pickering, T.R., 1999. Taphonomic interpretations of the Sterkfontein early hominid site (Gauteng, South Africa) reconsidered in light of recent evidence. Ph.D. dissertation, University of Wisconsin, MI.
Pickering, T.R., White, T.D., Toth, N., 2000. Cutmarks on a Plio-Pleistocene hominid from Sterkfontein, South Africa. American Journal of Physical Anthropology 111, 579–584.
Plummer, T., 2004. Flaked stone and old bones: biological and cultural evolution at the dawn of technology. Yearbook of Physical Anthropology 47, 118–164.
Plummer, T., Bishop, L.C., Ditchfield, P., Hicks, J., 1999. Research on Late Pliocene Oldowan sites at Kanjera South, Kenya. Journal of Human Evolution 36, 151–170.
Pobiner, B.L., 2007. Hominin-carnivore interactions: evidence from modern carnivore bone modification and Early Pleistocene archaeofaunas (Koobi Fora, Kenya; Olduvai Gorge, Tanzania). Ph.D. dissertation, Rutgers University.
Potts, R., 1988. Early Hominid Activities at Olduvai. Aldine de Gruyter, New York.
Potts, R., Shipman, P., 1981. Cut-marks made by stone tools on bones from Olduvai Gorge, Tanzania. Nature 291, 577–580.
Prat, S., Brugal, J.-P., Tiercelin, J.-J., Barrat, J.-A., Bohn, M., Delagnes, A., Harmand, S., Kimeu, K., Kibunjia, M., Texier, P.-J., Roche, H., 2005. First occurrence of early *Homo* in the Nachukui Formation (West Turkana, Kenya) at 2.3–2.4 Myr. Journal of Human Evolution 49, 230–240.
Pruetz, J.D., Bertolani, P., 2007. Savanna chimpanzees, *Pan troglodytes verus*, hunt with tools. Current Biology 17, 412–417.
Roche, H., 1989. Technological evolution in early hominids. Ossa 14, 97–98.
Roche, H., 2005. From simple flaking to shaping: stone knapping evolution among early hominins. In: Brill, B., Roux, V. (Eds.), Stone Knapping: the Necessary Conditions for a Uniquely Hominid Behaviour. Cambridge University Press, Cambridge, pp. 35–52.
Roche, H., Kibunjia, M., 1996. Contribution of the West Turkana Plio-Pleistocene sites to the archaeology of the lower Omo Turkana Basin. Kaupia 6, 27–30.
Roche, H., Tiercelin, J.-J., 1977. Découverte d'une industrie lithique ancienne in situ dans la Formation d'Hadar. Comptes Rendus de l'Acadamie des Sciences Paris, 284, série D, 1871–1874.
Roche, H., Tiercelin, J.-J., 1980. Industries lithiques de la Formation plio-pléistocène de Hadar, Ethiopie (campagne 1976). In: Leakey, R.E.F., Ogot, B.A. (Eds.), Proceedings of the Seventh Panafrican Congress of Prehistory and Quaternary Studies (Nairobi 1977). The International Louis Leakey Memorial Institute for African Prehistory, Nairobi, Kenya, pp. 194–199.
Roche, H., Delagnes, A., Brugal, J.-P., Feibel, C.S., Kibunja, M., Mourre, V., Texier, P.-J., 1999. Early hominid stone tool production and technical skill 2.34 Myr ago, in West Turkana, Kenya. Nature 399, 57–90.
Roche, H., Brugal, J.-P., Delagnes, A., Feibel, C., Harmand, S., Kibunjia, M., Prat, S., Texier, P.-J., 2003. Les sites archeologiques plio-pleistocenes de la formation de Nachukui, Ouest-Turkana, Kenya: bilan synthétique 1997–2001. Comptes Rendus Palevol 2, 663–673.
Sahnouni, M., 2005. Point des connaissances du Paléolithique ancien d'Afrique du Nord et la question de la première occupation humaine au Maghreb. In: Sahnouni, M. (Ed.), Les cultures paléolithique en Afrique. ArtCom-Errance, Paris, pp. 99–128.
Sahnouni, M., 2006. The North African Early Stone Age and the sites at Ain Hanech, Algeria. In: Toth, N.P., Schick, K. (Eds.), The Oldowan: Case Studies Into the Earliest Stone Age. Stone Age Institute Press, Bloomington, pp. 77–112.
Schick, K., Toth, N.P., 2006. An overview of the Oldowan industrial complex: the sites and the nature of the evidence. In: Toth, N.P., Schick, K. (Eds.), The Oldowan: Case Studies Into the Earliest Stone Age. Stone Age Institute Press, Bloomington, pp. 3–42.
Selvaggio, M.M., 1998. Evidence for a three-stage sequence of hominid and carnivore involvement with long bones at FLK *Zinjanthropus*, Olduvai Gorge, Tanzania. Journal of Archaeological Science 25, 191–202.
Semaw, S., 2000. The world's oldest stone Artefacts from Gona, Ethiopia: their implications for understanding stone technology and patterns of human evolution between 2.6-1.5 million years ago. Journal of Archaeological Science 27, 1197–1214.
Semaw, S., Renne, P., Harris, J.W.K., Feibel, C.S., Bernor, R.L., Fesseha, N., Mowbray, K., 1997. 2.5-million-year-old stone tools from Gona, Ethiopia. Science 284, 625–629.
Semaw, S., Rogers, M.J., Quade, J., Renne, P.R., Butler, R.F., Dominguez-Rodrigo, M., Stout, D., Hart, W.S., Pickering, T., Simpson, S.W., 2003. 2.6-million-year-old stone tools and associated bones from OGS-6 and OGS-7, Gona, Afar, Ethiopia. Journal of Human Evolution 45, 169–177.
Shea, J.J., 2006. Child's play: reflections on the invisibility of children in the paleolithic record. Evolutionary Anthropology 15, 212–216.
Shea, J.J., 2007. Lithic archaeology, or, what stone tools can (and can't) tell us about early hominin diets. In: Ungar, P.S. (Ed.), Evolution of the Human Diet: The Known, the Unknown and the Unknowable. Oxford University Press, Oxford, pp. 321–351.
Shipman, P., Walker, A., 1989. The costs of becoming a predator. Journal of Human Evolution 18, 373–392.
Spencer, L.M., 1997. Dietary adaptations of Plio-Pleistocene Bovidae: implications for hominid habitat use. Journal of Human Evolution 32, 201–228.
Stanford, C.B., Bunn, H.T. (Eds.), 2001. Meat Eating and Human Evolution. Oxford University Press, New York.

Stiles, D.N., Hay, R.L., Oneil, J.R., 1974. The MNK chert factory site, Olduvai Gorge, Tanzania. World Archaeology 5, 285–308.

Stout, D., Quade, J., Semaw, S., Rogers, M.J., Levin, N.E., 2005. Raw material selectivity of the earliest stone toolmakers at Gona, Afar, Ethiopia. Journal of Human Evolution 48, 365–380.

Susman, R.L., 1991. Who made the Oldowan tools? Fossil evidence for tool behavior in Plio-Pleistocene hominids. Journal of Anthropological Research 47, 129–151.

Tactikos, J.C., 2005. A Landscape Perspective on the Oldowan from Olduvai Gorge, Tanzania. Ph.D. dissertation, Rutgers University.

Theime, H., 1997. Lower Paleolithic hunting spears from Germany. Nature 385, 807–810.

Toth, N., Schick, K., Semaw, S., 1996. Further investigations of the Oldowan: lessons from the prehistoric record, experimental archaeology, and apes. In: Braun D., Hovers, E. (Orgs), Symposium: Interdisciplinary Approaches to Understanding the Oldowan, Annual Meeting of the Society for American Archaeology, San Juan, Puerto Rico.

Toth, N.P., 1987. Behavioral inferences from early stone artifact assemblages: an experimental model. Journal of Human Evolution 16, 763–787.

Toth, N.P., 1997. The artifact assemblages in the light of experimental studies. In: Isaac, G.L., Isaac, assisted by Isaac, B. (Eds.), Koobi Fora Research Project, Volume 5: Plio-Pleistocene Archaeology. Clarendon, Oxford, pp. 363–402.

Toth, N.P., Schick, K. (Eds.), 2006. The Oldowan: Case Studies Into the Earliest Stone Age. Stone Age Institute Press, Bloomington, IN.

Ungar, P. (Ed.), 2007. Evolution of the Human Diet: The Known, the Unknown, and the Unknowable. Oxford University Press, New York.

Walker, A., Leakey, R.E., Harris, J.M., Brown, F., 1986. 2.5-myr *Australopithecus boisei* from west of Lake Turkana, Kenya. Nature 322, 517 522.

Wang, W., Potts, R., Baoyin, Y., Huang, W., Cheng, H., Edwards, R.L., Ditchfield, P., 2007. Sequence of mammalian fossils, including hominoid teeth, from the Bubing Basin caves, South China. Journal of Human Evolution 52, 370–379.

Washburn, S.L., Lancaster, C.S., 1968. The evolution of hunting. In: Lee, R.B., DeVore, I. (Eds.), Man the Hunter. Aldine, Chicago, IL, pp. 293–303.

Werdelin, L., Lewis, M., 2005. Plio-Pleistocene Carnivora of eastern Africa: species richness and turnover patterns. Zoological Journal of the Linnean Society 144, 121–144.

Part V
Environmental and Ecological Perspectives

Chapter 13
Plio-Pleistocene East African Pulsed Climate Variability and Its Influence on Early Human Evolution

Mark A. Maslin and Martin H. Trauth

Keywords East Africa • tectonics • regional climate • global climate • paleo-lakes • precessional forcing • pulsed climatic variability

Introduction

Long-term climate change seems to be modulated primarily by tectonic changes at both the global and local scale (Maslin et al., 2001). Late Cenozoic global cooling has been ascribed to both the uplift of Tibet (Ruddiman and Raymo, 1988), and the closure of the Panama Isthmus (Haug and Tiedemann, 1998), although the exact role of atmospheric carbon dioxide is still unclear (Sundquist and Visser, 2004). In East Africa, long-term climate change is also controlled by local tectonics, especially the dynamic development of the branching East African Rift System (Sepulchre et al., 2006). Early hominin evolution in East Africa thus occurs at the same time as both long-term global cooling and extensive local tectonic changes. There is a compelling need to understand how these two environmental factors interact at the local scale and affect flora and fauna living in the East African Rift. The geologic record of the last 5 million years demonstrates that both local and global influences can lead to extremely rapid environmental change (Maslin and Christensen, 2007).

M.A. Maslin (✉)
Environmental Change Research Centre, Department of Geography, University College London, 26 Bedford Way, Pearson Building, Gower Street, London WC1E 6BT, UK
e-mail: m.maslin@geog.ucl.ac.uk

M.H. Trauth
Institut für Geowissenschaften, Universität Potsdam, POB 601553, D-14415 Potsdam, Germany
e-mail: trauth@geo.uni-potsdam.de

East African Tectonic History

The East African Rift System (EARS) is one of the most extensive geological features on the Earth's surface, running North-South for approximately 4,500 km from Syria through East Africa to Mozambique. Volcanism associated with the EARS began as early as 45–33 Ma in the Ethiopian Rift, by 33 Ma in northern Kenya, and by 15–8 Ma in the central and southern segments of the rift in Kenya and Tanzania (Fig. 13.1).

The early stages of rifting were characterized by updoming and downwarping, while subsequent faulting progressed from north to south (Fig. 13.1). Major faulting in Ethiopia between 20–14 Ma was followed by the generation of East dipping faults in northern Kenya between 12 and 7 Ma, and superseded by normal faulting on the western side of the central and southern Kenya Rift between 9 and 6 Ma (Baker et al., 1988, Strecker et al., 1990; Ebinger et al., 2000). These early half grabens were subsequently antithetically faulted between about 5.5 and 3.7 Ma, leading to a full-graben morphology (Baker et al., 1988, Strecker et al., 1990). This full-graben stage was preceded by the formation of the large Aberdare volcanic complex with elevation in excess of 4,000 m, forming an important orographic barrier in Kenya by ~5 Ma (Williams et al., 1983). By 2.6 Ma, the graben was further segmented in the central Kenya Rift by west-dipping faults, creating the 30 km wide intrarift Kinangop Plateau and the tectonically active 40-km-wide inner rift (Fig. 13.1) (Baker et al., 1988; Strecker et al., 1990; Bosworth and Strecker, 1997). In the Tanzanian sector of the rift, sedimentation in isolated basins began at ~5 Ma (Foster et al., 1997). A major phase of rift faulting occurred at 1.2 Ma and produced the present-day rift escarpments (Foster et al., 1997).

Plio-Pleistocene Variations in East African Moisture Availability

Figure 13.1 illustrates that late Cenozoic tectonic activity in the EARS led to the production of isolated basins within

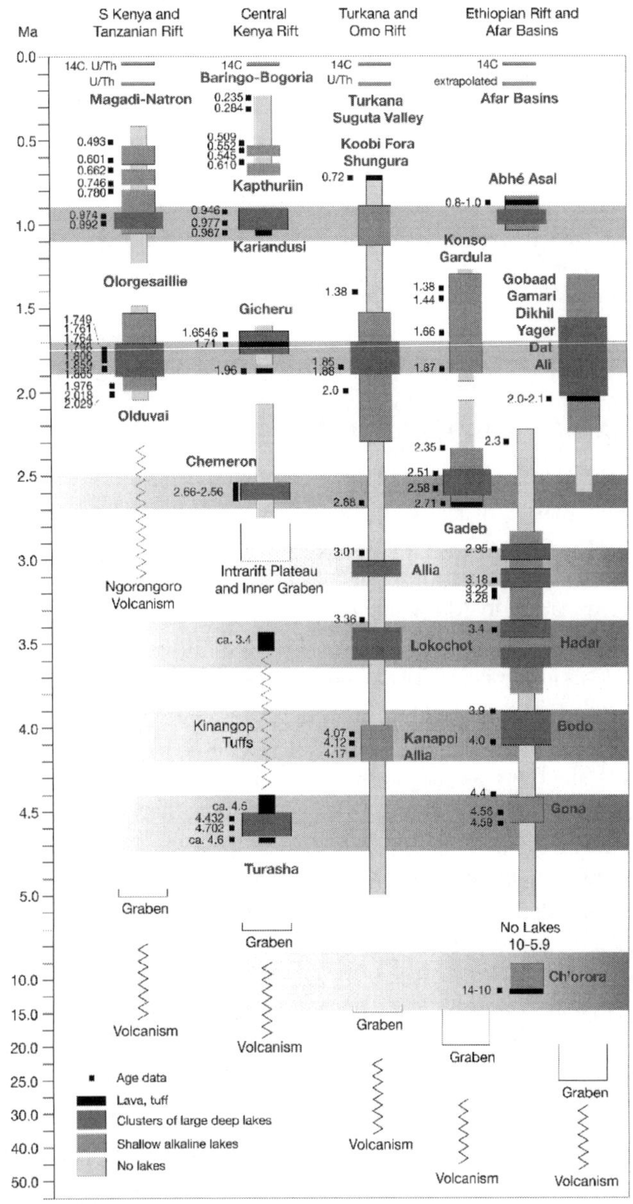

Fig. 13.1 Compilation of tectonic features and prominent lake periods for the eastern branch of the East African Rift System. Tectonic features and events complied from Baker et al. (1988), Strecker et al. (1990), Ebinger et al. (2000), Williams et al. (1983) and Foster et al. (1997). Paleoenvironmental and radiometric age data for the Olduvai Basin from Walter et al. (1991) and Ashley and Hay (2002); for the Magadi-Natron and Olorgesailie Basins from Potts (1998,1999), and Behrensmeyer et al. (2002). Natron has one persistent lacustrine interval (a Member of the Monik Formation, called the Moinik Clays) dated to 1.1–1.0 Ma (Deino, A., pers. comm., 2008). Paleoenvironmental and radiometric age data for the Gicheru Basin from Baker et al. (1988), Strecker (1991), Boven (1992) and this work; for the Naivasha Basin from Strecker et al. (1990) and Trauth et al. (2003, 2005); for the Nakuru-Elmenteita Basin from Evernden and Curtis (1965), Strecker (1991), Boven (1992) and Trauth et al. (2005); for the Baringo-Bogoria Basin from Owen et al. (2002) and Deino et al.(2006); for the Suguta Basin from Butzer et al.(1969), Hillarie-Marcel et al. (1986) and Sturchio et al. (1993); for the Omo-Turkana Basin from McDougall and Watkins (1988) and Brown and Feibel (1991); for the Ethiopian Rift from Williams et al. (1979), Gasse (1990) and WoldeGabriel et al. (2000); for the Afar Basin from Gasse (1990).

which lakes could form. Southward propagation of rifting and magmatic activity resulted in formation of lake basins first in the northern parts of the EARS. For example, the fluviolacustrine history of the Afar, Omo-Turkana and Baringo-Bogoria Basins in the north began in the Middle and Late Miocene, whereas the oldest lacustrine sequences in the central and southern segments of the rift in Kenya and Tanzania occur in the Early Pliocene (Tiercelin and Lezzar, 2002). In general, palaeo-lakes first appear in the EARS earlier in the north than in the south, due to the progressive formation of separate basins. If tectonics were the sole control over lake formation, then either a North to South or Northwest to Southeast temporal trend would be expected. However, what is observed is the appearance of large, deep lakes synchronously across large geographical areas at specific points in time (Trauth et al., 2005, 2007), suggesting that regional climatic control is operative.

Carbon isotope records from both soil carbonates (Levin et al., 2004; Wynn, 2004; Segalen et al., 2007) and biomarkers (*n*-alkanes) extracted from deep-sea sediments (Feakins et al., 2005) provide clear evidence a progressive vegetation shift from C_3 (~trees and shrubs) to C_4 (~tropical grasses) plants during the Plio-Pleistocene. This shift has been ascribed to increased aridity that arose from the progressive rifting of East Africa (deMenocal, 2004; Sepulchre et al., 2006). Superimposed on this regime of subdued moisture availability, three periods characterized by the occurrence of large and deep lakes have been broadly identified in East Africa at 2.7–2.5, 1.9–1.7 and 1.1–0.9 Ma (Trauth et al., 2005, 2007), indicating consistency in the moisture history of the Kenyan and Ethiopian Rifts. Although preservation of East African lake records prior to 2.7 Ma is patchy, there is limited evidence for lake phases at ~3.20–2.95, ~3.4–3.3, 4.0–3.9, and ~4.7–4.3 Ma (Fig. 13.1). The lake phases correspond to drops in the East Mediterranean marine dust abundance (Larrasoaña et al., 2003), which are thought to reflect the aridity of the eastern Algerian, Libyan, and western Egyptian lowlands located north of the central Saharan watershed (Fig. 13.2). The lake phases also correspond to an increased occurrence of sapropels in Mediterranean Sea, which are thought to be caused by increased Nile River discharge (Lourens et al., 2004). The correspondence of the Mediterranean marine records with lake records of East Africa suggest a consistent moisture record for a region encompassing much of central and northern Africa over the last 3–5 million years.

In contrast, these East African wet phases correlate with significant intermediate-term increases in the dust records from ocean sediment cores adjacent to West Africa and Arabia (deMenocal, 1995, 2004). While, at first, this seems contradictory, examination of these data in chronologic detail demonstrates that both the lake and dust records are responding to precessional forcing, and that they are in-phase.

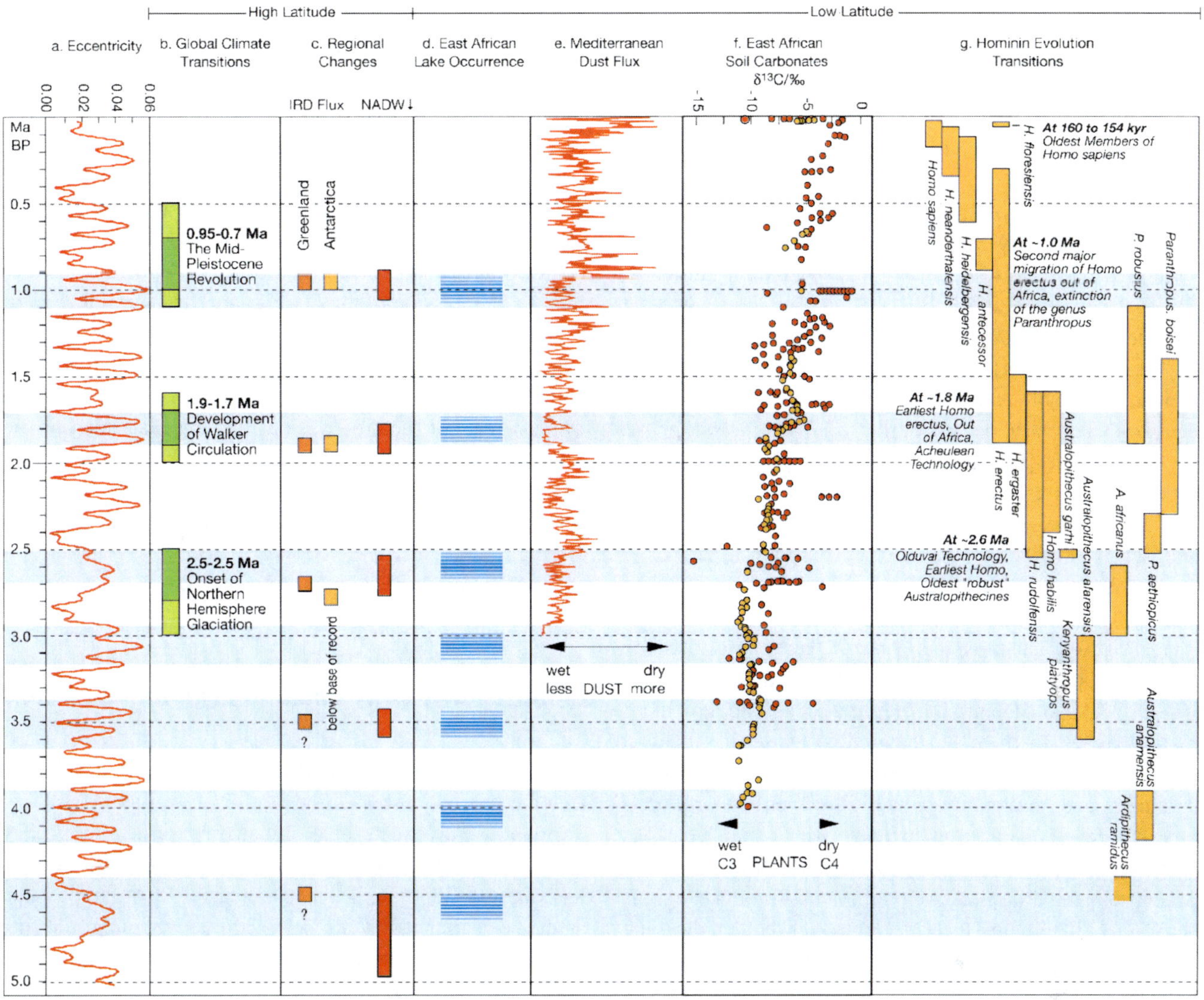

Fig. 13.2 Comparison of eccentricity variations (Berger and Loutre, 1991) with high latitude climate transitions (St John and Krissek, 2002; Cowan, 2001) and Mediterranean dust flux (Larrasoaña et al., 2003). Soil carbonate carbon isotopes: (yellow dots = Levin et al., 2004; red dots = Wynn et al., 2004). Data for East African lake occurrences from Trauth et al. (2005, 2007). Hominin species appearances and durations from Reed (1997), Dunsworth and Walker (2002), McHenry (2002), White (2002) and White et al. (2006).

Deino et al., (2006) and Kingston et al., (2007) found that the major lacustrine episode of the Baringo Basin between 2.7–2.55 Ma actually consisted of five paleo-lake phases separated by a precessional cyclicity of 23 kyr. The lakes occurrences are in-phase with increased freshwater discharge, and therefore sapropel formation in the Mediterranean Sea (Lourens et al., 2004), and are out of phase with the dust records from the Indian Ocean (deMenocal, 1995, 2004). Hence, the lake records from East Africa and the Indian Ocean dust records document extreme climate variability with precession-forced wet and dry phases. Precessional forcing of vegetation change also occurred at this time in southwest Africa, independent of glacial-interglacial cycles (Denison et al., 2005). There is also emerging evidence for precessional forcing of the 1.9–1.7 Ma lake phase in the KBS Member of the Koobi Fora Formation in the northeast Turkana Basin of Kenya (Lepre et al., 2007). During the same period, an oxygen isotope record from the Buffalo Cave flowstone (Makapansgat Valley, Limpopo Province, South Africa) shows clear evidence of precessionally-forced changes in rainfall in South Africa (Hopley et al., 2007).

Orbital-Forcing of African Climate

There is a growing body of evidence for precession-forcing of moisture availability in the tropics, in East Africa during the Pliocene (deMenocal, 1995, 2004; Deino et al., 2006; Kingston et al., 2007; Hopley et al., 2007; Lepre et al., 2007), and elsewhere in the tropics during the Pleistocene (Bush et al., 2002; Trauth et al., 2003; Cruz et al., 2005; Wang et al., 2004). The precessional control on tropical moisture has also been clearly illustrated by the climate modelling of Clement et al. (2004), which showed that a 180° shift in precession could change annual precipitation in the tropics by at least 180 mm/year and cause a significant shift in seasonality. Support for increased seasonality during these extreme periods of climate variability also comes from mammalian community structures (Reed, 1997; Bobe and Eck, 2001) and hominin paleo-diet reconstructions (Teaford and Ungar, 2000).

The late Cenozoic periods of extreme climate variability appear to correlate with maxima in the 400 kyr component of the earth's eccentricity cycle. Prior to 2.7 Ma the wet phases appear every 400 ka (see Fig. 13.1). After 2.7 Ma, however, the wet phases appear every 800 ka, with periods of precessional-forced extreme climate variability at 2.7–2.5, 1.9–1.7 and 1.1–0.9 Ma before present, whereas other periods of eccentricity maxima at ~2.2, ~1.4 and ~0.6 Ma are not associated with the alternating formation of large lakes or increased dust. The three late Cenozoic lake phases do, however, correlate with significant global climatic transitions as well as peaks in eccentricity. Hence after 2.7 Ma, global climate changes seem to be required to cause an increased regional climate sensitivity to precessional-forced insolation and increased seasonality, which allows either large deep lakes to develop or causes extreme aridity and large dust loads to the adjacent oceans. In contrast, prior to 2.7 Ma, eccentricity maxima alone were sufficient to produce regional sensitivity. It remains to be determined whether the long-term drying trend in East Africa, or the global cooling trend is responsible for this shift from a simple linear response to long-term eccentricity forcing.

Global Climate Transitions

The last three major Plio-Pleistocene lake phases correspond to global climate transitions. The lake phase at 2.7–2.5 Ma corresponds to intensification of the Northern Hemisphere Glaciation (INHG) (Haug and Tiedemann, 1998), that at 1.9–1.7 Ma to development and significant intensification of the Walker Circulation (Ravelo et al., 2004), and that at 1.1–0.9 Ma to initiation of the Mid-Pleistocene Revolution (Berger and Jansen, 1994). Each of these global climate transitions was accompanied by reduced North Atlantic Deep Water (NADW) formation (Haug and Tiedemann, 1998) and increased ice rafting from both Greenland and Antarctica (St. John and Krissek, 2002; Cowan, 2001). Ice expansion and cooling in either hemispheres would have significantly increased the Pole-Equator thermal gradient, leading to a northern and/or southern compression of the Intertropical Convergence Zone (ITCZ). A similar effect occurred during the Last Glacial Maximum, where a strong compression of the ITCZ is observed both in paleo-reconstructions of tropical hydrology (e.g., Peterson et al., 2000; Chiang et al., 2003; Wang et al., 2005), and *via* climate modelling (Lautenschlager and Herterich, 1990; Bush and Philander, 1999; Bush, 2001). Most important for East Africa moisture availability is the compression of the northern Hemisphere component of the ITCZ because it influences the strength of the SE Asian monsoons. Compression of the ITCZ is thus an essential component to increasing the sensitivity of East Africa to precessional forcing of moisture availability; otherwise moisture is transported north and south away from the Rift Valley. Along the whole length of the rift, without this high-latitude climate control, East Africa cannot receive enough rainfall to fill large deep freshwater lakes during positive precessional periods. Hence after 3 Ma, it seems that both global climate forcing and eccentricity maxima are required to generate episodes of extreme precessional forced climate.

Climate Variability and Early Human Evolution

On time scales of more than 100 kyr, rift-related volcano-tectonic processes shaped the landscape of East Africa and profoundly influenced local climate and surface hydrology through the development of relief. Through uplift of the Kenyan and Ethiopian Plateaus, changes in orography and associated rain shadow are believed to be the major driving force for increased variability of moisture availability throughout Eastern Africa. This increased sensitivity has resulted in a modern Rift Valley that hydrological modelling suggests could support lakes as deep as 150 m with an annual precipitation increase of only 15–30% (Bergner et al., 2003). Prior to the INHG there is a linear relationship between long-term eccentricity variations and the development of deep freshwater lakes in the East African Rift. From the ONHG onwards, global climate transitions, which resulted in an increased Pole-Equator gradients and compression both north and south boundaries of the ITCZ, were required to make East African moisture availability sensitive to maxima in eccentricity and thus changes in precession.

The alternating extreme wet and dry periods would have had a profound affect on the climate and vegetation of East Africa. The sinusoidal precessional forcing at the equator consists of periods of less than 2,000 years, during which 60% of total variation in daily insolation and seasonality occurs. These are followed by ~8,000 years when relatively little change in daily insolation occurred (Maslin et al., 2005; Maslin and Christensen, 2007). Hence, instead of precession being a smooth forcing, it combines rapid strong forcing with long periods of relatively weak forcing. Rapid stratigraphic transitions from deep lacustrine to fluvial deposition associated with the diatomite Pliocene lakes deposits in the Baringo Basin suggests that this sinusoidal precessional-forcing caused lakes to appear rapidly, remain part of the landscape for thousands of years, then disappear rapidly (Deino et al., 2006; Kingston et al., 2007). In fact, the absence of shallow-water diatom species from key Plio-Pleistocene lake deposits (Deino et al., 2006; Kingston et al., 2007) suggests that these lakes could have dried up in less than 500 years. This has important implications for the speciation and dispersal of mammals (including hominins) in East Africa. Figure 13.2 shows that between 5.0 and 0.5 Ma, the periods of highly variable East African climate – those oscillating from very wet to very dry (indicated by the striped blue boxes in Fig. 13.2d) – occupied less than a third of the total time. In contrast, 12 out of the 15 hominin species (~80%) first appeared in one of these extreme 'wet-dry' periods. In particular there seems to be a strong correspondence between these extreme climate periods and the appearance of appearance of such species as *Homo habilis*, *H. rudolfensis*, *H. erectus* and/or *H. ergaster* (See Fig. 13.2g). Even taking into the account the great difficulty in dating the first appearance of African hominins, and the problem of pseudo-speciation events (Vrba, 1993; Smith, 1994), this is compelling evidence for the preferential evolution of hominins during extreme climate periods. What we cannot rule out or account for is the possibility that the increased moisture availability during the periods increases the likely preservation of fossils and thus produces a false correlation.

Linking African Lake Variability and Theories of Human Evolution

We suggest that ephemeral lakes, expanding and contracting on precessional timescales, would have evoked wide-spread, regional-scale, rapid, and extreme environmental variability. However, the difficulty in invoking orbital forced changes in local hydrology arises not out of the question of scale, but of timing: what part of these climate variations may have influenced the speciation and extinction events? Figure 13.3 presents three different models of the lake response to local orbital forcing.

The first model suggests that there is a relatively smooth, gradual transition between periods with deep lakes and periods without lakes. If this 'smooth' model is correct, then there may have been prolonged periods of wet and arid conditions, which may invoke the Red Queen or the Turnover Pulse Hypothesis (TPH) as possible causes of evolution (Van Valen, 1973; Vrba, 1993). Alternatively, there may have been non-linear dynamic changes related to the complex interaction of precipitation, temperature, and seasonality patterns that produced threshold changes in the local vegetation which may have influenced evolution (Maslin, 2004).

The second model is a 'threshold' model, so, instead of a smooth gradual transition from wet and drier condition, the ephemeral lakes expanded and contracted extremely rapidly, producing a wide-spread, regional-scale, rapid, and extreme environmental variability, required by the Variability Selection Hypothesis of human evolution (Potts, 1998). Model three is a more extreme example of the threshold model, in which there is 'extreme climate variability' during the rapid transition from deep-lake to no-lake phases. This would provide extreme short-term variability that could influence speciation and extinction events, especially if this climate change occurred over a large geographic region.

There is, of course, a fourth possibility, namely that all three models contain prolonged extreme wet and dry periods, which would have provided prolonged periods of either extremely abundant or scarce water and food resources. The extreme dry periods would support a model such as the TPH. In contrast, the extreme wet periods, with very deep freshwater lakes, are rare events in the paleoclimatic history of East Africa. As such, speciation events may have occurred in the high energy/high competition environments provided by the wet periods. This would conform to the Red Queen Hypothesis.

At present, the preliminary data from Lake Baringo (Kingston et al., 2007) suggest the diatomites are typically bracketed by 20–30 cm of fine sand and silt horizons containing fish fossils. These grade into high-energy terrestrial facies, indicating relatively rapid cycling between deep lake and fully subaerial conditions. This suggests that for this region, at least, model three – extreme climate variability – is the most likely. What is now required are high resolution paleoclimate data with which to test the different models outlined above. There are also other methods for testing which of the three theoretical models is closest to reality. First, oxygen isotopes of the diatoms in the lake sediment can be analyzed, as these provide a measure of the evaporation-precipitation balance of the whole lake. This, in turn, provides an estimate of how quickly the lake was expanding and contracting. Second, one of us (MT) has already sampled the Late Glacial – Early Holocene paleo-lake in the Suguta Valley of northern Kenya, and its appearance and disappearance has been dated using radiocarbon (Garcin et al., 2009). This will yield an accurate estimate of how quickly recent lakes can vary providing an analog for the older material.

The Pulsed Climate Variability Hypothesis

In summary, new paleoclimate data suggest that the long-term drying trend in East Africa was punctuated by episodes

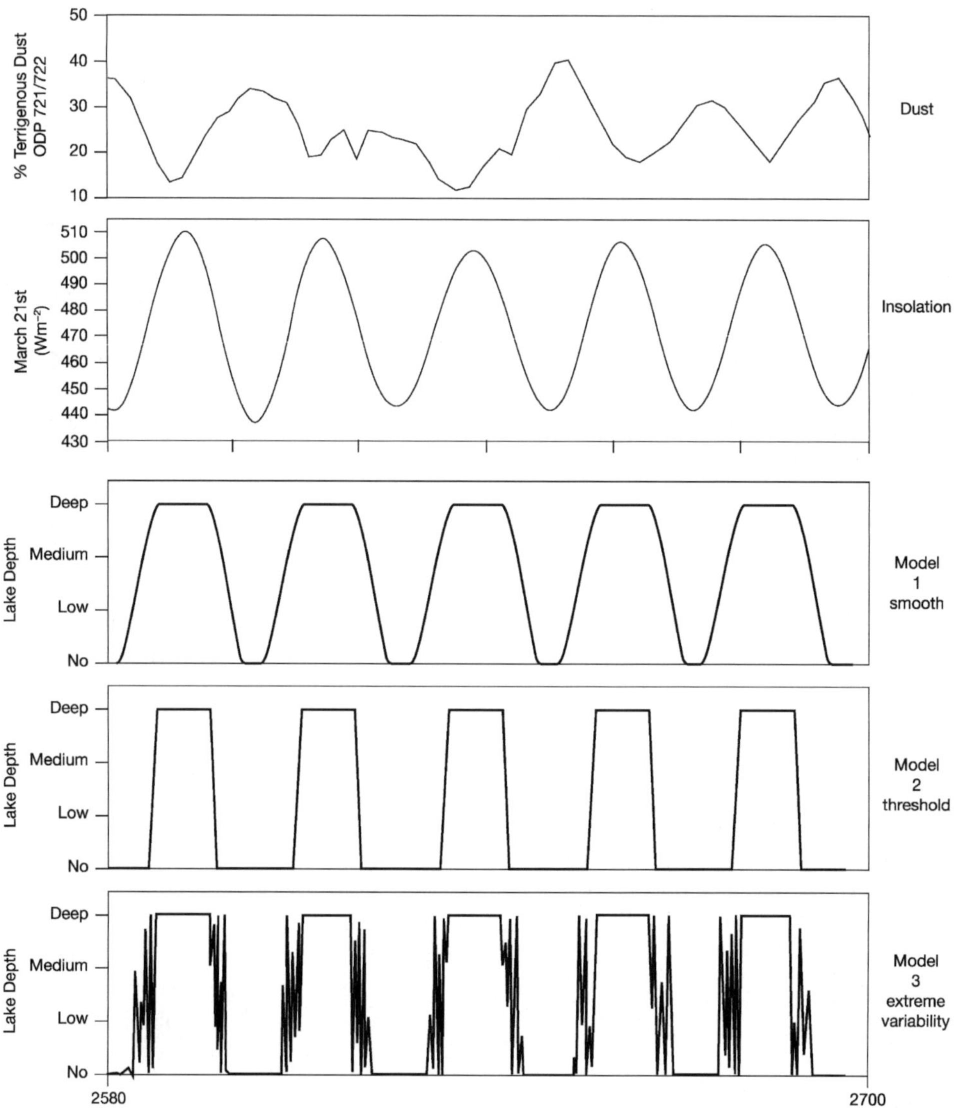

Fig. 13.3 Three theoretical models of lake changes in East Africa during the Plio-Pleistocene. Model 1: 'smooth' and relatively slow transitions from *deep* to *no* lake conditions, which would imply that either high energy wet conditions or prolonged aridity may have influenced human evolution. Model 2: 'threshold' rapid transitions from *deep* to *no* lake conditions, which would imply that rapid transition may have influenced human evolution, or the high energy wet conditions or prolonged aridity as suggested by Model 1. Model 3: 'extreme variability,' with high variability during the transitions between *deep* and *no* lake conditions, which implies variability influenced human evolution or, again, either high energy wet conditions or prolonged aridity.

of short, alternating periods of extreme humidity and aridity. These periods of 'pulsed climate variability' are characterized by the precession-forced appearance and disappearance of large, deep lakes in the East African Rift Valley, and are paralleled by low and high wind-driven dust loads reaching the adjacent ocean basins. During the last 3 million years, such periods only occur at the times of major global climatic transitions, such as the intensification of Northern Hemisphere Glaciation (2.7–2.5 Ma), development of the Walker circulation (1.9–1.7 Ma), and the Mid-Pleistocene Revolution (1.0–0.7 Ma). We suggest that high latitude forcing in both hemispheres is required to compress the Inter-Tropical Convergence Zone so that East Africa becomes locally sensitive to precessional forcing, resulting in rapid shifts from wet to dry conditions. Building on Potts' (1998) variability selection hypothesis, we suggest that these periods of pulsed climate variability may have provided a catalyst for evolutionary change, and driven key speciation and dispersal events amongst mammals and hominins in Africa. Hominin species, in particular the early species attributed to the genus *Homo*, seem to differentially originate and go extinct during periods of extreme climate variability. Results presented in this paper may represent the basis of a new theory of early human evolution in Africa.

Acknowledgments This project was funded by three grants to M.T. and Prof. Manfred Strecker by the German Research Foundation (DFG), and a UCL Graduate School grant to M.M. Moreover this paper could not have been written without the excellent fieldwork of the Potsdam research group. We also thank the government of Kenya and the Kenya Wildlife Service for research permits and support. The authors thank L. Aellio, A. Bergner, P. Blisniuk, B. Christensen, P. deMenocal, A. Deino, M. Dühnforth, F. Gasse, F. Grine, P. Hopley, J. Kingston, R. Leakey, A. Lister and G. Muchemi for inspiring discussions and the anonymous reviewers for their helpful suggestions. We also thank S. Higgins, S. Kabingu, T. Schlüter and M. Ibs-von Seht for logistical support. M.M. would like to dedicate this paper to memory of Nick Shackleton, friend and mentor.

References

Ashley, G.M., Hay, R.L., 2002. Sedimentary patterns in a Plio-Pleistocene volcaniclastic rift-margin basin, Olduvai Gorge, Tanzania. In: Renaut, R.W., Ashley, G.M. (Eds), Sedimentation in Continental Rifts. Volume 73 of SEPM (Society for Sedimentary Geology) Special Publication. SEPM, Tulsa, pp. 107–122.

Baker, B.H., Mitchell, J.G., Williams, L.A.J., 1988. Stratigraphy, geochronology and volcano-tectonic evolution of the Kedong-Naivasha-Kinangop region, Gregory Rift Valley, Kenya. Journal of the Geological Society of London 145, 107–116.

Behrensmeyer, A.K., Potts, R., Deino, A., Ditchfield, P., 2002. Olorgesailie, Kenya: a million years in the life of a rift basin. In: Renaut, R.W., Ashley, G.M. (Eds), Sedimentation in Continental Rifts. Volume 73 of SEPM (Society for Sedimentary Geology) Special Publication. SEPM, Tulsa, pp. 97–106.

Berger, W.H., Jansen, E., 1994. Mid-Pleistocene climate shift: the Nansen connection. In: Johannessen, O., Muench, R. Overland, J. (Eds), The Polar Oceans and Their Role in Shaping the Global Environment. Geophysical Monograph 85, American Geophysical Union, Washington, DC, pp. 295–311.

Berger, A., Loutre, M.F., 1991. Insolation values for the climate of the last 10 million years. Quaternary Science Reviews 10, 297–317.

Bergner, A.G.N., Trauth, M.H., Bookhagen, B., 2003. Magnitude of precipitation/evaporation changes in the Naivasha Basin (Kenya) during the last 150 kyrs. Global and Planetary Change 36, 117–135.

Bobe, R., Eck, G.G., 2001. Response of African bovids to Pliocene climatic change. Paleobiology 27, 1–47.

Bosworth, W., Strecker, M.R., 1997. Stress field changes in the Afro-Arabian Rift system during the Miocene to recent period. Tectonophysics 278, 47–62.

Boven, A., 1992. The applicability of the K-Ar and $^{40}Ar/^{39}Ar$ techniques to Quaternary volcanics: a fundamental and experimental investigation. Ph.D. thesis, Vrije Universiteit Brussels.

Brown, F.H., Feibel, C.S., 1991. Stratigraphy, depositional environments and palaeogeography of the Koobi Fora Formation. In: Harris, J.M. (Ed), Koobi Fora Research Project, Volume 3. Clarendon, Oxford, pp. 1–30.

Bush, A.B.G., 2001. Simulating climates of the Last Glacial Maximum and of the mid-Holocene: wind changes, atmosphere-ocean interactions, and the tropical thermocline. In: Seidov, D., Haupt, B.J., Maslin, M. (Eds), The Oceans and Rapid Climate Change: Past, Present, and Future. Geophysical Monograph Series, 126, American Geophysical Union, Washington, DC, pp. 135–144.

Bush, A.B.G., Philander, S.G.H., 1999. The climate of the Last Glacial Maximum: results from a coupled atmosphere-ocean general circulation model. Journal of Geophysical Research 104(D20), 24509–24525.

Bush, M.B., Moreno, E., de Oliveira, P.E., Colinvaux, P.A., 2002. Orbital-forcing signal in sediments of two Amazonian lakes. Journal of Paleolimnology 27, 341–352.

Butzer, K.W., Brown, F.W., Thruber, D.L., 1969. Horizontal sediments of the lower Omo Valley; the Kibish Formation. Quaternaria 11, 15–29.

Chiang, J.C.H., Biasutti, M., Battisti, D.S., 2003. Sensitivity of the Atlantic Intertropical Convergence Zone to Last Glacial Maximum boundary conditions. Paleoceanography 18, 345–358.

Clement, A.C., Hall., A., Broccoli, A.J., 2004. The importance of precessional signals in the tropical climate. Climate Dynamics 22, 327–341.

Cowan, E.A., 2001. Identification of the glacial signal from the Antarctic Peninsula since 3.0 Ma at Site 1011 in a continental rise sediment drift. In: Barker, P.F., Camerlenghi, A., Acton, G.D., Ramsay, A.T.S. (Eds), Proceedings of the Ocean Drilling Program, Scientific Results, Volume 178 [Online]: Available at <http://www.odp.tamu.edu/publications/178_SR/chap_10/chap_10.htm>.

Cruz, F.W., Burns, S.J., Karmann, I., Sharp, W.D., Vuille, M., Cardoso, A.O., Ferrari, J.A., Silva Dias, P.L., Viana, O., 2005. Insolation-driven changes in atmospheric circulation over the past 116,000 years in subtropical Brazil. Nature 434, 63–66.

Deino, A.L., Kingston, J.D., Glen, J.M., Edgar, R.K., Hill, A., 2006. Precessional forcing of lacustrine sedimentation in the Late Cenozoic Chemeron Basin, Central Kenya Rift, and calibration of the Gauss/Matuyama boundary. Earth and Planetary Science Letters 247, 41–60.

deMenocal, P., 1995. Plio-Pleistocene African climate. Science 270, 53–59.

deMenocal, P., 2004. African climate change and faunal evolution during the Pliocene-Pleistocene. Earth and Planetary Science Letters 220, 3–24.

Denison, S., Maslin, M.A., Boot, C., Pancost, R., Ettwein, V., 2005. Precession-forced changes in South West African vegetation during marine oxygen isotope stages 100 and 101. Palaeogeography, Palaeoclimatology, Palaeoecology 220, 375–386.

Dunsworth, H., Walker, A., 2002. Early genus *Homo*. In: Hartwig W.C. (Ed), The Primate Fossil Record. Cambridge University Press, New York, 401–406.

Ebinger, C.J., Temane, T., Harding, D., Tesfaye, S., Kelley, S., Rex, D., 2000. Rift deflection, migration, and propagation: linkage of the Ethiopian and Eastern rifts, Africa. Geological Society of America Bulletin 112, 163–176.

Evernden, J.G., Curtis, G.H., 1965. The potassium-argon dating of Late Cenozoic rocks in East Africa and Italy. Current Anthropology 6, 177–189.

Feakins, S.J., deMenocal, P.B., Eglinton, T.I., 2005. Biomarker records of Late Neogene changes in northeast African vegetation. Geology 33, 977–980.

Foster, A., Ebinger, C., Mbede, E., Rex, D., 1997. Tectonic development of the northern Tanzanian sector of the East African Rift System. Journal of the Geological Society 154, 689–700.

Garcin, Y., Junginger, A., Melnick, D., Olago, D., Strecker, M.R., Trauth, M.H., 2009. Late Pleistocene-Holocene rise and collapse of Lake Suguta, northern Kenya Rift. Quaternary Science Reviews, doi:10.1016/j.quascirev.2008.12.006.

Gasse, F., 1990, Tectonic and climatic controls on lake distribution and environments in Afar from Miocene to present. In: Katz, B.J. (Ed), Lacustrine Basin Exploration, Case Studies, and Modern Analogs, Volume 3. American Association of Petroleum Geologists Memoir, pp. 1–340.

Haug, G.H., Tiedemann, R., 1998. Effect of the formation of the Isthmus of Panama on Atlantic Ocean thermohaline circulation. Nature 393, 673–675.

Hillarie-Marcel, C., Carro, O., Casanova, J., 1986. 14C and Th/U dating of Pleistocene and Holocene stromatolites from East African paleolakes. Quaternary Research 25, 312–329.

Hopley P., Marshall, J., Weedon, G., Latham, A., Herries, A., Kuykendall, K., 2007. Orbital forcing and the spread of C4 grasses in the late Neogene: stable isotope evidence from South African speleothems. Journal of Human Evolution 53, 620–634.

Kingston J., Deino, A., Edgar, R., Hill, A., 2007. Astronomically forced climate change in the Kenyan Rift Valley 2.7 - 2.55 Ma: implications for the evolution of early hominin ecosystems. Journal of Human Evolution 53, 487–503.

Larrasoaña J. C., Roberts, A.P., Rohling, Winklhofer, M., Wehausen, R., 2003. Three million years of monsoon variability over the northern Sahara. Climate Dynamics 21, 689–698.

Lautenschlager, M., Herterich, K, 1990. Atmospheric response to ice age conditions - climatology near the earth's surface. Journal of Geophysical Research 95, 22547–22557.

Lepre, C.J., Quinn, R., Joordens, J., Swisher III, C., Feibel, C., 2007. Plio-Pleistocene facies environments from the KBS Member of the Koobi Fora Formation. Journal of Human Evolution 53, 504–514.

Levin, N.E., Quade, J., Simpson, S.W., Semaw, S., Rogers, M., 2004. Isotopic evidence for Plio-Pleistocene environmental change at Gona, Ethiopia. Earth and Planetary Science Letters 219, 93–110.

Lourens, L., Hilgen, F., Shackleton, N.J., Laskar, J., Wilson, D., 2004. The Neogene period. In: Gradstein, F., Ogg, J.G., Smith, G. (Eds), A Geologic Time Scale 2004. Cambridge University Press, Cambridge, pp. 409–440.

Maslin, M.A., 2004. Ecological verses climatic thresholds. Science, 306, 2197–2198.

Maslin, M.A., Christensen, B., 2007. Tectonics, orbital forcing, global climate change, and human evolution in Africa. Journal of Human Evolution 53, 443–464.

Maslin, M.A., Seidov, D., Lowe, J., 2001. Synthesis cf the nature and causes of sudden climate transitions during the Quaternary. In: Seidov, D., Haupt, B.J, Maslin, M. (Eds), The Oceans and Rapid Climate Change: Past, Present, and Future, Geophysical Monograph Series 126, American Geophysical Union, Washington, DC, pp. 9–52.

Maslin, M.A., Mahli, Y., Phillips O., Cowling S., 2005. New views on an old forest: assessing the longevity, resilience and future of the Amazon rainforest. Transactions of the Institute of British Geographers 30, 390–401.

McDougall, I., Watkins, R.T., 1988. Potassium-Argon ages of volcanic rocks from northeast of Lake Turkana, northern Kenya. Geological Magazine 125, 15–23.

McHenry, H.M., 2002. Introduction to the fossil record of human ancestry. In: Hartwig W.C. (Ed), The Primate Fossil Record. Cambridge University Press, New York, pp. 401–406.

Owen, R.B., 2002. Sedimentological characteristics and origins of diatomaceous deposits in the East African Rift System. Volume 73, SEPM (Society for Sedimentary Geology) Special Publication. SEPM, Tulsa, pp. 233–246.

Peterson, L.C., Haug, G., Hughen, K., Rohl, U., 2000. Rapid changes in the hydrologic cycle of the tropical Atlantic during the Last Glacial. Science 290, 1947–1951.

Potts, R., 1998. Environmental hypothesis of hominin evolution. Yearbook of Physical Anthropology 41, 93–136.

Potts, R., 1999. Paleolandscape variation and Early Pleistocene hominid activities: members 1 and 7, Olorgesailie Formation, Kenya. Journal of Human Evolution 37, 747–788.

Ravelo, C., Andreasen, D., Lyle, M, Lyle, A.O., Wara, M.W., 2004. Regional climate shifts caused by gradual global cooling in the Pliocene epoch. Nature 429, 263–267.

Reed, K.E., 1997. Early hominid evolution and ecological change through the African Plio-Pleistocene. Journal of Human Evolution 32, 289–322.

Ruddiman, W.F., Raymo, M.E., 1988. Northern Hemisphere climate regimes during the past 3 Ma: possible tectonic connections. Philosophical Transactions of the Royal Society London B 318, 411–430.

Segalen, L., Lee-Thorp, J.A., Cerling, T., 2007. Timing of C4 grasses expansion across sub-Saharan Africa. Journal of Human Evolution 53, 549–559.

Sepulchre P., Ramstein, G., Fluteau, F., Schuster, M., 2006. Tectonic uplift and eastern Africa aridification. Science 313, 1419–23.

Smith, A.B., 1994. Systematics and the Fossil Record. Blackwell, London.

St. John, K., Krissek, L., 2002. The Late Miocene to Pleistocene ice-rafting history of southeast Greenland. Boreas 41, 28–35.

Strecker, M.R., 1991. Das zentrale und südliche Kenia-Rift unter besonderer Berücksichtigung der neotektonischen Entwicklung. Habilitation thesis, University of Karlsruhe, Germany.

Strecker, M.R., Blisniuk, P.M., Eisbacher, G.H., 1990. Rotation of extension direction in the central Kenya Rift. Geology 18, 299–302.

Sturchio, N.C., Dunkley, P.N., Smith, M., 1993. Climate-driven variations in geothermal activity in the northern Kenya rift valley. Nature 362, 233–234.

Sundquist E.T., Visser K., 2004. The geologic history of the carbon cycle. In: Schlesinger, W.H. (Ed), Treatise on Geochemistry, Volume 8: Biogeochemistry. Elsevier, Amsterdam, pp. 425–472.

Teaford, M.F., Ungar, P.S., 2000. Diet and the evolution of the earliest human ancestors. Proceedings of the National Academy of Science USA 97, 13506–13511.

Tiercelin, J.J., Lezzar, K.E., 2002. A 300 million years history of rift lakes in Central and East Africa: an updated broad review. In: Odada, E.O., Olago, D.O., (Eds), The East African Great Lakes: Limnology, Paleolimnology and Biodiversity. Kluwer, Dordrecht, pp. 3–60.

Trauth, M.H., Deino, A., Bergner, A.G.N., Strecker, M.R., 2003. East African climate change and orbital forcing during the last 175 kyr BP. Earth and Planetary Science Letters 206, 297–313.

Trauth, M., Maslin, M.A., Deino, A., Strecker M., 2005. Late Cenozoic moisture history of East Africa. Science 309, 2051–2053.

Trauth, M.H., Maslin, M., Deino, A., Strecker, M.R., Bergner, A.G.N., Duhnforth, M., 2007. High and low latitude forcing of Plio-Pleistocene East African climate and human evolution. Journal of Human Evolution 53, 475–486.

Van Valen, L., 1973. A new evolutionary law. Evolutionary Theory 1, 1–30.

Vrba, E.S., 1993. Turnover-pulses, the Red Queen, and related topics. American Journal of Science 293A(S1), 418–452.

Walter, R.C., Manega, P.C., Hay, R.L., Drake, R.E., Curtis, G.H., 1991. Laser-fusion $^{40}Ar/^{39}Ar$ dating of Bed I, Olduvai Gorge, Tanzania. Nature 354, 145–194.

Wang, X., Auler, A.S., Edwards, R.L., Cheng, H., Cristalli, P.S., Smart, P.L., Richards, D.A., Shen, C.C., 2004. Wet periods in northeastern Brazil over the past 210kyr linked to distant climate anomalies. Nature 432, 740–743.

White, T., 2002. Earliest hominids. In: Hartwig W.C. (Ed), The Primate Fossil Record. Cambridge University Press, New York, pp. 406–417.

White, T.D., WoldeGabriel, G., and 20 other authors, 2006. Asa Issie, Aramis and the origin of *Australopithecus*. Nature 440, 883–889.

Williams, L.A.J., Macdonald, R., Leat, P.T., 1983. Magmatic and structural evolution of the central part of the Kenya Rift. In: Proceedings of Regional Seminar on Geothermal Energy in Eastern and Southern Africa, UNESCO/USAID, Nairobi, pp. 61–67.

Williams, M.A.J., Williams, F.M., Gasse, F., Curtis, G.H., Adamson, D.A., 1979. Pliocene-Pleistocene environments at Gadeb prehistoric site, Ethiopia. Nature 282, 29–33.

WoldeGabriel, G., Aronson, J., Walter, R., 2000. Volcanism, tectonism, sedimentation, and the paleoanthropological record in the Ethiopian Rift System. Geological Society of America Special Paper 345, 83–99.

Wynn, J.G, 2004. Influence of Plio-Pleistocene aridification on human evolution: evidence from paleosols of the Turkana Basin, Kenya. American Journal of Physical Anthropology 123, 106–118.

Chapter 14
Tracking Ecological Change in Relation to the Emergence of *Homo* Near the Plio-Pleistocene Boundary

Kaye E. Reed and Samantha M. Russak

Keywords Mammalian community structure • *Homo habilis* • *Homo erectus* • East Africa • South Africa

Introduction

The emergence of species included in the genus *Homo* at ~2.4 Ma, as well as appearances of other hominin taxa between 3.0 and 1.8 Ma have often been attributed to global climate change (deMenocal, 1995; deMenocal and Bloemendal, 1995; Vrba, 1995), but the precise details of these climatic transitions, including causal factors, are still debated. In addition, the Plio-Pleistocene boundary (~1.8 Ma; Pasini and Colalongo, 1997) is an important benchmark in human evolution as many hominin species (e.g., *Homo erectus/ergaster, Paranthropus robustus*) have first appearance data (FADs) close to this date, while other taxa (e.g., *Homo rudolfensis, Paranthropus aethiopicus*) appear to have last appearance data (LADs) prior to this time. There are also some species that are found on both sides of the boundary (e.g., *Paranthropus boisei, Homo habilis*) (Kimbel, 1995; Wood and Richmond, 2000; Cameron, 2003; Spoor et al., 2007).

Perhaps the most frequently cited relationship between climate and faunal change is Vrba's turnover pulse hypothesis (Vrba, 1995; Potts, 1998a; Behrensmeyer, 2006) which refers to an intensity of turnover (speciation and extinction events) during a brief period of time as a result of environmental change. More specifically, synchronous change in multiple groups was said to have occurred at 2.5 Ma due to a shift from a moist, warm habitat to drier, cooler and more open conditions. More recently, this hypothesis has been countered by one positing a prolonged and gradual period of turnover between 2.5 and 1.8 Ma (Behrensmeyer et al., 1997), although this study is specific to the Turkana Basin. Examinations of individual site patterns in East Africa have shown that turnovers in some mammalian lineages occur during the 3.0–2.0 Ma period, but that the exact timing appears to fluctuate depending upon site and type of analysis (Bobe and Eck, 2001; Alemseged, 2003). It has become apparent that the picture of hominin evolution is thus contextually diverse. Strategic analyses in different basins across both time and space are necessary to further our understanding of evolution during this time period (Behrensmeyer, 2006).

East and South Africa underwent major climatic and subsequent environmental changes from the Late Pliocene to Early Pleistocene. Marine records show marked shifts in amplitude and variation at around 2.8, 1.7, and 1.0 Ma, characterizing the climate during these time periods as a continuum of alternating wet and dry conditions (deMenocal, 1995). Analyses of paleosol carbonates from the Turkana and Olduvai basins indicate that open savanna grasslands gradually replaced woodland environments between 3.0 and 1.0 Ma and that there was a particular increase in aridity between 1.8–1.6 Ma (Cerling, 1992; deMenocal, 1995, 2004; Potts, 1998b; Wynn, 2004). Faunal analyses of particular sites also indicate gradual change toward more open woodland habitats until about 1.8 Ma when there is a great increase in arid and grazing adapted mammals (Reed, 1997; Spencer, 1997; Bobe et al., 2002; Bobe and Behrensmeyer, 2004; Bobe and Leakey, 2009).

Here we explore pan-African habitats at various hominin sites across three time periods. The first block of time is centered on 3.0 Ma (3.31–2.85 Ma) to examine habitats before the emergence of *Homo* in the fossil record. The second time period brackets 2.4 Ma (2.52–2.33 Ma) to represent the appearance of *Homo* species (as well as other hominins). We use this as a marker considering that the first stone tools, often attributed to *Homo* species, appear at 2.58 Ma (Semaw et al., 2003) and the first well-dated *Homo* cf. *H. habilis* is

K.E. Reed (✉)
School of Human Evolution and Social Change,
Institute of Human Origins, Arizona State University,
P.O. Box 874101, Tempe, AZ 85287-4101, USA
e-mail: kreed@asu.edu

S.M. Russak
School of Human Evolution and Social Change,
Institute of Human Origins, Arizona State University,
P.O. Box 872402, Tempe, AZ 85287-2402, USA
e-mail: Samantha.russak@asu.edu

dated to 2.33 Ma (Kimbel et al., 1996). Finally, we examine the time period surrounding the Plio-Pleistocene boundary at 1.8 Ma (2.0–1.6 Ma) to explore environmental changes that might be associated with the appearance of *H. ergaster/erectus* in the fossil record. We ask three questions: (1) What were the broad based habitats that characterized African fossil sites directly before the appearance of *Homo*? (2) Was there a change of habitat associated with the emergence *Homo* species? (3) Was there a discernable change of habitat across the Plio-Pleistocene boundary after which specimens of *Homo erectus* increase substantially?

To address these questions we reconstruct the habitats of several East and South African hominin sites through ecological analyses of the associated large mammal fauna. The early sites include stratigraphic members from Hadar, Koobi Fora, West Turkana, and Makapansgat. Hadar, Koobi Fora, West Turkana, Shungura, and Sterkfontein represent the middle time period. Although many of the postcranial fossils attributed to *Homo* at Olduvai and the Turkana Basin may not be referable to species, we expect that the increase of *Homo erectus* specimens occurred sometime during the late period. To explore this time bracket, we use the Turkana and Olduvai basins in the broad sense, as they are well dated and span this full range of time. In addition, we use Swartkrans (~1.8 Ma) for a broader perspective on *Homo erectus*. We will reconstruct the habitats of these localities using a subset of large mammal adaptations that have proven to be useful in that endeavor (Reed, 1998, 2002, 2008). General trends in the changes of composition of the faunal communities, including hominin species, at these sites can possibly be correlated with larger environmental changes. In addition, the amount of dissimilarity between faunal assemblages from which earlier hominins, *Homo habilis* and those from which *H. erectus* has been recovered, can indirectly indicate the degree of changes occurring in species representations at this time.

Materials and Methods

Modern Habitat Sites

Table 14.1 lists the modern areas from which large mammal communities were derived (Swynnerton, 1958; Lamprey, 1962; Child, 1964; Vesey-Fitzgerald, 1964; Sheppe and Osborne, 1971; Smithers, 1971; Ansell, 1978; Rautenbach, 1976, 1978a, b; Behrensmeyer et al., 1979; Delany and Happold, 1979; Bremen and de Wit, 1983; Happold, 1987; Lanjouw, 1987; Emmons et al., 1983; Ansell and Dowsett, 1988; Skinner and Smithers, 1990). The habitat sample includes forests, closed woodlands, bushland, open woodlands, shrublands, ecotones, and grasslands. The ecotone habitat refers to disparate habitats that adjoin abruptly with no transitional habitat, e.g., forest and grassland or forest and mountain heath. In a sense, all African habitats today are either ecotone or mosaic for several reasons. First, rivers and lakes provide underground water that effect the landscape close to their courses and thus one can have a riverine forest within an overall shrubland habitat, e.g., modern Hadar. Second, geomorphology, drainage patterns, and soil types provide different environments within which plants grow, thus interspersing bushland and woodland habitats with patches of grasslands or forests. Here, we classify grasslands as both secondary and edaphic. Edaphic grassland in this instance refers to grasses that are water-logged for much of the year or are formed due to overbank flood deposits providing rich soils in which grasses can rapidly grow. Thus, there would also be great seasonal diversity in habitat structure in these regions. Secondary grasslands are a relatively new phenomenon in Africa and are due to regular burning and/or overgrazing (White, 1983; Pratt and Gwynne, 1977; Spencer, 1997). For more information on habitat types see Reed and Rector (2007) and White (1983).

Hominin Fossil Sites

This study focuses on three fossiliferous sedimentary accumulations in the Turkana Basin (Shungura Formation, Omo Valley, Ethiopia; Koobi Fora Formation, East Turkana, Kenya; and Nachukui Formation, West Turkana, Kenya), as well as the Hadar Formation (Ethiopia), Olduvai Gorge (Tanzania), and cave deposits in South Africa (Makapansgat, Sterkfontein and Swartkrans). Areas along the Rift Valley in East Africa were chosen because the fossil assemblages and Plio-Pleistocene strata of each of these sites are well documented. Further, several of these formations offer an almost continuous sampling of time between 1.6 and 2.0 Ma (Table 14.2).

The Hadar Formation provides a continuous sequence of deposits from 3.6 through 2.94 Ma (Alemseged et al., 2005; Campisano, 2007). Each member has a volcanic tephra at its base, and each member has been divided into sub-members using various marker beds. The Hadar site was dominated by a large meandering river with ephemeral tributaries, with occasional transgressions of paleolake Hadar from the east throughout the sequence (Campisano, 2007). For this study, we use the mammals recovered from the Kada Hadar Member (3.18–2.94 Ma) as representative of the early time period and the Makaamitalu region (2.33 Ma) as representative of the middle time period.

Olduvai Gorge is a steep-sided ravine located in northern Tanzania that is well known for its hominin-bearing deposits, particularly those associated with *H. habilis* and *Paranthropus boisei*. Bed I is composed mainly of coarse volcanic material, deposited in lacustrine sedimentary environments, and is dated from about 2.0–1.75 Ma (Walter et al., 1991; Fernandez-Jalvo

Table 14.1 Modern African sites

Locality	Code	General habitat	Annual rainfall (mm)
Aberdares NP, Kenya	Ab	Montane heath/forest (ecotone)	750
Amboseli NP, Kenya	Amb	Shrubland/bushland/grassland	510
Chobe NP, Botswana[a]	Ch	Scrub woodland	650
Congo Basin, DRC	Co	Edaphic rainforest	1,832
East of Cross River, Nigeria	Ec	Rainforest	1,550
East of Niger River, Nigeria	En	Rainforest	1,596
Gemsbok NP, South Africa	Gn	Shrubland	140
Golden Gate NP, South Africa	Gg	Grassland/woodland	450
Guinea Woodland, Nigeria	Gw	Closed woodland	1,000
Hadar, Ethiopia	Ha	Shrubland/gallery forest	150
Hluhluwe NP, South Africa	Hl	Woodland/floodplains	750
Kafue Flats, Zambia[a]	Kf	Edaphic grasslands (flood plain)	821
Kafue NP, Zambia[a]	Kfn	Edaphic grassland (flood plain)/woodland	821
Kapama NP, South Africa	Kap	Woodland	400
Karoo (Nama), South Africa	Kar	Shrubland	300
Kidepo NP, Uganda	Kid	Bushland/grassland	625
Kilimanjaro, Tanzania	Kil	Montane forest	1,050
Kruger NP, South Africa	Kru	Woodland/bushland	675
Lake Mweru, Zambia[a]	Lm	Bushland/woodland	750
Lake Nkuru, Kenya	Ln	Woodland/open grassland (ecotone)	1,000
Linyanti Swamp, Botswana[a]	Ls	Marsh	650
Liwonde NP, Malawi	Li	Edaphic grassland (flood plain)/woodland	750
Makakou, Gabon	Ma	Rainforest	1,800
Masai Mara, Kenya	Mm	Forest/grassland (ecotone)	1,000
Miambo Woodland, Angola	Ang	Woodland	850
Namib Desert, Namibia	Na	Desert	<150
Natal Woodland, South Africa	Nw	Closed woodland	875
Nyika NP, Malawi	Ny	Forest/grassland (ecotone)	1,200
Okavango Delta, Botswana[a]	Ok	Shrubland/woodland/edaphic grassland	600
Rukwa Valley, Tanzania	Rk	Bushland/woodland	700
Rwenzori NP, Uganda	Rw	Closed woodland	900
Sahel Savanna, Nigeria	Sa	Shrubland/grasses	450
Serengeti Bushland, Tanzania	Sb	Bushland	803
Serengeti NP, Tanzania	Snp	Bushland/woodland/grassland	750
Serengeti Plains, Tanzania	Sp	Grassland	500
SS Grasslands, South Africa	Ssg	Grassland	500
Sudan Woodland, Nigeria	Sw	Woodland	689
Tai Forest, Ivory Coast	Ti	Rainforest	1,900
Tarangire NP, Tanzania	Ta	Scrub woodland	600
Tongwe NP, Tanzania	To	Forest/woodland (ecotone)	1,012
Tsavo NP, Kenya	Ts	Bushland/woodland	500
West Lunga NP, Zambia[a]	Wl	Bushland/woodland	875
West of Niger River, Nigeria	Wn	Rainforest	1,600

[a] Sites that have non-rainfall abundant water.

et al., 1998). Bed II is characterized by lacustrine and fluvial sediments in the lower series, aeolian sands and weathered clays in the middle, and once again by fluvial and lacustrine deposits in the upper series; it has been dated to 1.75–1.2 Ma (Curtis and Hay, 1972; Walter et al., 1991). The paleoenvironments of both Beds I and II have been previously reconstructed as grasslands (Reed and Rector, 2007). Beds I and II are here included in part of the later time period.

The Shungura Formation, part of the Omo Group deposits in Ethiopia, is one of the best-dated sequences, with 12 members each marked by a volcanic tephra at its base (Feibel et al., 1989; Brown, 1995). For most of the Late Pliocene, the lower Omo Basin was dominated by a large meandering river, with a lacustrine transgression at about 2.1 Ma, possibly as a result of tectonic uplift in the area (Brown, 1995; Bobe et al., 2002). Therefore, most of the fossils from Members A to lower G are from fluvial deposits, while those from upper Member G were deposited in fine-grained lacustrine sediments (Howell et al., 1987). Members C through G are used in this study as part of the middle time-period.

The deposits of West Turkana comprise the Nachukui Formation, which is divided into eight members, ranging from greater than 4–0.7 Ma. Much like the Shungura Formation, early deposits are characterized by fluvial sediments, shifting towards a lacustrine depositional environment at around 2.0 Ma (Brown and Feibel, 1991; Bruga et al., 2003). We use the middle and upper Lomekwi Member as part of the early time-period, the mammals from the "Black Skull" site and Kalochoro Member to represent the middle time-period, and the Katio and Natoo Members as the later time-period.

The Koobi Fora Formation in East Turkana is also divided into eight members, ranging in age from 4.2 Ma (Lonyumun Member) to 0.7 Ma (Chari Member) and representing five major depositional environments: fluvial channel, fluvial floodplain, lake margin, and lake basin (Brown and Feibel, 1991; McDougall and Brown, 2006). Here we include the Tulu Bor Member in the early time period, and the Upper Burgi, KBS and Okote Members in the late time period.

The three South African localities are cave sites that have been accumulated by various carnivores or possibly hill wash (Maguire et al., 1980). Makapansgat Member 3 is a fairly homogenous deposit (Reed, 1998) and is included in the early time period. Sterkfontein Member 4 has been roughly estimated to range in age from 2.8 to 2.3 Ma, and will be used here in the middle time period. Finally, Swartkrans Member 1, from which both *P. robustus* and *Homo* have been recovered, will represent the later time period.

Table 14.2 Fossil localities used in analyses

Formation	Member or site	Bracketing dates in Ma
Hadar	Makaamitalu (HMAK)	>2.33
	Kada Hadar 2 (KH2)	3[a]–2.95
	Kada Hadar 1 (KH1)	3.18–3.0[a]
Koobi Fora	Okote (OK)	1.6
	KBS (KBS)	1.88–1.6
	Upper Burgi (UB)	2.0–1.88
	Tulu Bor (TB)	3.4–2.95
Shungura	Shungura G (SHG)	2.33–1.90
	Shungura F (SHF)	2.40–2.36
	Shungura D (SHD)	2.52–2.40
	Shungura C (SHC)	2.85–2.52
South Africa	Swartkrans 1 (SK1)	1.8[a] (2.0–1.6)
	Sterkfontein 4 (ST4)	2.7[a] (2.8–2.4)
	Makapansgat 3 (M3)	3.0[a] (3.2–2.8)
West Turkana	Natoo (NT)	1.60–1.34
	Kaito (KAI)	1.88–1.60
	Kalachoro (KC)	2.35–1.88
	WT17000 (W17)	2.50–2.35
	Upper Lomekwi (UL)	2.94–2.52
	Lower-middle Lomekwi (ML)	3.36–2.94
Olduvai	Olduvai I (OLDI)	2.0–1.75
	Olduvai II (OLDII)	1.75–1.2

[a] Dates are estimated.

Modern Mammal Communities

Mammals greater than 4 kg were included in species lists from each modern habitat locality. These species were assigned to trophic and substrate categories (Table 14.3) based on observations, stomach contents (Hoffman and Stewart, 1972), and mammal guides (Dorst and Dandelot, 1969; Kingdon, 1974a, b, 1977, 1979, 1982a, b; Skinner and

Table 14.3 Adaptations of extant and fossil mammals (From Reed, 2008)

Adaptation	Description	Example
Substrate		
Arboreal (A)		*Colobus guereza*
Aquatic (AQ)		*Hippopotamus amphibius*
Fossorial (F)		*Mellivora capensis*
Terrestrial/arboreal (TA)		*Panthera pardus*
Terrestrial (T)		*Connochaetes taurinus*
Trophic		
Browser (B)	Dicot bushes, trees	*Tragelaphus strepsiceros*
Meat eater (C)	Flesh	*Panthera leo*
Meat/bone eater (CB)	Flesh, bone	*Crocuta crocuta*
Meat/invertebrates (CI)	Flesh, insects	*Genetta genetta*
Fresh grass grazer (FG)	Floodplain/wetland grasses	*Kobus vardoni*
Frugivore (FL)	Fruit with leaves or insects	*Chlorocebus aethiops*
Grazer (G)	Grass	*Alcelaphus buselaphus*
Insectivore	Insects	*Orycteropus afer*
Browser/grazer (MF)	Mixed grass and leaves	*Gazella granti*
Omnivore (OM)	No preference	*Mellivora capensis*
Roots, tubers (R)	USOs	*Hystrix africaustralis*

Boldface indicates those adaptations that are significantly different among more than one pair of habitats.

Smithers, 1990; Kitchener, 1991). The numbers of species in each trophic and substrate category for each community were tabulated. For some analyses the percentages of each of the trophic and substrate adaptations were calculated for each community. Previous studies have shown that particular adaptations could be used to separate habitat types from one another (Reed 1997, 1998; Reed and Rector, 2007). The modern sites were analyzed with correspondence analysis using the seven mammalian adaptations that were found to be significantly different among habitats: arboreal, terrestrial, and aquatic substrate use and fruit-eating, grazing, fresh grass grazing and mixed feeding in the trophic categories.

Fossil Mammal Communities

Faunal lists for each site were taken from the published literature (Leakey, 1967; Harris, 1991; Watson, 1993; Turner et al., 1999), or from first-hand examination of the mammalian fossils (Hadar, Makapansgat, Sterkfontein, Swartkrans). This study focuses exclusively on macromammals and those taxa identified to the species level. Genera that were not represented by particular species were included if specimens could be measured as to trophic adaptation and their substrate use was evident, e.g., all bovid species use a terrestrial substrate. Each taxon was assigned a trophic and substrate adaptation (Table 14.3) after Reed (1998), and relied upon morphological analysis, published information, and, at times, the principle of taxonomic uniformitarianism. Taxonomic uniformitarianism is sometimes useful for attributing trophic and substrate characteristics to fossil taxa when fossil specimens cannot be measured (Reed, 1998).

Analyses

The numbers of species in each trophic and substrate category were calculated for each stratigraphic member, submember or cave site in order to be compared to the modern data base. Those adaptations that have been shown to be highly significant between habitats using ANOVA and Tukey's Honestly Significant Difference posthoc test were used to reconstruct environments (Reed and Rector, 2007; Reed, 2008). These data for the fossil sites were then included in correspondence analyses with the modern communities. Correspondence analysis is an exploratory technique that allows examination of the patterns of both the sites and the adaptations. These were conducted with Statistica Version 7.1. Dice/Sorenson Similarity Indices, which are based on the presence and absence of species, were calculated between all pairs of fossil localities using SPSS 14.0 to produce a similarity matrix. This formula is:

$$C_s = 2j / (a + b) \tag{14.1}$$

where j is the number of species in common to both fossil localities, a is the number of species in fossil site A, and b the number of species in fossil site B (Magurran, 1988). The shared species are weighted more heavily, thus minimizing the effect of sample size differences especially in the case of fossil assemblages where the sample sizes may not be indicative of lower diversity but of taphonomic processes. The matrices were then used to produce a tree diagram of a cluster analysis using Statisitica Version 7.1. The cluster analysis was used to look for patterns of species change within and between each of the time periods and geographic regions under review.

Results

The means of the significantly different adaptations (arboreal, terrestrial, and aquatic substrate use; fruit-eating, grazing, fresh grass grazing and mixed feeding) show trends across habitats such that rainforests have high percentages of fruit-eating and arboreality with no grazing adaptations, whereas grasslands have almost no fruit-eating, no arboreality and high percentages of grazing within their communities (Reed, 2008). These data can be seen in Fig. 14.1 where the means for each habitat group are presented. Similarly, the correspondence analysis of modern African sites and these significantly different adaptations produce clusters of similar habitats (Fig. 14.2). Forests and grasslands, at opposite ends of the x-axis are most distinct from woodland and bushland habitats. The latter habitats have less tree and bush cover than forests, but are not as open as grasslands. Thus the habitats align on the x-axis from the left with those that have more annual rainfall and are more closed (i.e., forests) to those habitats that are drier with more seasonal rainfall and open. This axis encompasses 74.15% of the difference from the expected chi-squared values and as such accounts for much of the variation among habitats. The y-axis, accounting for 14.45% of the variation among habitats, separates those sites in which there is abundant water (that is not necessarily based on local rains) from those that do not. The wet sites include the Linyanti Swamp (woodland), Kafue Flats (grassland), and the Okavango Delta (shrubland), all of which have a broad based habitats that are different from one another. They group towards the bottom of the graph with other sites that have lacustrine and/or fluvial systems, and as such, influence the types of mammals that utilize them.

Figure 14.3 shows the fossil localities in chronologically order from oldest to youngest with the grazing adaptation from each locality plotted. Grazing increases to 30% or

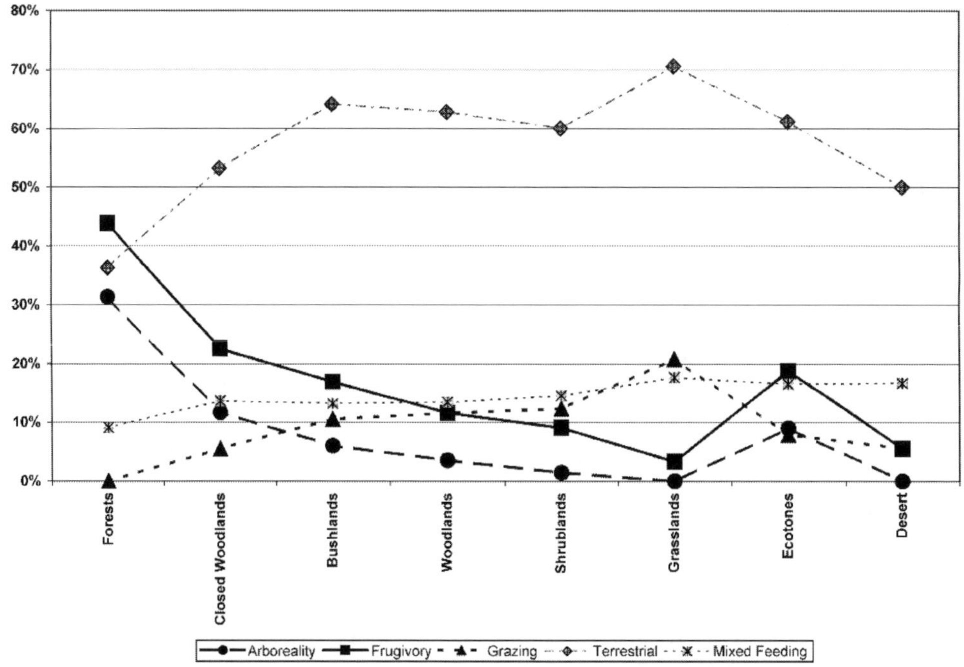

Fig. 14.1 The distribution of five significantly different adaptations across modern habitat types is depicted through means of each habitat type except desert where there is only one locality. Arboreal substrate use and fruit and leaf eating adaptations drop in abundance from forest through grasslands, whereas grazing adaptations rise across the same span. Aquatic substrate use and fresh grass grazing are not pictured as they separate all habitats on different criteria.

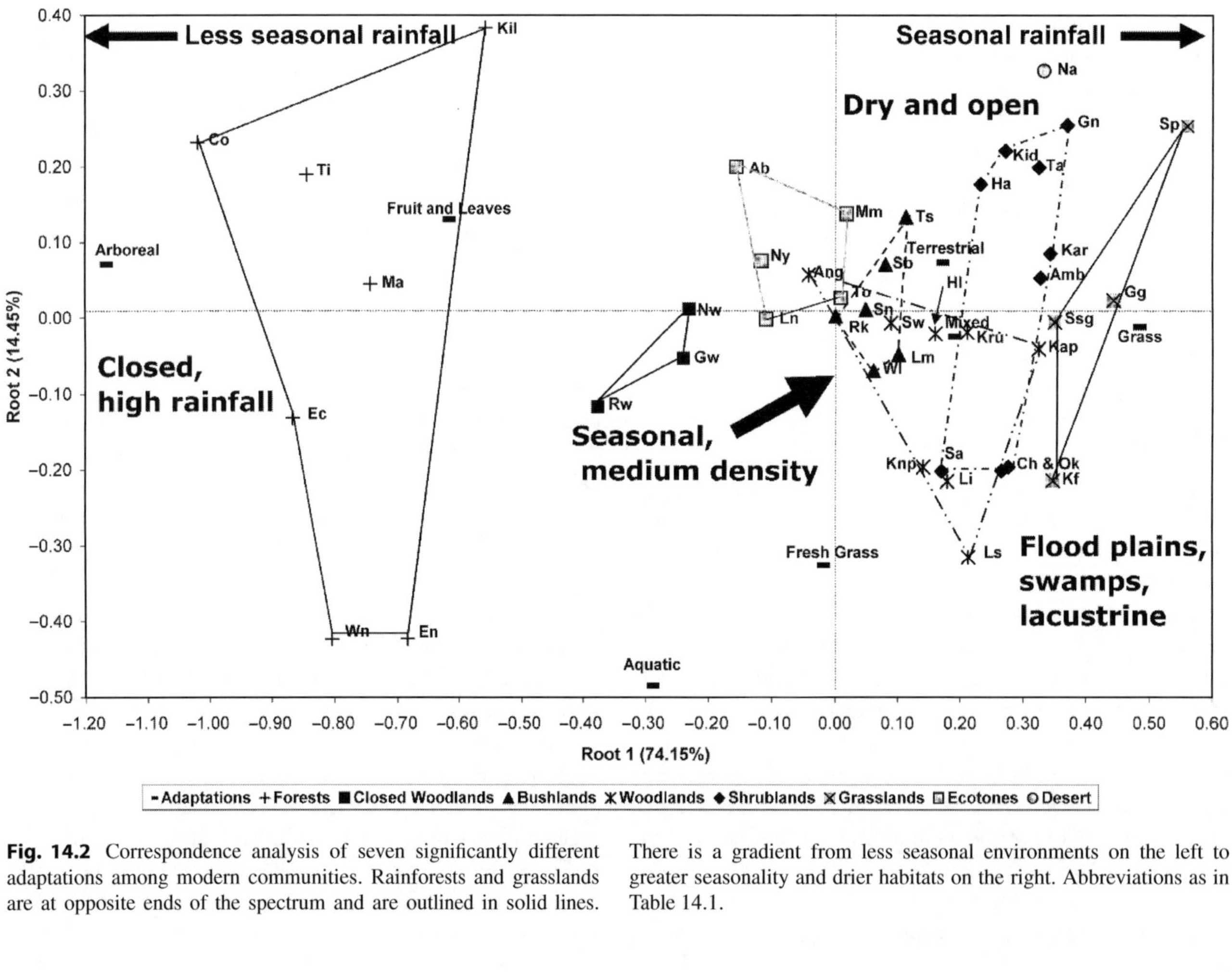

Fig. 14.2 Correspondence analysis of seven significantly different adaptations among modern communities. Rainforests and grasslands are at opposite ends of the spectrum and are outlined in solid lines. There is a gradient from less seasonal environments on the left to greater seasonality and drier habitats on the right. Abbreviations as in Table 14.1.

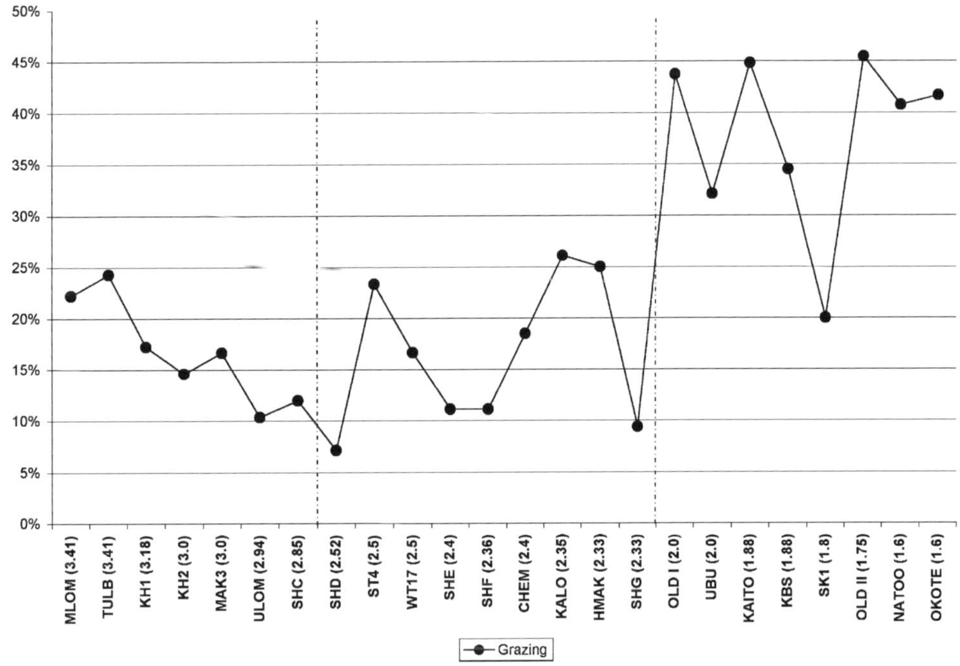

Fig. 14.3 The distribution of grazing across the fossil sites used in this study. The fossil sites are arranged in chronological order and there is no consistent trend toward more grazing, although there is a jump in the percentage of grazers at ~2.0 Ma (Olduvai Bed I).

greater in East African localities from approximately 2.0 Ma. This contrasts with all modern sites whose percentage of grazing mammals never exceeds 25%. Figure 14.4 shows the patterns fruit-eating, mixed feeding, and arboreality in the fossil sites. Overtime, there is no apparent trend in any one of these adaptations. If each region is separately examined in chronological order (Fig. 14.5) there are few direct trends within each. Although the end points in time for each location tend to have high grazing percentages and low percentages of fruit-eating and arboreality, none of the sites have a direct trend to that end. In relation to the hominins that have been recovered from the middle and late periods, there appears to be a larger change in mammal adaptations compared to what came before. That is, there is more change in mammal adaptations and habitats with the appearance of *Homo erectus*.

The correspondence analyses of each time period identified the habitat type of each fossil site. Figure 14.6 displays the early period centered around 3.0 Ma. The fossil sites at this time are positioned in the right half of the graph with seasonal rainfall and woodland bushland extant sites. This contrasts with pre 3.0 Ma when many fossil localities are in the midrange of the graph indicating more closed habitats (Reed and Rector, 2007). The mammals of Middle Lomekwi Member suggest a lacustrine or deltaic component as does the depositional environment. The Tulu Bor, Kada Hadar 1 and 2, and the Upper Lomekwi Members, are positioned with open woodlands although the depositional environment is also fluvial or lacustrine. The habitat of Makapansgat 3 is in the drier open woodland – scrub woodland range.

Figure 14.7 shows the placement of the middle time period sites. The WT-17000 site from which *P. aethiopicus* has been recovered, is reconstructed as bushland/medium density woodland. The fauna from Shungura C and F, as well as the Maakamitalu from Hadar indicates wooded grasslands, while Shungura D and E are drier shrubland habitats. The fauna from Shungura G and the Kalochoro Members cause these sites to be located to the left of the graph within modern grassland habitats. Sterkfontein 4 is positioned with the most arid of the modern habitats. The latest fossil sites from Koobi Fora and West Turkana (~1.6 Ma) fall in the range of modern habitats that include flood plains, swamps or lacustrine environs (Fig. 14.8). This is contrasted with the KBS and Upper Burgi Members of Koobi Fora, which are reconstructed as not as wet, and the Olduvai sites and Swartkrans 1, which are even more dry and open. The Olduvai mammals produce habitat reconstructions that are outside of the range of modern habitats, including the Serengeti Plains.

Finally, the cluster analysis provides a look at differences among species compositions at all sites (Fig. 14.9). Rather than grouping clusters by time, the first major break is between South and East Africa. This indicates that the fauna within each region is distinctly associated with the area. Sequential pan-African species turnovers, which would be represented by time clusters, are not evident. Within South African sites, each site is almost equidistant from the others, likely indicating similar amounts of species differences between these three time periods. Therefore, there appears to be significant species change every 500 ka in South Africa,

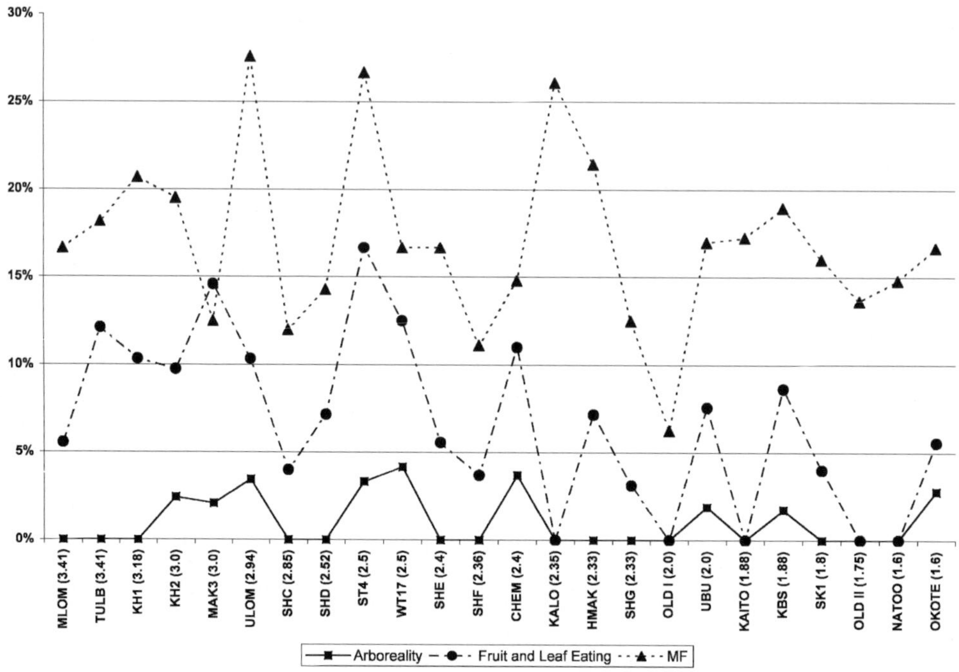

Fig. 14.4 The distribution of arboreality, fruit and leaf eating, and mixed feeding across the fossil sites. The fossil sites are arranged in chronological order and again there is no consistent trend towards less arboreality and fruit and leaf eating, nor an increase in mixed feeding.

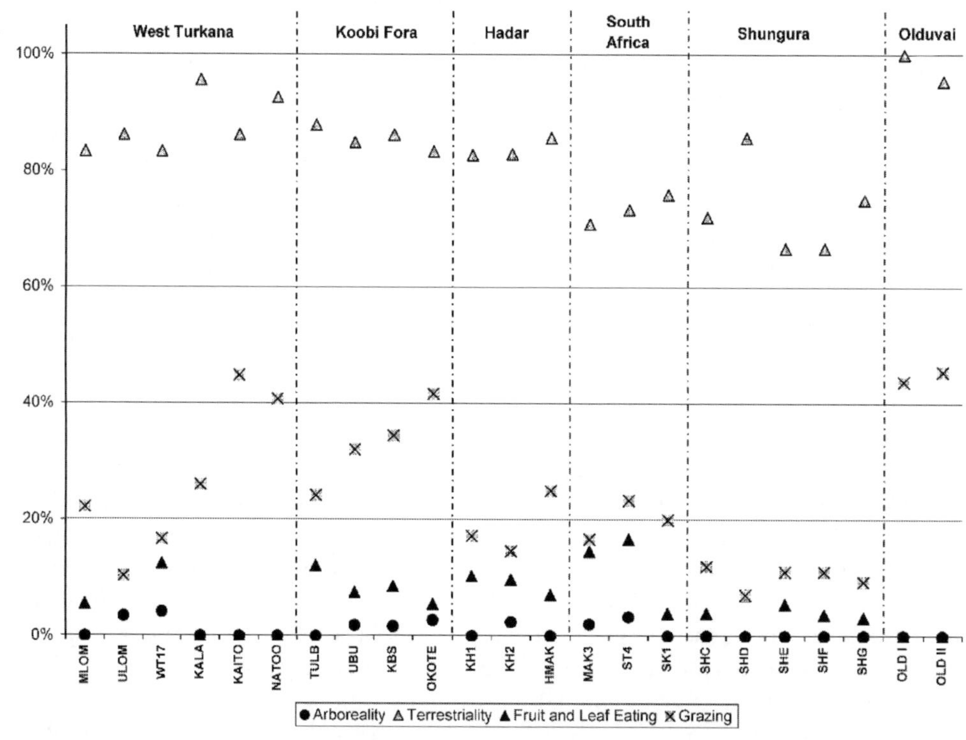

Fig. 14.5 The distribution of four significantly different adaptations across fossil sites arranged by formation or geographic region. In general there are slight trends in each area toward mammalian adaptation percentages that indicate more open, seasonal habitats especially at Koobi Fora and West Turkana.

although without having sites in the missing time periods we cannot say if the changes were gradual, or if they involved migration or speciation events at specific intervals.

The East African sites are grouped by both region and time. First the early Hadar sites from which *A. afarensis* have been recovered are the most unique of this large cluster. This is probably because of its distance from the other more geographically restricted localities (Reed, 2008). The later A.L. 666 *Homo* cf. *H. habilis* site at Hadar (Makaamitalu), on the other hand, is positioned between the Olduvai and Turkana Basin Formations. The Olduvai Beds group further from this main cluster also likely due to distance from the greater Turkana Basin in northern Kenya and southern Ethiopia. It is not surprising that the Shungura

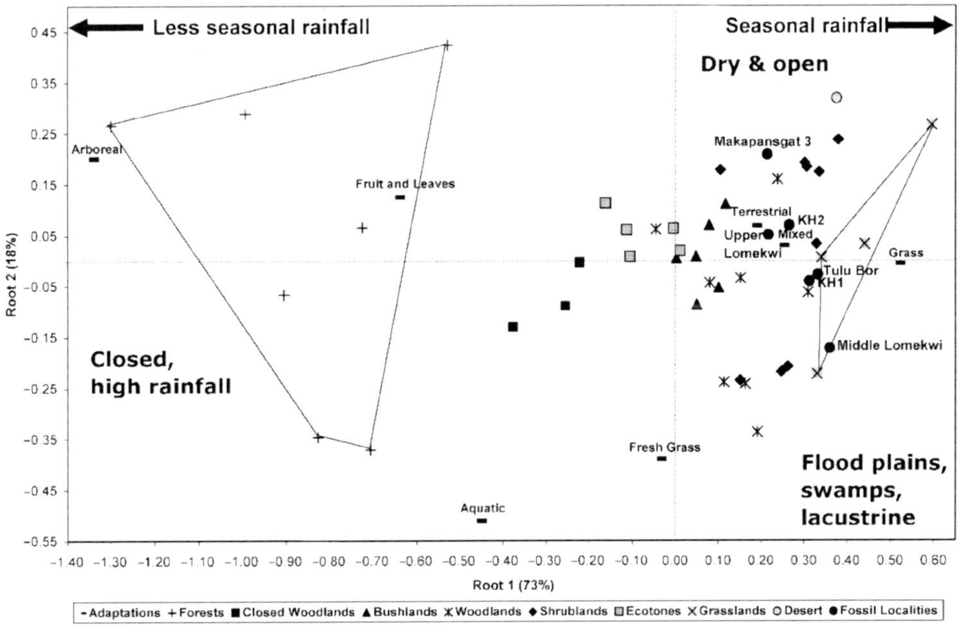

Fig. 14.6 Correspondence analysis of the adaptations with the modern localities and early time period fossil sites. Rainforests and grasslands are at opposite ends of the spectrum and outlined in solid lines. The fossil sites are distributed within medium density (wood, bush, or scrub), seasonal modern habitats.

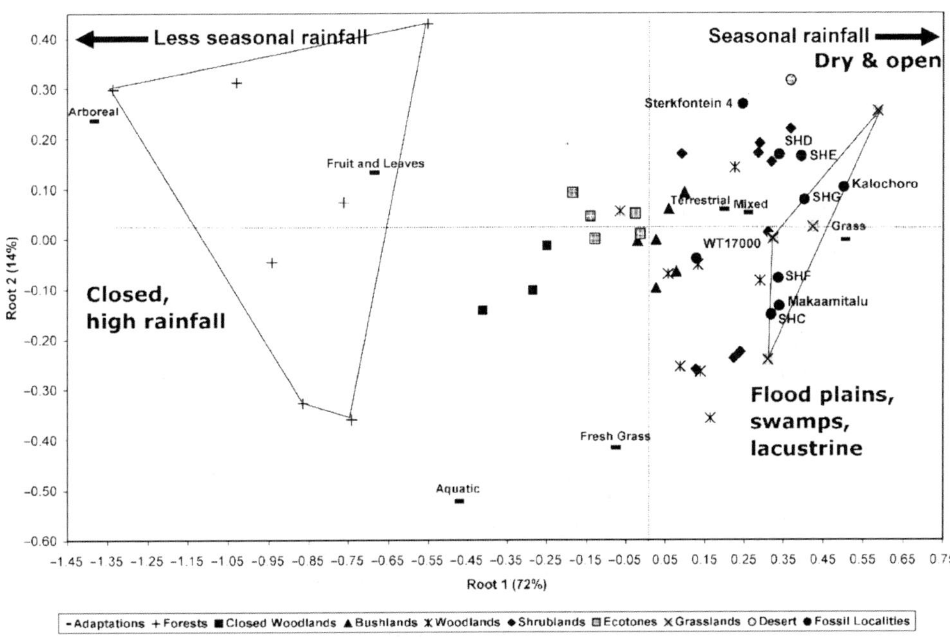

Fig. 14.7 Correspondence analysis of the adaptations with the modern localities and the middle time period fossil sites. Rainforests and grasslands are at opposite ends of the spectrum and outlined in solid lines. The fossil sites are generally distributed within low density, seasonal modern habitats such as scrub woodlands, shrublands, and grasslands. The *Paranthropus aethiopicus* site of WT17000 is more closed than all of the others of this time period. Shungura Members C and F, and the Makaamitalu from Hadar are depicted as habitats with greater proportions of wetlands or floodplains whereas the remaining sites are reconstructed as drier shrublands and grasslands.

sites group together irrespective of time within the large cluster because the region is proposed to have been a refugia against some of the major pan-African drying trends (Vrba, 1995). What is interesting is that the Shungura mammals more closely resemble those from the earlier Turkana Basin sites from the west side of the lake, while the sites later in time from the surrounding lake area (from ~1.8 Ma and younger) group together. In fact, all sites in this later group have had a fairly large species change from the previous time periods.

Discussion

The results show a major difference in both habitat structure and species composition between the earliest and latest time periods. Habitats are more open at 2.0–1.6 Ma as evinced by greater percentages of grazing fauna and the presence of few arboreal and frugivorous taxa. This is not supported by directed change from closed and wet to dry and open in any of the time successive members of the basins examined here,

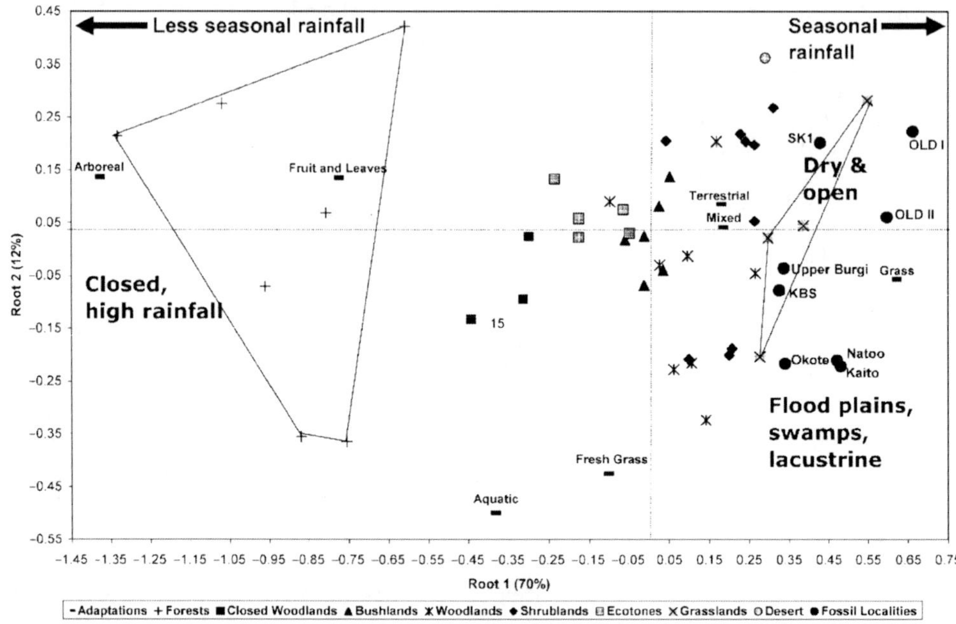

Fig. 14.8 Correspondence analysis of the adaptations with the modern localities and the late time period fossil sites. Rainforests and grasslands are at opposite ends of the spectrum and outlined in solid lines. The fossil sites are distributed near open, highly seasonal modern habitats, i.e., grasslands. Although only the Upper Burgi and KBS collection areas fall within modern secondary grassland ranges. The Olduvai sites have extremely large numbers of grazers compared with all extant localities. It is perhaps expected due to depositional environments that some of the Turkana Basin sites are more like the Kafue Flats with abundant wetland grasslands and flood plains.

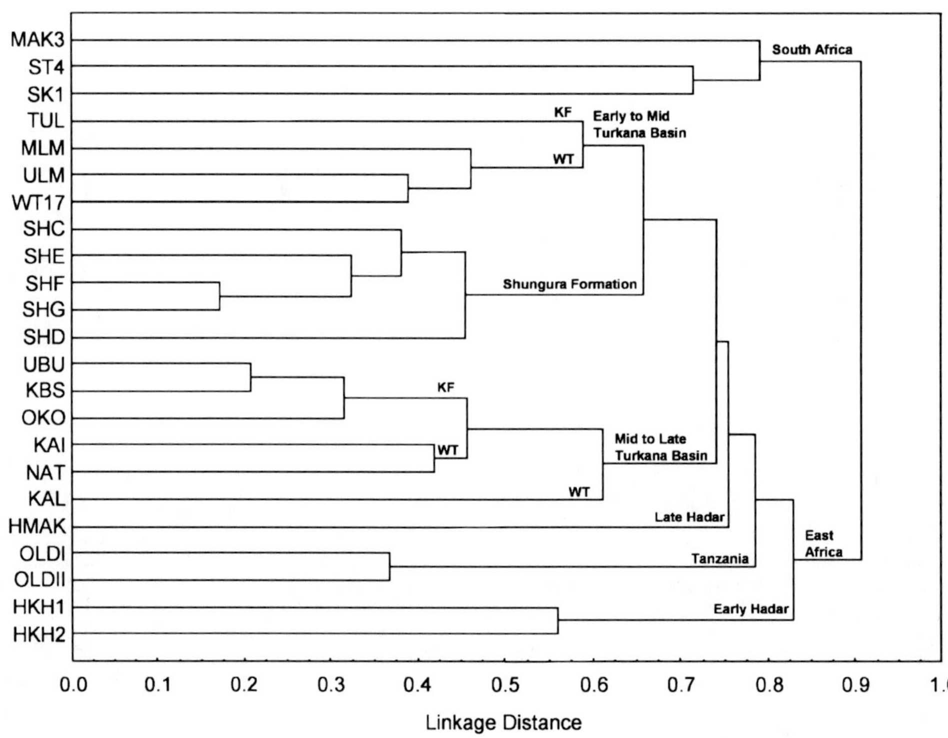

Fig. 14.9 Tree diagram of a cluster analysis based on the Dice/Sorenson Index of species similarity among fossil sites used in this study. The major difference among the sites is geographical rather than chronological (South Africa vs. East Africa). However, the mid-to late Turkana Basin sites from which *Homo erectus/ergaster* has been recovered form a smaller cluster away from the Kalochoro Member and all other localities. The early Turkana Basin sites represent *Australopithecus* and *Paranthropus* sites.

although overall it is apparent that environments became more seasonal and dry. This correlates to other findings (e.g., Behrensmeyer et al., 1997; Spencer, 1997; Reed, 1997, 1998; Bobe and Behrensmeyer, 2004; Bobe and Leakey, 2009) of increased numbers of taxa adapted for grazing and as an extension suggests fewer trees in more open grasslands. However, the increase in aquatic and fresh grass grazing fauna, particularly at West Turkana and Hadar in the later time period also indicates that while habitats were relatively treeless, they were wet, in contrast to other areas of Africa that were drying out. These data support findings based on the analysis of lacustrine sediments indicating the presence of a large lake in the Turkana basin fed by the proto-Omo River between 1.9 and 1.7 Ma and a sizeable lake in the Olduvai basin between 1.92 and 1.7 Ma (Potts, 1998a; Trauth et al., 2005). Other slight differences in adaptive patterns between formations demonstrate the importance of accounting for local environmental influences on the structure and composition of faunal assemblages and not assuming that all East African habitats were changing at the same rates or in the same directions.

While the results of this study apply to a relatively small range of time, it can be concluded that the effects of climate change as indicated by deMenocal (1995) at 2.8 Ma can be observed in fauna before and after this time period in a variety of sites. However, this did not occur at the same time at each locality. Southern African localities, as one might expect from temperate sites, appear to be more seasonal and perhaps drier than penecontemporaneous East African fossil sites. This does not mean the South African sites were accumulated in more open environments, however. While the East African sites are fluvial or lacustrine deposits, the South African deposits have been accumulated by carnivores in caves. Despite this, the South African sites have aquatic taxa represented and the difference in habitats seems to be more a result of latitude rather than taphonomy. Additionally, a different faunal community in the two regions is indicated by the primates alone in that there are more colobine taxa in East Africa traditionally thought to inhabit more closed environments.

Based on our results it seems likely that habitats in both east and south Africa were gradually changing but, perhaps more importantly, fluctuating, over long periods of time, resulting in different species compositions in mammal communities dependant on geographic region. Wynn (2004) has suggested that mosaic environments consisting of drier and more open habitats on floodplains, in contrast to wetter and more closed refuge habitats near rivers and lakes across this time period resulted in an increased number of localized ecological niches for species to fill. This premise could also account for species similarities being higher among sites and regions in close proximity (e.g., Olduvai, early Hadar, and the Shungura Formation).

There appear to be only slight environmental changes that were associated with the emergence of species of the genus *Homo*, and indeed also with species of *Paranthropus* that appeared at roughly the same time. Habitats tended to become slightly more xeric and open, but not at the same time in different areas. More open habitats are also an indication of greater seasonality in rainfall as higher levels of annual rainfall in less seasonal environments support greater woody vegetation. Rainforests have perhaps a 3 month "dry" season whereas open woodlands can have dry seasons lasting up to 8 months in either a unimodal or bimodal pattern. There are increases in grazing mammals at 2.35 Ma (Kalochoro), 2.0 Ma (Upper Burgi), 2.33 Ma (Hadar), and 2.4 Ma (Shungura F). As Olduvai Bed I is the first in that sequence, there is nothing with which to compare to what came before in that region, but the grazing adaptations as represented by the fossil assemblage in Olduvai Bed I times exceeds 40%. Sterkfontein 4 has high numbers of grazing animals, but lianas have also been recovered from the site (Bamford, 1999). Thus, the mammal community gives evidence of drier climate, but a probable ecotone habitat in which grasslands were contiguous with forests. Although the presence of *Homo* from Sterkfontein 4 is debated, there is an increase in mammals indicating more arid habitats compared with Makapansgat 3.

We may have evidence of a new hominin taxon appearing in the fossil record as early as 2.6 Ma with the manufacture of stone tools. Thus, habitats in the middle time period that are associated with *Homo* and other hominins are likely more seasonal and more open than earlier sites from which *Australopithecus afarensis* has been recovered in East Africa. The key to understanding climatic factors that are associated with the emergence of *Homo* and other taxa such as *A. garhi* and *P. aethiopicus*, both recovered from sediments of 2.5 Ma, is to comprehend the causes of the extinction of earlier australopithecines. This is a complex issue but with more research in individual basins, an understanding of the patterns of localized tectonic influenced changes as well as climate changes may shed light on this issue. We know that during the period from ~3.0 to 2.0 Ma, fossil and other evidence reveal habitat changes and species turnovers. This does not appear to be a pan-African event where species changed in all regions at once. For example, there is evidence at Hadar of a mammal species turnover at ~3.0 Ma that did not involve *A. afarensis* (Reed, 2008). On the other hand, it has been reported that there are morphological differences in the mandibles of *A. afarensis* before and after this faunal shift (Lockwood et al., 2000).

While there were no stark divergences in habitat between members in which *H. erectus* fossils are found (upper Burgi, KBS, Kaito, Olduvai Bed II, Okote, Natoo) compared to those from which *H. erectus* has not been recovered (Kalochoro, Member G, Olduvai Bed I), it is clear that the composition of mammalian assemblages differ, both with respect to time and location. While the Olduvai Beds group together in the cluster analysis, they are not similar to one another. Furthermore, there was no abrupt change in fauna (in terms of speciation, extinction or migration events) around the time in which *H. erectus* first appears (~1.9 Ma). It is interesting however, that *H. erectus* has been recovered from all members of the middle and late Turkana Basin cluster (*sensu lato*) in Fig. 14.8 except for the Kalochoro Member, which is the most distant site in the cluster.

The results of this study emphasize prolonged turnover of fauna in the late Pliocene to early Pleistocene, correlated with gradual and fluctuating shifts in paleoclimate and thus habitat representation. It is clear that while environments across Africa were generally becoming drier and more open, large mammals (including hominins) existed in a variety of habitats. And as the Hadar record shows, individual hominin species were able to live in a variety of environments (Reed, 2008).

Conclusions

The broad based habitats that characterized African fossil localities before the appearance of the genus *Homo* in the fossil record were bushlands and woodlands with a density of trees and bushes midway between forests and grasslands. East African sites, not surprisingly, contain fauna that indicates lacustrine or fluvial settings. The sites from which early *Homo* has been recovered indicate slightly drier and more open density woodland and wooded grasslands compared with fossil assemblages recovered with earlier hominins. Finally, there is not a dramatic change in habitat from early *Homo* to *H. erectus/ergaster*. There are other species turnovers associated with this change such that there is an increase in grazing animals and a drop in frugivorous animals. These changes, however, occur at different times in different basins or localities.

References

Alemseged, Z., 2003. An integrated approach to taphonomy and faunal change in the Shungura Formation (Ethiopia) and its implications for hominid evolution. Journal of Human Evolution 44, 451–478.

Alemseged, Z., Wynn, J., Kimbel, W.K., Reed, D., Geraads, D., Bobe, R., 2005. A new hominin from the Basal Member of the Hadar Formation, Dikika, Ethiopia, and its geological context. Journal of Human Evolution 49, 499–514.

Ansell, W.F.H., 1978. The Mammals of Zambia. National Parks and Wildlife Service, Chilanga.

Ansell, W.F.H., Dowsett, R.J., 1988. The Mammals of Malawi: An Annotated Checklist and Atlas. Trendrine Press, St. Ives.

Bamford, M., 1999. Pliocene fossil woods from an early hominid cave deposit, Sterkfontein, South Africa. South African Journal of Science 95, 231–237.

Behrensmeyer, A.K., 2006. Climate change and human evolution. Science 311, 476–478.

Behrensmeyer, A.K., Todd, N.E., Potts, R., McBrinn, G.E., 1997. Late Pliocene faunal turnover in the Turkana Basin, Kenya and Ethiopia. Science 278, 1589–1594.

Bobe, R., Behrensmeyer, A.K., 2004. The expansion of grassland ecosystems in Africa in relation to mammalian evolution and the origin of the genus *Homo*. Palaeogeography, Palaeoclimatology, Palaeoecology 207, 399–420.

Bobe, R., Eck, G.G., 2001. Responses of African bovids to Pliocene climatic change. Paleobiology 27, 1–48.

Bobe, R., Leakey, M.G., 2009. Ecology of Plio-Pleistocene mammals in the Omo-Turkana Basin and the emergence of *Homo*. In: Grine, F.E., Fleagle, J.G., Leakey, R.E. (Eds), The First Humans: Origin and Early Evolution of the Genus *Homo*. Springer, Dordrecht, pp. 173–184.

Bobe, R., Behrensmeyer, A.K., Chapman, R.E., 2002. Faunal change, environmental variability and late Pliocene hominin evolution. Journal of Human Evolution 42, 475–497.

Bremen, H., de Wit, C. T., 1983. Rangeland productivity and exploitation in the Sahel. Science 221, 1341–1347.

Brown, F.H., 1995. The potential of the Turkana Basin for paleoclimatic reconstruction in East Africa. In: Vrba, E.S., Denton, G.H., Partridge, T.C., Burckle, L.H. (Eds), Paleoclimate and Evolution, with Emphasis on Human Origins. Yale University Press, New Haven, CT, pp. 319–330.

Brown, F.H., Feibel, C.S., 1991. Stratigraphy, depositional environments, and paleogeography of the Koobi Fora Formation. In: Harris, J.M. (Ed), Koobi Fora Research Project. Clarendon, Oxford, pp. 1–30.

Bruga, J.P., Roche, H., Kibunjia, M., 2003. Faunes et paleoenvironements des principaux sites archeologiques plio-pleistocenes de la formation de Nachukui (Ouest-Turkana, Kenya). Comptes Rendus Palevol 2, 675–684.

Cameron, D.W., 2003. Early hominin speciation at the Plio/Pleistocene transition. Homo 54, 1–28.

Campisano, C.J., 2007. Tephrostratigraphy and hominin paleoenvironments of the Hadar Formation, Afar Depression, Ethiopia. Ph.D. dissertation, Rutgers, The State University of New Jersey.

Cerling, T.E., 1992., Development of grasslands and savannas in East Africa during the Neogene. Palaeogeography, Palaeoclimatology, Palaeoecology 97, 241–247.

Child, G.S., 1964. Some notes on the mammals of Kilimanjaro. Tanganyika Notes and Records. 64, 77–89.

Curtis, G.H., Hay, R.L., 1972. Further geologic studies and K-Ar dating of Olduvai Gorge and Ngorongoro Crater. In: Bishop, W.W., Miller, J.A. (Eds), Calibration of Hominoid Evolution. Scottish Academic Press, Edinburgh, pp. 289–301.

Delany, M., Happold, D.C.D., 1979. Ecology of African Mammals. Longman, London.

deMenocal, P.B., 1995. Plio-Pleistocene African climate. Science 270, 53–59.

deMenocal, P.B., 2004. African climate change and faunal evolution during the Pliocene-Pleistocene. Earth and Planetary Science Letters 220, 3–24.

deMenocal, P.B., Bloemendal, J., 1995. Plio-Pleistocene climatic variability in subtropical African and the paleoenvironment of Hominid evolution: a combined data-model approach. In: Vrba, E.S., Denton, G.H., Partridge, T.C., Burckle, L.H. (Eds), Paleoclimate and Evolution, with Emphasis on Human Origins. Yale University Press, New Haven, CT, pp. 262–288.

Dorst, J., Dandelot, P., 1969. A Field Guide to the Larger Mammals of Africa. Collins, London.

Emmons, L.H., Gautier-Hion, A., Dubost, G., 1983. Community structure of the frugivorous-folivorous forest mammals of Gabon. Journal of Zoology 199, 209–222.

Feibel, C.S., Brown, F.H., McDougall, I., 1989. Stratigraphic context of fossil hominids from the Omo Group deposits: northern Turkana Basin, Kenya and Ethiopia. American Journal of Physical Anthropology 78, 595–622.

Fernandez-Jalvo Y., Denys, C., Andrews, P., Williams, T., Dauphin, Y., Humphrey, L., 1998. Taphonomy and palaeoecology of Olduvai Bed I (Pleistocene, Tanzania). Journal of Human Evolution 34, 137–172.

Happold, D.C.D., 1987. The Mammals of Nigeria. Clarendon, Oxford.

Harris, J.M., 1991. Koobi Fora Research Project. Clarendon, Oxford.

Hoffman, R.R., Stewart, D.R.M., 1972. Grazer or browser: a classification based on the stomach-structure and feeding habits of East African ruminants. Mammalia 36, 226–240.

Howell, F.C., Haesaerts, P., de Heinzelin, J., 1987. Depositional environments, archeological occurrences and hominids from Members E and F of the Shungura Formation (Omo Basin, Ethiopia). Journal of Human Evolution 16, 665–700.

Kimbel, W.H., 1995. Hominid speciation and Pliocene climate change. In: Vrba, E.S., Denton, G.H., Partridge, T.C., Burckle, L.H. (Eds), Paleoclimate and Evolution with Emphasis on Human Origins. Yale University Press, New Haven, CT, pp. 425–437.

Kimbel, W.H., Walter, R.C., Johanson. D.C., Reed, K.E., Aronson, J.L., Assefa, Z., Marean, C.W., Eck, G.G., Bobe, R., Hovers, E., Rak, Y., Vondra, C., Yemane, T., York, D., Chen, Y., Evensen, N.M., Smith, P.E., 1996. Late Pliocene *Homo* and Oldowan tools from the Hadar Formation (Kada Hadar Member), Ethiopia. Journal of Human Evolution 31, 549–561.

Kingdon, J., 1974a. East African Mammals. Volume 1. University of Chicago Press, Chicago, IL.

Kingdon, J., 1974b. East African Mammals: Hares and Rodents. Volume IIB. University of Chicago Press, Chicago, IL.

Kingdon, J., 1977. East African Mammals: Carnivores. Volume IIIA. University of Chicago Press, Chicago, IL.

Kingdon, J., 1979. East African Mammals: Large Mammals. Volume IIIB. University of Chicago Press, Chicago, IL.

Kingdon, J., 1982a. East African Mammals: Bovids. Volume IIIC. University of Chicago Press, Chicago, IL.

Kingdon, J., 1982b. East African Mammals: Bovids. Volume IIID. University of Chicago Press, Chicago, IL.

Kitchener, A., 1991. The Natural History of the Wild Cats. Comstock Publishing Associates, Ithaca, NY.

Lamprey, H.F., 1962. The Tarangire game reserve. Tanganyika Notes and Records. 60, 10–22.

Lanjouw, A., 1987. Data review on the central Congo swamp and flood-plain forest ecosystem. Unpublished report. Royal Tropical Institute Rural Development Program.

Leakey, L.S.B., 1967. Olduvai Gorge, 1951–61. Cambridge University Press, Cambridge.

Lockwood, C.A., Kimbel, W.H. Johanson, D.C., 2000. Temporal trends and metric variation in the mandibles and dentition of *Australopithecus afarensis*. Journal of Human Evolution 39, 23–55.

Maguire, J.M., Pemberton, D., Collett, M.H., 1980. The Makapansgat Limeworks grey breccia: hominids, hyenas or hillwash? Palaeontologia Africana 23, 75–98.

Magurran A., 1988. Ecological Diversity and Its Measurement. Princeton University Press, Princeton, NJ.

McDougall, I., Brown, F.H., 2006. Precise $^{40}Ar/^{39}Ar$ geochronology for the upper Koobi Fora Formation, Turkana Basin, northern Kenya. Journal of the Geological Society, London 163, 205–220.

Pasini, G., Colalongo, M.L., 1997. The Pliocene-Pleistocene boundary stratotype at Vrica, Italy, In: Van Couvering, J.A. (Ed), The Pleistocene Boundary and the Beginning of the Quaternary. World and Regional Geology 9. Cambridge University Press, Cambridge, pp. 15–45.

Potts, R., 1998a. Environmental hypotheses of hominin evolution. Yearbook of Physical Anthropology, 93–136.

Potts, R., 1998b. Variability selection in hominid evolution. Evolutionary Anthropology 7, 81–96.

Pratt, D.J., Gwynne, M.D., 1977. Rangeland Management and Ecology in East Africa. Hodder and Stoughton, London.

Rautenbach, I.L., 1976. A survey of the mammals occurring in the Golden Gate Highlands National Park. Koedoe 19, 133–144.

Rautenbach, I.L., 1978a. A numerical re-appraisal of southern African biotic zones. Bulletin of the Carnegie Museum of Natural History 6, 175–187.

Rautenbach, I.L., 1978b. Ecological distribution of the mammals of the Transvaal (Vertebrata: Mammalia). Annals of the Transvaal Museum 31, 131–153.

Reed, K.E., 1997. Early hominid evolution and ecological change through the African Plio-Pleistocene. Journal of Human Evolution 32, 289–322.

Reed, K.E., 1998. Using large mammal communities to examine ecological and taxonomic structure and predict vegetation in extant and extinct assemblages. Paleobiology 24, 384–408.

Reed, K.E., 2002. The use of paleocommunity and taphonomic studies in reconstructing primate behavior. In: Plavcan, J.M., Kay, R.F., Jungers, W.L., van Schaik, C.P. (Eds), Reconstructing Behavior in the Primate Fossil Record. Plenum, New York, pp. 217–259.

Reed, K.E., 2008. Paleoecological patterns at the Hadar hominin site, Afar Regional State, Ethiopia. Journal of Human Evolution 54, 743–768.

Reed, K.E., Rector, A.L., 2007. African Pliocene paleoecology: hominin habitats, resources, and diets. In: Ungar, P.S. (Ed), Evolution of the Human Diet. Oxford University Press, Oxford, pp. 262–288.

Semaw, S., Rogers, M.J., Quade, J., Renne, P., Butler, R.F., Dominguez-Rodrigo, M., Stout, D., Hart, H.S., Pickering, T., Simpson, S.W., 2003. 2.6-million-year-old stone tools and associated bones from OGS-6 and OGS-7, Gona, Afar, Ethiopia. Journal of Human Evolution 45, 169–177.

Sheppe, W., Osborne, T., 1971. Patterns of use of a flood plain by Zambian mammals. Ecoogical Monographs 41, 179–205.

Skinner, J.D., Smithers, R.H.N., 1990. The Mammals of the Southern African Subregion. University of Pretoria Press, Pretoria.

Smithers, R.H.N., 1971. A Checklist of the Mammals of Botswana. Trustees of the National Museum of Rhodesia, Salisbury.

Spencer, L.M., 1997. Dietary adaptations of Plio-Pleistocene Bovidae: implications for hominid habitat use. Journal of Human Evolution 32, 201–228.

Spoor, F., Leakey, M. G., Gathogo, P.N., Brown, F.H., Antón, S.C., McDougall, I., Kiarie, C., Manthi, F.K., Leakey, L.N., 2007. Implications of new early *Homo* fossils from Ileret, east of Lake Turkana, Kenya. Nature 448, 688–691.

Swynnerton, G.H., 1958. Fauna of the Serengeti National Park. Mammalia 22, 435–450.

Trauth, M.H., Maslin, M.A., Deino, A., Strecker, M.R., 2005. Late Cenozoic moisture history of East Africa. Science 309, 2051–2053.

Turner, A., Bishop, L.C., Denys, C., McKee, J.K., 1999. Appendix: a locality-based listing of African Plio-Pleistocene mammals. In: Bromage, T.G., Schrenk, F. (Eds), African Biogeography, Climate Change, and Human Evolution. Oxford University Press, Oxford, pp. 369–399.

Vesey-Fitzgerald, D.F., 1964. Mammals of the Rukwa Valley. Tanganyika Notes and Records 62, 61–72.

Vrba, E.S., 1995. On the connections between paleoclimate and evolution. In: Vrba, E.S., Denton, G.H., Partridge, T.C., Burckle, L.C. (Eds), Paleoclimate and Evolution with Emphasis on Human Origins. Yale University Press, New Haven, CT, pp. 24–45.

Walter, R.C., Manenga, P.C., Hay, R.L., Drake, R.E., Curtis, G.H., 1991. Laser-fusion $^{40}Ar/^{39}Ar$ dating of Bed I, Olduvai Gorge, Tanzania. Nature 354, 145–149.

Watson, V., 1993. Composition of the Swartkrans bone accumulations, in terms of skeletal parts and animals represented. In: Brain, C.K. (Ed.), Swartkrans: A Cave's Chronicle of Early Man. Transvaal Museum, Transvaal Museum Monograph No. 8. Pretoria, pp. 35–74.

White, F., 1983. The Vegetation of Africa: A Descriptive Memoir to Accompany UNESCO/ AETFAT/UNSO Vegetation Maps of Africa. UNESCO, Paris.

Wood, B., Richmond, B.G., 2000. Human evolution: taxonomy and paleobiology. Journal of Anatomy 196, 19–60.

Wynn, J.G., 2004. Influence of Plio-Pleistocene aridification on human evolution: evidence from paleosols of the Turkana Basin, Kenya. American Journal of Physical Anthropology 123, 106–118.

Chapter 15
Ecology of Plio-Pleistocene Mammals in the Omo–Turkana Basin and the Emergence of *Homo*

René Bobe and Meave G. Leakey

Keywords Paleoenvironment • paleoecology • Shungura Formation • Koobi Fora Formation • Nachukui Formation • Pliocene • Pleistocene • mammals • primates • Suidae • Bovidae

Introduction

Understanding the origin of the genus *Homo* in Africa remains one of the central problems in paleoanthropology. Considerable empirical evidence relevant to this issue derives from the Omo–Turkana Basin, which includes areas surrounding Lake Turkana in northern Kenya and the lower Omo River Valley in southern Ethiopia (Fig. 15.1). This basin is well known for it rich record of Late Cenozoic vertebrates (Coppens et al., 1976; Harris, 1983, 1991; Harris et al., 1988; Harris and Leakey, 2003; Leakey and Harris, 2003; Leakey and Leakey, 1978). It also includes some of the earliest specimens attributed to the genus *Homo* (Prat et al., 2005; Suwa et al., 1996), as well as some of the best specimens of the genus from near the Pliocene/Pleistocene boundary (Leakey, 1973, 1976; Walker and Leakey, 1993; Wood, 1985, 1991).

Indirect evidence of *Homo* in the form of lithic artifacts in the Omo–Turkana Basin dates to about 2.4 Ma (Howell et al., 1987; Merrick, 1976; Roche et al., 2003), and well documented archeological sites were common during the Early Pleistocene (Isaac, 1997; Kibunjia et al., 1992; Rogers et al., 1994). Although lithic artifacts cannot be exclusively attributed to *Homo*, stone tools appear in the archeological record broadly contemporaneously with earliest *Homo* in the Omo–Turkana Basin and elsewhere in eastern Africa (Kimbel et al., 1996; Roche et al., 2009). In addition to its rich paleontological and archeological record, the basin has a well-established series of radiometric dates that facilitates correlation among different areas, and constrains the timing and rate of evolutionary processes (Brown, 1994; Brown and Feibel, 1991; Brown et al., 2006; Feibel et al., 1989; McDougall and Brown, 2006). Table 15.1 provides a summary of dating and correlations among the three main areas of the basin: the Omo, East Turkana, and West Turkana (Table 15.1 is adapted from Bobe et al., 2007: see references therein). With its unparalleled record of well-dated and abundant fossils, the Omo–Turkana Basin provides one of the best archives to study the emergence of the genus *Homo* in the context of mammalian evolution in eastern Africa. In this contribution, we focus on two key questions: (1) When does *Homo* first appear in the region? (2) What does the mammalian fossil record tell us about the environmental and ecological conditions associated with the emergence of *Homo*?

Materials

The primary data analyzed in this study derive from collections housed at the National Museums of Kenya and the National Museum of Ethiopia. The published record of fossil mammals from the Turkana Basin has been compiled into a database that uses FileMaker Pro software (Bobe and Behrensmeyer, 2007). The Turkana Basin Paleontology Database currently has about 16,500 records of fossil mammals from geological formations on both sides of Lake Turkana that span in time from the late Miocene to the early Pleistocene. The database includes all published mammals from the Nawata (Lothagam), Kanapoi, Nachukui (West Turkana), and Koobi Fora (East Turkana) Formations. The Turkana database is scheduled to be posted online by the National Museums of Kenya and the Smithsonian Institution. Data from the lower Omo Valley were first computerized by

R. Bobe (✉)
Department of Anthropology, University of Georgia, Athens, GA 30602, USA
e-mail: renebobe@uga.edu

M.G. Leakey
National Museums of Kenya, Department of Earth Sciences (Palaeontology), Nairobi 00502, Kenya; Turkana Basin Institute, Stony Brook University, Stony Brook, NY 11794-4364, USA
e-mail: meave.leakey@stonybrook.edu

F.E. Grine et al. (eds.), *The First Humans: Origin and Early Evolution of the Genus* Homo, Vertebrate Paleobiology and Paleoanthropology, © Springer Science + Business Media B.V. 2009

Fig. 15.1 Schematic map of the Omo–Turkana Basin, including Lake Turkana and the lower Omo Valley. Grey shading depicts geological formations discussed in the text. Inset shows the Turkana Basin in the context of eastern Africa (Redrawn from Feibel et al., 1989).

Gerald Eck working with the American contingent of the International Omo Research Expedition under the direction of F. Clark Howell (Eck, 2007). The Omo database is also in FileMaker Pro and includes nearly 24,000 records from the Mursi, Usno, and Shungura Formations. A separate Shungura database is maintained by Zeresenay Alemseged and includes records collected by the French contingent of the International Omo Research Expedition (Alemseged et al., 2007). The analyses in this study use data from the Turkana Basin Paleontology Database and the American Shungura database. Each record includes museum catalogue information as well as details regarding locality, age, publication, and various attributes of the specimens themselves, for the major groups of fossil mammals. Table 15.2 provides a summary of abundance among the main bovid tribes in the Plio-Pleistocene Omo–Turkana Basin.

When Does *Homo* Appear in the Omo–Turkana Basin?

An answer to this question involves several interrelated and very complex issues. One of the major problems is that *Homo* itself remains a poorly defined genus, and recognizing this taxon in the fossil record is thus highly problematic (Wood and Collard, 1999; Wood, 2009). The earliest specimens attributed to *Homo* on the basis of dental characters occur in Member E of the Shungura Formation, about 2.4 Ma (Suwa et al., 1996), and the lower Kalochoro Member of the Nachukui Formation, about 2.3 Ma (Prat et al., 2005). One of the problems with this first appearance datum (FAD) is that most of these specimens are isolated teeth, and therefore provide little information on most of the traits frequently used to define the genus (e.g., large endocranial volume). Another type of uncertainty relates to the fact that hominins were rare elements of Plio-Pleistocene mammalian faunas (Bobe et al., 2002). The Shungura fossil record collected by the American contingent of the International Omo Research expedition includes 22,335 specimens, of which 147 are identified as hominin. These numbers would indicate that hominins made up about 0.6% of the paleofauna (Fig. 15.2), but numerous collection and taphonomic factors can bias the number of collected specimens in relation to the original living faunas. A major portion of the Shungura record was collected by Gerald Eck using a well-defined and systematic methodology (Bobe and Eck, 2001; Eck, 2007). For example, he set out to collect all mammalian astragali along the Shungura deposits: a total of 601 astragali were collected, including 10 primates, *but not a single astragalus was identified as hominin*. This fact alone indicates that hominins were rare elements of the fauna near the environments of deposition of the paleo-Omo River. Although rare, hominins were clearly present in the paleo-Omo landscape. Mammalian mandibles were also systematically collected by Eck's team: 773 belong to identifiable mammalian families, and only 5 of these are hominin. Systematically collected mandibles thus indicate that hominins made up about 0.65% of the specimens on the surface of the Shungura Formation deposits (Fig. 15.2). All five of these hominin mandibles are identified as *Paranthropus*; no mandible attributed to *Homo* was found in these systematic surveys. In contrast, there are 147 cercopithecid mandibles (19% of the mammalian sample), and 96 of these (12.4% of the mammalian sample) can be identified to the genus *Theropithecus*, a primate that is taphonomically comparable to hominins (in terms of body size and morphology). These numbers do not necessarily indicate the actual abundance of hominins (or primates) in the Plio-Pleistocene of the lower Omo Valley, but they do indicate that hominins were rare and that *Homo* was particularly rare. Assessments of hominin paleobiology, as well as estimates of first and last appearances need to take this fact into consideration. Large samples are needed to detect the presence of rare taxa in a given landscape. When it comes to rare species such as those of early *Homo*, absence of evidence is not evidence of absence.

We can estimate the uncertainty associated with a first appearance datum (FAD) by considering the abundance of a

Table 15.1 Dating and correlations of Omo group deposits (Modified from Bobe et al., 2007; see references therein)

	Omo		West Turkana		East Turkana	GPTS	Age (Ma)	Error	Dated unit (Fm)	Method
				Chari	top Chari Mb		0.70	0.05		Stratigraphic scaling
			Silbo Tuff		Silbo Tuff		0.74	0.01	Silbo Tuff (KF)	K-Ar, Ar-Ar
							0.780			Matuyama-Brunhes
			Top Nariokotome Mb				0.980	0.05		Stratigraphic scaling
							0.990			GPTS Jaramillo top
L	L9 (top)						1.05	0.12		Stratigraphic scaling
							1.070			GPTS Jaramillo base
	L5/L6						1.12	0.02	L5/6 (Sh)	GPTS Cobb Mtn (N)
					Gele Tuff		1.25	0.02	Gele Tuff (KF)	K-Ar
		Nariokotome	L. Nariokotome Tuff				1.33	0.03	Nariokotome (Nk)	K-Ar
	Tuff L-3		Naito Tuff							
	Tuff L		Chari Tuff		Chari Tuff		1.39	0.02	Chari Tuff (KF)	K-Ar, Ar-Ar
K	Tuff K1				Koobi Fora Tuff					
					MFB		1.49	0.05	MFB (KF)	Stratigraphic scaling
	Tuff K (alpha)		Upper Okote Tuff		U. Okote Tuff		1.53	0.03	Tuff K (Sh)	Stratigraphic scaling
J	Tuff J7-1		Black Pumice Tuff		Black Pumice		1.55	0.03	BPT (KF)	Stratigraphic scaling
	Tuff J6-2	Natoo	L. Koobi Fora Tuff	Okote	L. Koobi Fora Tuff		1.62	0.02	LKF Tuff (KF)	Interpolation
					Okote Tuff					
					L. Okote Tuff		1.62	0.02	L. Okote Tuff (KF)	Stratigraphic scaling
	Tuff J4		Morutot Tuff		Morutot Tuff		1.65	0.03	J4 (Sh)	K-Ar
					A6		1.68	0.05	A6 (KF)	Stratigraphic scaling
					White Tuff		1.70	0.03	WT (KF)	Stratigraphic scaling
					C6		1.73	0.05	C6 (KF)	Stratigraphic scaling
	Tuff J						1.74	0.03	Tuff J (Sh)	Stratigraphic scaling
H	Unit H7	Kaitio					1.770		H7 (Sh)	GPTS Olduvai (top)
				KBS	A2		1.78	0.05	A2 (KF)	Stratigraphic scaling
							1.796			
	Unit H5				C4		1.86	0.05	C4 (KF)	Stratigraphic scaling
	Tuff H4		Malbe Tuff		Malbe Tuff		1.86	0.02	Malbe Tuff (KF)	K-Ar, Ar-Ar
	Tuff H2		KBS Tuff		KBS Tuff		1.88	0.02	KBS Tuff (KF)	K-Ar, Ar-Ar
	Tuff H						1.90	0.03	Tuff H (Sh)	Stratigraphic scaling
G					Lorenyang Tuff		1.90	0.05	Lorenyang (KF)	Stratigraphic scaling
	Unit G27						1.950		Unit G27 (Sh)	GPTS Olduvai base
	Unit G14						2.11	0.04	Unit G14 (Sh)	GPTS Reunion II top
	Unit G12						2.15	0.04	Unit G 12 (Sh)	GPTS Reunion II base
	Unit G9						2.19	0.04	Unit G9 (Sh)	GPTS Reunion I top
	Unit G4	Kalochoro		Burgi			2.27	0.04	Unit G4 (Sh)	GPTS Reunion I base
	Tuff G						2.33	0.03	Tuff G (Sh)	K-Ar
	Units F3/4						2.37		Units F3/4 (Sh)	GPTS X(N) subchron
	Tuff F1		Ekalalei Tuff							
F	Tuff F		Kalochoro Tuff				2.36	0.05	Tuff F (Sh)	K-Ar
E	Tuff E		Kokiselei Tuff				2.40	0.05	Tuff E (Sh)	Stratigraphic scaling

(continued)

Table 15.1 (continued)

	Omo		West Turkana		East Turkana	GPTS	Age (Ma)	Error	Dated unit (Fm)	Method
D	Tuff D	Lokalalei	Lokalalei Tuff		Lokalalei Tuff		2.52	0.05	Tuff D (Sh)	K-Ar, Ar-Ar
C	Tuff C9		Emekwi Tuff				2.581		Emekwi (Nk)	GPTS Gauss-Matuyama
	Unit C9	Lomekwi			Basal Burgi Mb		2.581		Unit C9 (Sh)	GPTS Gauss-Matuyama
					Burgi Tuff		2.64	0.05	Burgi Tuff (KF)	K-Ar
	Tuff C4				Ingumwai Tuff		2.74	0.08	Tuff C4 (Sh)	Stratigraphic scaling
	Tuff C (alpha)				Hasuma Tuff		2.85	0.08	Tuff C (Sh)	Stratigraphic scaling
B	Tuff B10						2.95	0.05	Tuff B-10 (Sh)	K-Ar
							3.04			GPTS Kaena top
					Ninikaa Tuff		3.08	0.03	Ninikaa (KF)	K-Ar, Ar-Ar
							3.11			GPTS Kaena base
	U14				Karo		3.2		U14 (Us)	
					Allia Tuff		3.22	0.05	Allia Tuff (KF)	Stratigraphic scaling
	Unit B2 (U12)			Tulu Bor			3.22		Unit B2 (Sh)	GPTS Mammoth top
	Unit B1						3.33		Unit B1 (Sh)	GPTS Mammoth base
					Toroto Tuff		3.32	0.02	Toroto (KF)	K-Ar, Ar-Ar
	Tuff B-beta		Tulu Bor Tuff (beta)		Tulu Bor Tuff		3.40	0.03	SHT (Hadar)	Ar-Ar
A			Burrowed bed				3.53			Interpolation
	Tuff A (U6)		Lokochot Tuff	Lokochot	Lokochot Tuff		3.594		Tuff A (Sh)	GPTS Gilbert-Gauss
					Wargolo Tuff		3.77		VT-3 (Maka)	
	Usno 1	Kataboi	Moiti Tuff	Moiti	Moiti Tuff		3.94	0.04	Moiti Tuff (Nk)	Ar-Ar
			Topernawi Tuff		Topernawi Tuff		3.96	0.03	Topernawi (Nk)	Ar-Ar
			Kataboi Basalt				4.05	0.06	Basalt (Nk)	K-Ar
			Kanapoi Tuff				4.07	0.02	KT (Kp)	Ar-Ar
	Usno Basalt						4.10	0.06	Basalt (Us)	K-Ar, Ar-Ar
	Mursi Basalt			Lonyumun			4.2	0.2	Basalt (Mu)	K-Ar

Abbreviations

GPTS = Geomagnetic Polarity Time Scale

Nk Nachukui Formation

Kp Kanapoi Formation

Nw Nawata Formation

KF Koobi Fora Formation

Sh Shungura Formation

Us Usno Formation

Mu Mursi Formation

Ages given of submembers reflect the age at the base of the unit, unless noted otherwise.

Colors are used to separate adjacent members within each formation.

Table 15.2 Number of specimens among bovid tribes in geological members of the Nachukui, Koobi Fora, Usno, and Shungura Formations (Nachukui data from Harris et al., 1988)

		Bovid tribes									
Formation	Member or unit	Aepycerotini	Alcelaphini	Antilopini	Bovini	Hippotragini	Reduncini	Tragelaphini	Total	%Alcelaphini –Antilopini	Time (midpoint)
Nachukui	Kataboi	4	2	2	0	0	3	3	14	28.6	3.7
Nachukui	Lomekwi	55	58	12	8	0	68	15	216	32.4	3
Nachukui	Lokalalei	3	5	1	4	0	2	1	16	37.5	2.5
Nachukui	Kalochoro	6	15	12	4	0	36	7	80	33.8	2.15
Nachukui	Kaitio	11	20	1	6	0	11	26	75	28.0	1.75
Nachukui	Natoo	4	10	10	7	0	13	4	48	41.7	1.5
Nachukui	Nariokotome	3	4	0	2	0	6	2	17	23.5	1.3
Koobi Fora	Lokochot	16	10	4	4	0	10	25	69	20.3	3.5
Koobi Fora	Tulu Bor	4	14	7	25	0	96	39	185	11.4	3.1
Koobi Fora	Upper Burgi – KBS	114	231	128	78	19	512	186	1,268	28.3	1.9
Koobi Fora	Okote	6	35	7	15	0	71	33	167	25.1	1.5
Usno	B(L)	128	4	3	28	0	4	50	217	3.2	3.3
Shungura	B(U)	49	7	1	24	0	63	32	176	4.5	2.9
Shungura	C	94	5	3	73	3	74	178	430	1.9	2.7
Shungura	D	56	6	1	15	0	30	66	174	4.0	2.5
Shungura	E	91	9	0	13	1	56	117	287	3.1	2.4
Shungura	F	141	25	5	15	0	69	87	342	8.8	2.36
Shungura	G(L)	461	51	4	38	2	684	375	1,615	3.4	2.2
Shungura	G(U)	24	13	9	2	0	29	2	79	27.8	2
Shungura	H	7	12	3	2	0	125	4	153	9.8	1.8
Shungura	J	6	8	0	1	0	39	0	54	14.8	1.65
Shungura	K	5	15	2	1	0	27	1	51	33.3	1.5
Shungura	L	4	11	1	2	0	42	2	62	19.4	1.35
	Total	1,292	570	216	367	25	2,070	1,255	5,795	13.6	

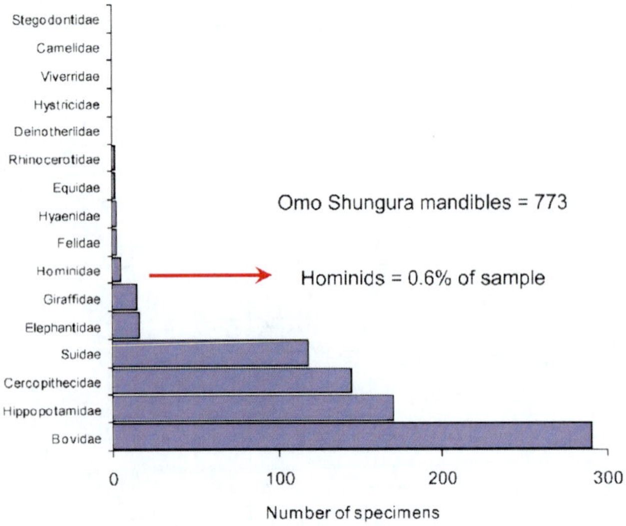

Fig. 15.2 Abundance (number of specimens) of mammalian mandibles systematically collected from the Shungura Formation deposits. This distribution provides an estimate of the relative abundance of mammalian families in the Shungura deposits. There are five hominid mandibles in a sample of 773 mammalian specimens. Thus, hominids make up about 0.65% of the sample.

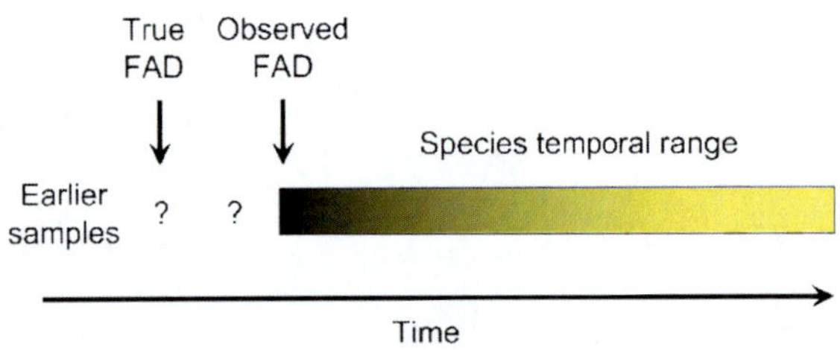

Fig. 15.3 The earliest occurrence of a species in the fossil record provides an upper limit to the timing of its true origination or migration into the area being sampled. The first appearance datum (FAD) should always be evaluated in relation to earlier samples. If the fossil record prior to the species' FAD is abundant and continuous, then the FAD has a high likelihood of representing a true origination or migration event. If the fossil record prior to the species' FAD is poor and discontinuous, then the FAD may be an artifact of sampling.

taxon once it appears in the fossil record, and the abundance and distribution of samples prior to that FAD. An observed FAD in the fossil record provides an upper limit to the species time of origination, but the true origination may be significantly earlier than the observed event (Fig. 15.3).

A first appearance preceded by a very sparse fossil record could simply indicate that the taxon has not been found in the sparse samples, even though it may have been present in the landscape. A first appearance of an abundant taxon preceded by a series of large samples without evidence of that taxon would provide a high level of confidence that the FAD truly reflects an origination or migration event, not an artifact of sampling. It should be noted that a FAD could indicate either *in situ* evolution or *migration* into the area being sampled. The only way to distinguish between these two phenomena would be to obtain samples from large areas, regional or continental in scale, that include the likely sources of the species in question. The current distribution of the fossil record in Africa, with a disproportionate contribution from the East African Rift Valley, precludes firm conclusions regarding immigration or speciation events. Nevertheless, we can use methods designed to evaluate the uncertainty of FADs depending on the distribution of samples within a region and the abundance of the species of interest. Here we use the methods developed by Koch and colleagues to place 95% confidence intervals on the FADs of *Homo* (Barry et al., 2002; Koch, 1987; Koch and Morgan, 1988). We estimate the probability (*Pi*) that *Homo* originated or migrated to the area being sampled by

$$P_i = 1 - (1 - n/m)^r \qquad (15.1)$$

where *n* is the abundance of *Homo* over its range (number of specimens), *m* is the total number of fossil mammals in the collection over the same range, and *r* is the number of specimens in successive intervals prior to the first occurrence (FAD) of the genus. Thus, the ratio n/m provides a measure of relative abundance. Following Barry et al. (2002), we use a P_i value of at least 0.8 to determine the 95% confidence interval. The same method can be applied to last occurrence data (LADs), but the focus here is on origination rather than extinction.

In the sample from the Shungura Formation, *Homo* appears in Member E, at about 2.4 Ma, represented by specimen L. 26-1, a right lower M1 (Suwa et al., 1996). Member E is preceded by relatively large samples of fossil mammals in Member D (with 946 specimens dated from 2.52 to 2.4 Ma), Member C (with 3301 specimens dated from 2.85 to 2.52 Ma), and Member B (with 1997 specimens derived mostly from the uppermost units dated to about 2.95 Ma). These earlier Shungura samples in the time range from 2.4 to nearly 3.0 Ma, and totaling 6,244 fossil mammals have not provided evidence of *Homo*. Thus, the earliest record of *Homo* at 2.4 Ma in the Shungura Formation may be very close in time to the true origination or migration event (in Fig. 15.4a the 95% confidence interval is narrow). But could we recognize *Homo* on the basis of isolated teeth prior to 2.4 Ma? Defining and recognizing *Homo* on the basis of an incomplete fossil record remains a daunting task. Nevertheless, we know that the *Homo* and *Paranthropus* lineages diverged sometime in the Pliocene. The earliest specimen assigned to *Paranthropus* is a right lower M2 (specimen L. 62-17) from Shungura Member C-5 (Suwa et al., 1996). Tuff C4 is dated to 2.74 Ma (Feibel et al., 1989) and the *Paranthropus* specimen derives from the sedimentary unit above it. Thus, *Homo* and *Paranthropus* clearly had separate evolutionary trajectories by 2.7 Ma, and possibly considerably earlier. If this is the case, the earliest record of *Homo* in

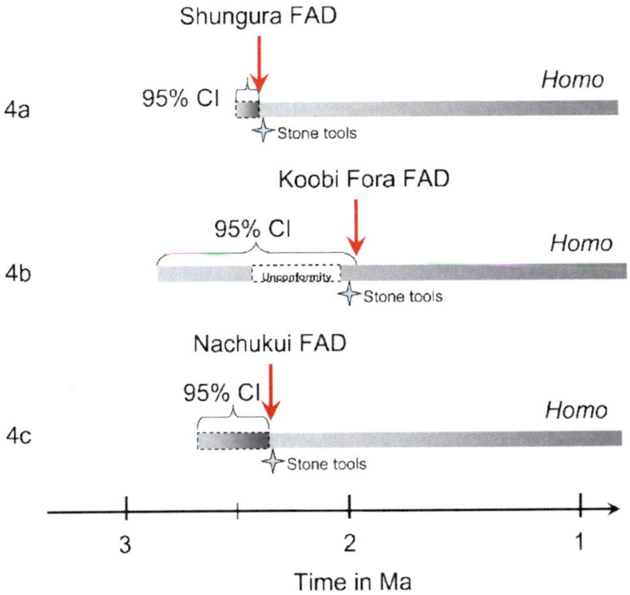

Fig. 15.4 First appearance datum (FAD indicated by arrow) of *Homo* in the (**a**) Shungura, (**b**) Koobi Fora, and (**c**) Nachukui Formations. Horizontal brackets denote the 95% confidence interval associated with the *Homo* FADs. See text for details.

the Shungura Formation at about 2.4 Ma would represent the migration of the genus into the lower Omo Valley.

At Koobi Fora there are exceptional specimens of unambiguous early *Homo*: KNM-ER 1470, KNM-ER 1813 and KNM-ER 3733 among the best known (Wood, 1991). The earliest of these are from the Upper Burgi Member of the Koobi Fora Formation, and date to slightly less than 2 Ma (Feibel et al., 1989; Spoor et al., 2007). They are roughly coincident with the earliest stone artifacts at East Turkana (Fig. 15.4b). But at Koobi Fora, we encounter the problem that there is a major unconformity spanning a few hundred thousand years below the Upper Burgi Member. Sediments from the *Lower* Burgi Member have yielded few fossils, and in the upper part of the Tulu Bor Member there is a sample of only 204 fossil mammals. Thus, prior to the FAD of *Homo* at Koobi Fora there is a long time interval, almost 1 Myr, with a relatively sparse fossil record. Thus, at Koobi Fora, the FAD of *Homo* at near 2 Ma has a large margin of error because of geological circumstances and a sparse fossil record prior to that first appearance (this is therefore depicted by a wide confidence interval in Fig. 15.4b).

The west side of Lake Turkana has produced spectacular hominins such as KNM-WT 15000, KNM-WT 17000 and KNM-WT 40000 (Brown et al., 1985; Leakey et al., 2001; Walker et al., 1986), but the earliest published specimen attributed to *Homo* is an isolated right lower M1 (KNM-WT 42718) from the base of the Kalochoro Member and dated to about 2.3 Ma (Prat et al., 2005). Compared to the Shungura Formation, the number of fossils from the Nachukui Formation is relatively small and therefore the earliest appearance of *Homo* at near 2.3 Ma in West Turkana has a wider 95% confidence interval (Fig. 15.4c). However, it is noteworthy that the archeological record at West Turkana is very close in age to the earliest *Homo* specimens, as is the case in East Turkana and the lower Omo Valley (Fig. 15.4). The approach taken here could be modified by archeologists to address confidence intervals associated with the earliest occurrences of lithic technologies, but the focus of this paper remains on the fossil bones.

The data discussed here indicate that the critical time period for understanding the origin and dispersal of *Homo* is between about 3 and 2.4 Ma. By 2.4 Ma there is evidence of *Homo* in the Omo–Turkana Basin and elsewhere in East Africa (Hill et al., 1992; Kimbel et al., 1997; Suwa et al., 1996). But profound changes in hominid adaptation seem to occur with the emergence of *Homo erectus* (*sensu lato*) after 2 Ma (Bramble and Lieberman, 2004; McHenry and Coffing, 2000; Wood and Collard, 1999; Wood, 2009). What was the environmental context in which these evolutionary processes took place? The mammalian fauna from the Omo–Turkana Basin provides a rich source of information on the environmental and ecological conditions that existed during the time when *Homo* first appeared and the subsequent emergence and dispersal of *Homo erectus*.

What Does the Mammalian Fossil Record Tell Us About the Environmental and Ecological Conditions Associated with the Emergence of *Homo*?

Multiple lines of evidence indicate that East African Plio-Pleistocene environments were complex and dynamic, having been composed of varying proportions of forest, woodland, bush, and grassland (Bobe, 2006; Bonnefille, 1995; Kingston, 2007; Kingston et al., 1994; Reed, 1997; Wynn, 2004). On broad scales, Pliocene African climate fluctuated between wet and dry conditions with an apparent overall trend toward greater aridity (deMenocal, 1995, 2004). The record of paleosol carbon isotopes shows a shift toward C_4 grasslands during the latest Pliocene and Early Pleistocene (Levin et al., 2004; Wynn, 2004). This Plio-Pleistocene shift is also indicated by data from biomarkers recovered from site 231 in the Gulf of Aden (Feakins et al., 2005), although the two records are not entirely congruent (Fig. 15.5). While the paleosol record shows the shift to C_4 vegetation at around the Pliocene/Pleistocene boundary, the biomarker record from the Gulf of Aden shows an earlier shift between about 3.4 and 3.0 Ma (Feakins et al., 2005). The explanation of this discrepancy is likely due to the larger area sampled by the wind-blown biomarker data. These residues of terrestrial

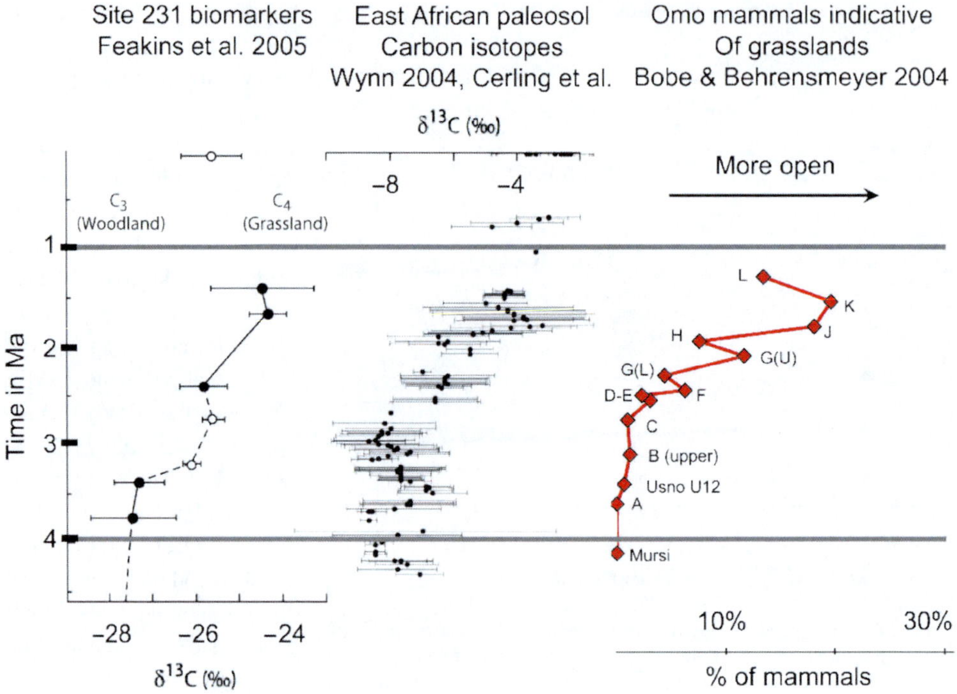

Fig. 15.5 (**a**) Plant biomarker data from the Gulf of Aden (site 231) depicting an increase in C4 vegetation at about 3.2 Ma followed by a further shift after 2 Ma (from Feakins et al., 2005). (**b**) East African paleosol carbonate data showing an increase in C4 biomass beginning about 2 Ma (data from Wynn, 2004). (**c**) Relative abundance of mammals indicative of seasonally arid grasslands in the lower Omo Valley showing a moderate increase in grassland-adapted mammals at about 2.5–2.4 Ma and more pronounced peaks after about 2 Ma (Data from Bobe and Behrensmeyer, 2004).

plants were recovered from marine records in the Gulf of Aden, and their source may have spanned an area including much of northeastern Africa as well as the Arabian Peninsula, while the paleosol data clearly derived from localized terrestrial sequences at hominin-bearing sites. An important conclusion to be drawn from these carbon isotope studies is that eastern African Pliocene/Pleistocene vegetation was complex and included an increasing component of C_4 grasslands, but this expansion of grassland habitats likely occurred in different parts of East Africa at different times.

Analyses of fossil mammals present a pattern of change that supplements the carbon isotope record. The record from the Shungura Formation, for example, shows remarkable faunal stability during much of the Late Pliocene, but taxa indicative of open grasslands become slightly more abundant at about 2.5 Ma and more significantly so after 2 Ma (Bobe and Behrensmeyer, 2004) (Fig. 15.5). The Omo–Turkana data also show that different parts of the basin had different habitats. Thus, bovids indicative of seasonally arid grasslands were consistently more abundant in the East and West Turkana areas (Koobi Fora and Nachukui Formations) than in the Omo area (Shungura Formation) (Bobe et al., 2007). The diversity of habitats in the Pliocene–Pleistocene Turkana Basin is demonstrated here by comparing the relative abundance of Alcelaphini and Antilopini across different areas (Fig. 15.6). These antelopes are associated with grassland and bushland, seasonally arid habitats (Greenacre and Vrba, 1984). Omo, West Turkana, and East Turkana show very different percentages of alcelaphines and antilopines during the late Pliocene, but display some convergence during the latest

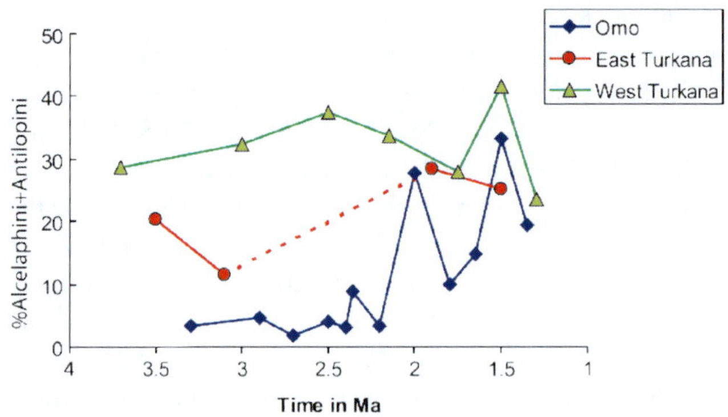

Fig. 15.6 Relative abundance of Alcelaphini + Antilopini (as a percentage of all Bovidae) in the three main regions of the Turkana Basin. High values suggest open, seasonally arid environments dominated by grasslands. The Omo appears to have been consistently more closed than other parts of the Turkana Basin, but all three regions show increases in grasslands, although with fluctuations, after 2 Ma.

Pliocene and early Pleistocene, beginning at about 2 Ma. This would indicate grasslands and bushland were becoming more prominent throughout the basin beginning at about 2 Ma, although woodlands and forest fringing the paleo-Omo River would have remained an important part of the vegetational landscape. Previous analyses controlling for taphonomic factors, depositional environments, and collection biases in the Omo region indicate that these faunal changes represent true biological phenomena (Bobe and Eck, 2001).

Among suids, species of *Nyanzachoerus* and *Notochoerus* decline precipitously toward the Pliocene/Pleistocene boundary, when species of *Metridiochoerus* and *Kolpochoerus* become numerically dominant (Fig. 15.7). Although the *Notochoerus* lineage shows increasing adaptation to a diet of grass over time (Harris and Cerling, 2002), it may have succumbed to competition from a large array of grazers that became widespread during the earliest Pleistocene. Among suids, *Metridiochoerus hopwoodi*, *M. compactus*, and *M. modestus* are all highly hypsodont, and have their first appearances near the Pliocene/Pleistocene boundary. Likewise, species of *Kolpochoerus*, which are characterized by significant elongation of the third molars, have their first appearances during the early Pleistocene (e.g., *K. majus*, *K. phacochoeroides*) or become dominant elements of the early Pleistocene fauna (e.g., *K. heseloni*) (Cooke, 1997; Harris and Cerling, 2002; White, 1995).

Other mammalian indicators of grasslands show a similar pattern of increasing dominance during the latest Pliocene or earliest Pleistocene. The genus *Equus*, for example, a quintessential grazer of the African savanna, first appears in the Omo record at about 2.3 Ma (at the base of Member G), where it comprises about 30% of all specimens of the family Equidae (other species belong to the genus *Eurygnathohippus*). After about 1.8 Ma, *Equus* makes up about half or more of all equid specimens (Fig. 15.8).

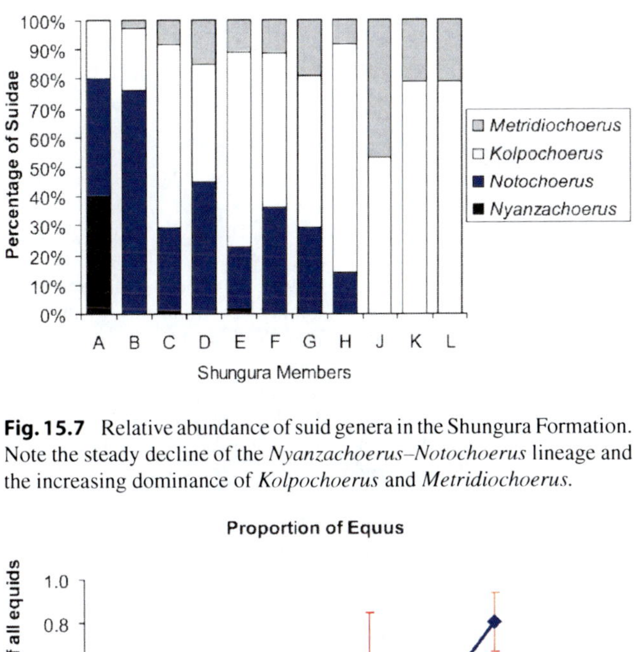

Fig. 15.7 Relative abundance of suid genera in the Shungura Formation. Note the steady decline of the *Nyanzachoerus–Notochoerus* lineage and the increasing dominance of *Kolpochoerus* and *Metridiochoerus*.

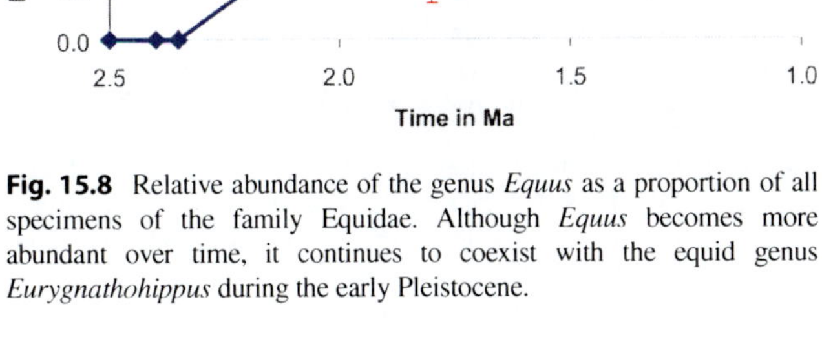

Fig. 15.8 Relative abundance of the genus *Equus* as a proportion of all specimens of the family Equidae. Although *Equus* becomes more abundant over time, it continues to coexist with the equid genus *Eurygnathohippus* during the early Pleistocene.

Among cercopithecids, *Theropithecus brumpti* is the dominant monkey in the Omo from about 3 to 2.4 Ma (Members B through D). It is also numerically dominant at Koobi Fora during Tulu Bor times (after 3.4 Ma), along with large bodied colobines such as *Rhinocolobus turkanaensis*, *Paracolobus mutiwa*, *Cercopithecoides kimeui*, and *C. williamsi* (Jablonski and Leakey, 2008). In Shungura Member E, at 2.4 Ma, *T. oswaldi* begins to replace *T. brumpti*, which becomes extinct by 2 Ma (Eck et al., 1987). The extinction of *T. brumpti* is followed by the decline of large bodied colobines in the Koobi Fora Formation by 1.5 Ma, but during this time *T. oswaldi* continues to thrive (Leakey et al., 2008).

These results point to consistent trends in the fauna over time. They do not necessarily provide environmental reconstructions of particular time periods, but they do document complex patterns of ecological change that likely influenced the behavior of hominin populations during the Plio-Pleistocene.

Discussion and Conclusions

This contribution addresses two key questions: (1) when does *Homo* first appear in the Omo–Turkana Basin? and (2) what does the mammalian fauna tell us about the ecological conditions at the time of the first appearance of *Homo*? We have highlighted quantitatively a fact long-known to field paleontologists: namely that hominins were rare elements of Pliocene–Pleistocene faunas. The consistently collected samples from the Shungura Formation suggest that hominins made up at most 0.6% of the fossils representing large mammals on the surface of the exposures. This percentage does not translate directly into the actual relative abundance of hominins in the Pliocene and Pleistocene ecosystems of eastern Africa, but it does indicate that hominins were relatively rare mammals. Clearly, if we were to consider the entire mammalian fauna from the Plio-Pleistocene, including small mammals that are rarely found in fossil collections, the relative abundance of hominins would be significantly lower than 0.6%. Nevertheless, this estimate indicates that the first and last appearances of hominin species may have more to do with sampling issues than with actual origination or migration events. Here we have placed 95% confidence intervals on the earliest records of *Homo* in the Omo–Turkana Basin, and have emphasized that we need to understand not just the point in time of these earliest records, but also a span of time preceding them. In this regard, it is noteworthy that earliest *Homo* and lithic artifacts first appear at about the same time in both the Shungura and Nachukui Formations. Another event of significance in human evolution is the appearance of *Homo erectus* at about 1.9–1.8 Ma. *Homo erectus* shows a novel adaptive complex relative to earlier species of the genus (Wood and Collard, 1999; Wood, 2009), but it is also clear that much remains to be discovered about the paleobiology of Plio-Pleistocene *Homo*.

The faunal evidence reviewed here indicates that the Omo–Turkana Basin consisted of complex and diverse habitats during the late Pliocene and early Pleistocene. These included varying proportions of forests, woodlands, grasslands, and bush. However, multiple lines of evidence indicate that grasslands became more prominent in the landscape during the time interval when *Homo* first appeared in the fossil record (2.5–2.4 Ma). The appearance of *Homo* and lithic artifacts in the Omo–Turkana Basin at about 2.4 Ma, which corresponds to an increase in grassland fauna at 2.5–2.4 Ma, suggests that a mix of grasslands and woodlands provided an opportunity for *Homo* to migrate into the region (Figs. 15.4–15.6).

Although mosaic habitats characterized the landscapes associated with the earliest appearance of *Homo*, the evidence shows that from about 2 Ma, the Omo–Turkana Basin underwent pronounced faunal changes that likely reflected a significant increase in the extent and persistence of grasslands. This pulse of grassland expansion may have allowed *Homo erectus* to migrate to the Omo–Turkana Basin from a source area still unknown (Rightmire and Lordkipanidze, 2009). The bovids *Connochaetes gentryi*, *Megalotragus isaaci*, *Beatragus antiquus*, *Pelorovis oldowayensis*, *P. turkanensis*, and *Tragelaphus strepsiceros* are likely immigrants into the region at a time when *Homo erectus* also makes its first appearance (Bobe and Behrensmeyer, 2004). These bovids, with the exception of *T. strepsiceros*, represent a community of highly hypsodont grazers.

Both the archeological and fossil records indicate that early *Homo* migrated to the Omo–Turkana region at about 2.4 Ma, and that the genus likely originated elsewhere at an earlier time. The Omo fauna, which seems to track broader environmental changes rather well (Fig. 15.5), indicates that the latest Pliocene was a time of increasing aridity and environmental instability, and that grasslands were becoming more important components of the vegetation. Further faunal changes near the Pliocene/Pleistocene boundary likely reflect even greater expansion of grassland ecosystems. *Homo erectus* appeared in the Turkana Basin at the same time as several highly hypsodont bovids and suids that likely were exploiting these grasses. The specific adaptations of *Homo* species to these ecological conditions remain to be fully discerned. Nevertheless, the emergence of *Homo* and the spread of *Homo erectus* can be seen as processes within the larger context of environmental change toward increased areas of grassland and the evolution of more grassland-adapted mammals in eastern Africa in the late Pliocene and earliest Pleistocene.

Acknowledgments We would like to thank the volume editors, Fred Grine, John Fleagle and Richard Leakey for inviting us to participate in this discussion of the first humans. The editors and the two anonymous reviewers provided valuable feedback and suggestions that greatly improved the manuscript. The Turkana Basin Database has been compiled with support from the National Science Foundation (NSF BCS 0137235).

References

Alemseged, Z., Bobe, R., Geraads, D., 2007. Comparability of fossil data and its significance for the interpretation of hominin environments: a case study in the lower Omo valley, Ethiopia. In: Bobe, R., Alemseged, Z., Behrensmeyer, A.K. (Eds), Hominin Environments in the East African Pliocene: An Assessment of the Faunal Evidence. Springer, Dordrecht, pp. 159–181.

Barry, J.C., Morgan, M.E., Flynn, L.J., Pilbeam, D.R., Behrensmeyer, A.K., Raza, S.M., Khan, I.A., Badgley, C., Hicks, J., Kelley, J., 2002. Faunal and environmental change in the Late Miocene Siwaliks of northern Pakistan. Paleobiology Memoirs, Paleobiology 28, 1–71.

Bobe, R., 2006. The evolution of arid environments in eastern Africa. Journal of Arid Environments 66, 564–584.

Bobe, R., Behrensmeyer, A.K., 2004. The expansion of grassland ecosystems in Africa in relation to mammalian evolution and the origin of the genus *Homo*. Palaeogeography, Palaeoclimatology, Palaeoecology 207, 399–420.

Bobe, R., Behrensmeyer, A.K., 2007. The Turkana Basin Paleontological Database: An Archive of Vertebrate Evolution in Eastern Africa. East African Association for Paleoanthropology and Paleontology, Nairobi, pp. 22–23.

Bobe, R., Eck, G.G., 2001. Responses of African bovids to Pliocene climatic change. Paleobiology Memoirs, Paleobiology 27(Supplement to number 2), 1–47.

Bobe, R., Behrensmeyer, A.K., Chapman, R.E., 2002. Faunal change, environmental variability and late Pliocene hominin evolution. Journal of Human Evolution 42, 475–497.

Bobe, R., Behrensmeyer, A.K., Eck, G.G., Harris, J.M., 2007. Patterns of abundance and diversity in late Cenozoic bovids from the Turkana and Hadar Basins, Kenya and Ethiopia. In: Bobe, R., Alemseged, Z., Behrensmeyer, A.K. (Eds), Hominin Environments in the East African Pliocene: an Assessment of the Faunal Evidence. Springer, Dordrecht, pp. 129–157.

Bonnefille, R., 1995. A reassessment of the Plio-Pleistocene pollen record of East Africa. In: Vrba, E.S., Denton, G.H., Partridge, T.C., Burckle, L.H. (Eds), Paleoclimate and Evolution with Emphasis on Human Origins. Yale University Press, New Haven, CT, pp. 299–310.

Bramble, D.M., Lieberman, D.E., 2004. Endurance running and the evolution of *Homo*. Nature 432, 345–352.

Brown, F.H., 1994. Development of Pliocene and Pleistocene chronology of the Turkana Basin, East Africa, and its relation to other sites. In: Corruccini, R.S., Ciochon, R.L. (Eds), Integrative Paths to the Past. Prentice-Hall, Englewood Cliffs, NJ, pp. 285–312.

Brown, F.H., Feibel, C.S., 1991. Stratigraphy, depositional environments and palaeogeography of the Koobi Fora Formation. In: Harris, J.M. (Ed), Koobi Fora Research Project, Volume 3: The Fossil Ungulates: Geology, Fossil Artiodactyls, and Paleoenvironments. Clarendon, Oxford, pp. 1–30.

Brown, F.H., Harris, J.M., Leakey, R.E., Walker, A., 1985. Early *Homo erectus* skeleton from west Lake Turkana, Kenya. Nature 316, 788–792.

Brown, F.H., Haileab, B., McDougall, I., 2006. Sequence of tuffs between the KBS Tuff and the Chari Tuff in the Turkana Basin, Kenya and Ethiopia. Journal of the Geological Society of London 163, 185–204.

Cooke, H.B.S., 1997. The status of the African fossil suids *Kolpochoerus limnetes* (Hopwood, 1926), *K. phacochoeroides* (Thomas, 1884) and *"K." afarensis* (Cooke, 1978). Geobios 30, 121–126.

Coppens, Y., Howell, F.C., Isaac, G.L., Leakey, R.E. (Eds), 1976. Earliest Man and Environments in the Lake Rudolf Basin: Stratigraphy, Paleoecology, and Evolution. University of Chicago Press, Chicago, IL.

deMenocal, P.B., 1995. Plio-Pleistocene African climate. Science 270, 53–59.

deMenocal, P.B., 2004. African climate change and faunal evolution during the Plio-Pleistocene. Earth and Planetary Science Letters 220, 3–24.

Eck, G.G., 2007. The effects of collection strategy and effort on faunal recovery: a case study of the American and French collections from the Shungura Formation, Ethiopia. In: Bobe, R., Alemseged, Z., Behrensmeyer, A.K. (Eds), Hominin Environments in the East African Pliocene: An Assessment of the Faunal Evidence. Springer, Dordrecht, pp. 183–215.

Eck, G.G., Jablonski, N.G., Leakey, M.G. (Eds), 1987. Les faunes Plio-Pléistocène de la basse vallée de l'Omo (Éthiopie). Tome 3: Cercopithecidae de la Formation de Shungura. CNRS, Paris.

Feakins, S.J., deMenocal, P.B., Eglinton, T.I., 2005. Biomarker records of late Neogene changes in northeast African vegetation. Geology 33, 977–980.

Feibel, C.S., Brown, F.H., McDougall, I., 1989. Stratigraphic context of fossil hominids from the Omo Group deposits: Northern Turkana Basin, Kenya and Ethiopia. American Journal of Physical Anthropology 78, 595–622.

Greenacre, M.J., Vrba, E.S., 1984. Graphical display and interpretation of antelope census data in African wildlife areas, using correspondence analysis. Ecology 65, 984–997.

Harris, J.M. (Ed), 1983. Koobi Fora Research Project, Volume 2: The Fossil Ungulates: Proboscidea, Perissodactyla, and Suidae. Clarendon Press, Oxford.

Harris, J.M. (Ed), 1991. Koobi Fora Research Project, Volume 3: The Fossil Ungulates: Geology, Fossil Artiodactyls, and Paleoenvironments. Clarendon Press, Oxford.

Harris, J.M., Cerling, T.E., 2002. Dietary adaptations of extant and Neogene African suids. Journal of Zoology 256, 45–54.

Harris, J.M., Leakey, M.G. (Eds), 2003. Geology and Vertebrate Paleontology of the Early Pliocene Site of Kanapoi, Northern Kenya. Natural History Museum of Los Angeles County, Los Angeles.

Harris, J.M., Brown, F.H., Leakey, M.G., 1988. Stratigraphy and Paleontology of Pliocene and Pleistocene Localities West of Lake Turkana, Kenya. Natural History Museum of Los Angeles County, Los Angeles.

Hill, A., Ward, S.C., Deino, A., Curtis, G.H., Drake, R., 1992. Earliest *Homo*. Nature 355, 719–722.

Howell, F.C., Haesaerts, P., de Heinzelin, J., 1987. Depositional environments, archeological occurrences and hominids from Members E and F of the Shungura Formation (Omo Basin, Ethiopia). Journal of Human Evolution 16, 665–700.

Isaac, G.L. (Ed.), 1997. Koobi Fora Research Project Volume 5: Plio-Pleistocene Archaeology. Clarendon Press, Oxford.

Jablonski, N.G., Leakey, M.G. (Eds), 2008. The Fossil Monkeys. Koobi Fora Research Project Volume 6. California Academy of Sciences Press, San Francisco.

Kibunjia, M., Roche, H., Brown, F.H., Leakey, R.E., 1992. Pliocene and Pleistocene archaeological sites, west of Lake Turkana, Kenya. Journal of Human Evolution 23, 432–438.

Kimbel, W.H., Walter, R.C., Johanson, D.C., Reed, K.E., Aronson, J.E., Assefa, Z., Marean, C.W., Eck, G.G., Bobe, R., Hovers, E., Rak, Y., Vondra, C.F., Yemane, T., York, D., Chen, Y., Evenssen, N., Smith, P., 1996. Late Pliocene *Homo* and Oldowan tools from the Hadar Formation (Kada Hadar Member), Ethiopia. Journal of Human Evolution 31, 549–561.

Kimbel, W.H., Johanson, D.C., Rak, Y., 1997. Systematic assessment of a maxilla of *Homo* from Hadar, Ethiopia. American Journal of Physical Anthropology 103, 235–262.

Kingston, J.D., 2007. Shifting adaptive landscapes: progress and challenges in reconstructing early hominid environments. American Journal of Physical Anthropology 134, 20–58.

Kingston, J.D., Marino, B.D., Hill, A., 1994. Isotopic evidence for Neogene hominid paleoenvironments in the Kenya Rift Valley. Science 264, 955–959.

Koch, C.F., 1987. Prediction of sample size effects on the measured temporal and geographic distribution patterns of species. Paleobiology 13, 100–107.

Koch, C.F., Morgan, J.P., 1988. On the expected distribution of species' ranges. Paleobiology 14, 126–138.

Leakey, M.G., Harris, J.M. (Eds), 2003. Lothagam: The Dawn of Humanity in Eastern Africa. Columbia University Press, New York.

Leakey, M.G., Leakey, R.E. (Eds), 1978. Koobi Fora Research Project, Volume 1: The Fossil Hominids and an Introduction to Their Context, 1968–1974. Clarendon Press, Oxford.

Leakey, M.G., Gathogo, P.N., Jablonski, N.G., 2008. Geological background, geochronology and stratigraphic analysis. In: Jablonski, N.G., Leakey, M.G. (Eds), The Fossil Monkeys. Koobi Fora Research Project Volume 6. California Academy of Sciences Press, San Francisco, pp. 359–396.

Leakey, M.G., Spoor, F., Brown, F.H., Gathogo, P.N., Kiarie, C., Leakey, L.N., McDougall, I., 2001. New hominin genus from eastern Africa shows diverse middle Pliocene lineages. Nature 410, 433–440.

Leakey, R.E., 1973. Evidence for an advanced Plio-Pleistocene hominid from East Rudolf, Kenya. Nature 242, 447–450.

Leakey, R.E., 1976. An overview of the Hominidae from East Rudolf, Kenya. In: Coppens, Y., Howell, F.C., Isaac, G.L., Leakey, R.E. (Eds), Earliest Man and Environments in the Lake Rudolf Basin. University of Chicago Press, Chicago, IL, pp. 476–483.

Levin, N., Quade, J., Simpson, S.W., Semaw, S., Rogers, M., 2004. Isotopic evidence for Plio-Pleistocene environmental change at Gona, Ethiopia. Earth and Planetary Science Letters 219, 93–110.

McDougall, I., Brown, F.H., 2006. Precise 40Ar/39Ar geochronology for the upper Koobi Fora Formation, Turkana Basin, northern Kenya. Journal of the Geological Society of London 163, 205–220.

McHenry, H.M., Coffing, K., 2000. *Australopithecus* to *Homo*: transformations in body and mind. Annual Review of Anthropology 29, 125–146.

Merrick, H.V., 1976. Recent archaeological research in the Plio-Pleistocene deposits of the lower Omo valley, southwestern Ethiopia. In: Isaac, G.L., McCown, E.R. (Eds), Human Origins. W.A. Benjamin, Menlo Park, CA, pp. 461–481.

Prat, S., Brugal, J.-P., Tiercelin, J.-J., Barrat, J.-A., Bohn, M., Delagnes, A., Harmand, S., Kimeu, K., Kibunjia, M., Texier, P.-J., Roche, H., 2005. First occurrence of early *Homo* in the Nachukui Formation (West Turkana, Kenya) at 2.3–2.4 Myr. Journal of Human Evolution 49, 230–240.

Reed, K.E., 1997. Early hominid evolution and ecological change through the African Plio-Pleistocene. Journal of Human Evolution 32, 289–322.

Rightmire, G.P., Lordkipanidze, D., 2009. Comparisons of early Pleistocene skulls from East Africa and the Georgian Caucasus: evidence bearing on the origin and systematics of genus *Homo*. In: Grine, F.E., Fleagle, J.G., Leakey, R.E. (Eds), The First Humans: Origin and Early Evolution of the Genus *Homo*. Springer, Dordrecht, pp. 39–48.

Roche, H., Brugal, J.-P., Delagnes, A., Feibel, C., Harmand, S., Kibunjia, M., Prat, S., Texier, P.-J., 2003. Les sites archeologiques plio-pleistocenes de la formation de Nachukui, Ouest-Turkana, Kenya: bilan synthetique 1997–2001. Comptes Rendus Palevol 2, 663–673.

Roche, H., Blumenschine, R.J., Shea, J.J., 2009. Origins and adaptations of early *Homo*: what archeology tells us. In: Grine, F.E., Fleagle, J.G., Leakey, R.E. (Eds), The First Humans: Origin and early Evolution of the Genus *Homo*. Springer, Dordrecht, pp. 135–147.
Rogers, M.J., Harris, J.W.K., Feibel, C.S., 1994. Changing patterns of land use by Plio-Pleistocene hominids in the Lake Turkana Basin. Journal of Human Evolution 27, 139–158.
Spoor, F., Leakey, M.G., Gathogo, P.N., Brown, F.H., Anton, S.C., McDougall, I., Kiarie, C., Manthi, F.K., Leakey, L.N., 2007. Implications of new early *Homo* fossils from Ileret, east of Lake Turkana, Kenya. Nature 448, 688–691.
Suwa, G., White, T.D., Howell, F.C., 1996. Mandibular postcanine dentition from the Shungura Formation, Ethiopia: crown morphology, taxonomic allocations, and Plio-Pleistocene hominid evolution. American Journal of Physical Anthropology 101, 247–282.
Walker, A., Leakey, R.E. (Eds), 1993. The Nariokotome *Homo erectus* Skeleton. Harvard University Press, Cambridge, MA.
Walker, A., Leakey, R.E., Harris, J.M., Brown, F.H., 1986. 2.5-Myr *Australopithecus boisei* from west of Lake Turkana, Kenya. Nature 322, 517–522.
White, T.D., 1995. African omnivores: global climatic change and Plio-Pleistocene hominids and suids. In: Vrba, E.S., Denton, G.H., Partridge, T.C., Burckle, L.H. (Eds), Paleoclimate and Evolution with Emphasis on Human Origins. Yale University Press, New Haven, CT, pp. 369–384.
Wood, B.A., 1985. Early *Homo* in Kenya, and its systematic relationships. In: Delson, E. (Ed), Ancestors: The Hard Evidence. Alan R. Liss, New York, pp. 206–214.
Wood, B.A., 1991. Koobi Fora Research Project Volume 4: Hominid Cranial Remains. Clarendon Press, Oxford.
Wood, B.A., 2009. Where does the genus *Homo* begin, and how would we know? In: Grine, F.E., Fleagle, J.G., Leakey, R.E. (Eds), The First Humans: Origin and Early Evolution of the Genus *Homo*. Springer, Dordrecht, pp. 17–28.
Wood, B.A., Collard, M., 1999. The human genus. Science 284, 65–71.
Wynn, J.G., 2004. Influence of Plio-Pleistocene aridification on human evolution: evidence from paleosols of the Turkana Basin, Kenya. American Journal of Physical Anthropology 123, 106–118.

Chapter 16
Biogeochemical Evidence for the Environments of Early *Homo* in South Africa

Matt Sponheimer and Julia A. Lee-Thorp

Keywords Stable isotopes • carbon isotopes • oxygen isotopes • herbivore enamel • *Homo habilis* • *Homo erectus* • South Africa • *Australopithecus* • Sterkfontein • Swartkrans • Makapansgat

Introduction

It has long been observed that early members of the genus *Homo* are associated with more open and arid-adapted fauna than antecedent australopith species (e.g., Vrba, 1980; Wesselman, 1985; Reed, 1997; Bobe and Behrensmeyer, 2004). Moreover, the first appearances of earliest *Homo* (~2.5 Ma; Hill et al., 1992; Schrenk et al., 1993; Kimbel et al., 1996) and *Homo ergaster* (~1.8 Ma; Feibel et al., 1989) are broadly coincident with a series of global climate changes that included the onset of northern hemisphere glaciation and the Walker circulation respectively (Trauth et al., 2005). Together, these observations suggest the possibility of some linkage between the origin and subsequent speciation of *Homo* and climatic/environmental change; yet, there are several reasons to be cautious about accepting this linkage at face value. For one, large-scale climate processes are not discrete "events," and thus are very difficult to tie to specific speciation and/or extinction events in the fragmentary, and often chronologically ambiguous, terrestrial fossil record. For another, the paleoclimatic data that are often used to posit such linkages are usually global in nature (e.g., Prentice and Denton, 1988; deMenocal, 1995), and thus their relevance for animals in particular basins and around specific sites is unclear. This is particularly problematic with respect to southern Africa, where climate change may be anti-phased or at least partially decoupled with respect to the better-known East African record (e.g., Partridge et al., 2004). In addition, paleoanthropologists working in southern Africa have fewer proxies with which to explore paleoclimate and environmental trends. For instance, the absence of pedogenic carbonates (from relevant ages and places), which have been used to investigate long-term environmental change (e.g., Cerling, 1992; Quade and Cerling, 1995; Wynn, 2004) and document vegetational diversity across paleolandscapes (Sikes, 1994), has made it difficult to cross-check both regional and local paleoenvironmental reconstructions in southern Africa. Thus, researchers in southern Africa have relied perforce upon global or regional datasets that may or may not be relevant, or have relied on the taxonomic or morphological character of hominin-associated fauna.

Here, we draw on stable isotope data from fossil herbivores to augment the taxonomic and ecomorphological studies that have investigated the habitats of *Australopithecus africanus* and later *Homo* in South Africa (e.g., Cooke, 1978; Vrba, 1980, 1985; Reed, 1997; Spencer, 1997). We explore the degree to which faunal isotopic proxies bear on two principle questions: (1) Did early members of the genus *Homo* inhabit more open environments than their predecessor (*A. africanus*) in South Africa, and (2) If so, was the trend to open environments accompanied by aridification, so that *Homo* lived in drier conditions than *A. africanus*? In applying and evaluating the isotopic data, we will also touch on several methodological and theoretical considerations.

Using Herbivore Carbon Isotope Ratios to Investigate Paleoenvironments

The basis for using carbon isotope data from herbivores to investigate paleoenvironments lies in the different photosynthetic pathways that are utilized by most tropical trees, bushes, shrubs, and forbs (C_3 plants) on the one hand, and tropical grasses (C_4 plants) on the other. Biochemical and anatomical differences between C_3 and C_4 plants result in their having very different, non-overlapping carbon isotope

M. Sponheimer (✉)
Department of Anthropology, University of Colorado at Boulder, Boulder, CO 80309, USA
e-mail: msponheimer@gmail.com

J.A. Lee-Thorp
Division of Archeological, Geographical and Environmental Sciences, University of Bradford, Bradford BD7 1DP, UK
e-mail: J.A.Lee-thorp@Bradford.ac.uk

ratios (Smith and Epstein, 1971; see Codron et al., 2005 for a large African dataset). Herbivores incorporate plant carbon into their tissues, and thus one can readily distinguish an animal eating vegetation from trees, bushes, shrubs or forbs (e.g., leaves, fruits) from one eating tropical grasses, based on the carbon isotope ratios of their tissues (Vogel, 1978). Furthermore, this distinction is readily preserved in dental enamel from the Pliocene (Lee-Thorp and van der Merwe, 1987; Cerling et al., 1997). Therefore, the rationale for using carbon isotope data from herbivores to investigate paleoenvironments is that, at a very general level, a relationship exists between the number of C_4 grass consumers at a site and the availability of palatable grasses in the local environment. For instance, if virtually every animal at a site is found to have been a C_4 grass-consumer, it is a reasonable assumption that the area was dominated by grasses (we discuss potential collection biases below). Conversely, if all of the animals at a site consumed C_3 vegetation, it would be fair to say that there was little if any C_4 grass available for consumption.

Figure 16.1 shows the percentage of C_3 and C_4 consumers at 15 African game reserves, and as can be seen, this method generally distinguishes between "closed" and "open" habitats (as defined by Vrba, 1980), with the latter being isolated in the bottom right corner (few C_3 specialists and many C_4 specialists) (from Sponheimer and Lee-Thorp, 2003). The only "closed" area from Vrba's survey that clusters with "open" habitats is Lake Manyara, where buffaloes comprise 66% of the total bovid population. However, Lake Manyara also contains many "open" areas, so this result is at least partly a matter of definition. These data strongly suggest that when an area has fewer than 20% C_3 consumers and more than 35% C_4 consumers, it is likely to be an "open" environment. To a significant extent such information could be provided

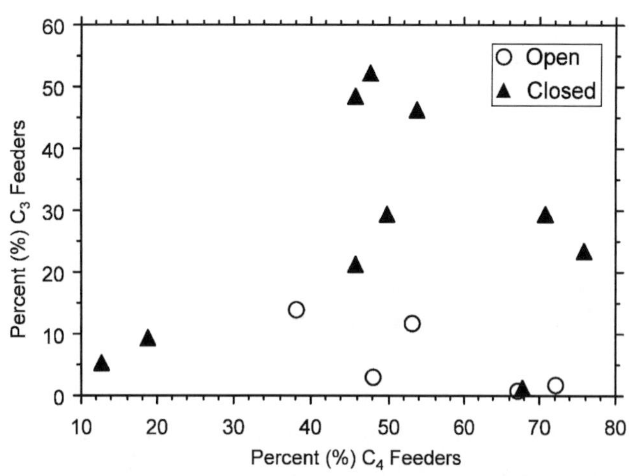

Fig. 16.1 Bivariate plot showing the percentages of C_3 and C_4 specialists at a variety of African game reserves (see Sponheimer and Lee-Thorp, 2003). The relatively "open" environments (as defined by Vrba, 1980) are confined to the lower right hand corner of the graph.

using ecomorphology (see Reed, 1996). However, there is a decided advantage to supplementing ecomorphological with stable carbon isotope data (or other techniques that provide non-genetic information such as dental microwear analysis; see Schubert et al., 2006), as they are direct indicators of diet, while the former tells us more about the diets that challenged an animal's ancestors.

We have also applied this technique to fossil herbivores from the sites Makapansgat Limeworks (Member 3), Sterkfontein (Members 4 and 5), and Swartkrans (Members 1 and 2) (data in Lee-Thorp et al., 2007; Fig. 16.2). These are all karst sites within the Malmani dolomite formation. Conventionally, the deposits of each site have been divided into a series of Members that are believed to form a sequence from older to younger on the basis of lithostratigraphy and biostratigraphy; however, their stratigraphy is complex as a result of multiple depositional and erosional events, and as a result some of these divisions are contested (Brain, 1981, 1993; Maguire, 1985; Berger et al., 2002). The ages of the Members are based largely on biostratigraphic comparisons with well-dated sites in East Africa. For the purposes of this study we use ages of ~3 Ma for Makapansgat Member 3, ~2.2–2.5 Ma for Sterkfontein Member 4, ~1.5–2 Ma for Sterkfontein Member 5, and ~1.6–1.8 Ma and ~1.2–1.6 Ma for Swartkrans Member 1 and 2 respectively (based on McFadden et al., 1979; Vrba, 1982; Brain, 1993; Kuman and Clarke, 2000; Partridge, 2000; Clarke and Partridge, 2002).

We divided data for all large herbivore specimens into three broad trophic categories: C_3 consumers ($\delta^{13}C < -9.0‰$), C_4 consumers ($\delta^{13}C > -3.0‰$), and mixed feeders ($\delta^{13}C$ values between these two extremes). Figure 16.2 shows a general decline in the proportions of C_3 consumers (browsers) through the sequence, and a concomitant rise in the proportions of C_4 consumers (grazers). The two *Australopithecus*-bearing members, Makapansgat M3 and Sterkfontein M4, contain more than 30% C_3 consumers and fewer than 40% C_4 consumers. In contrast, all of the members which contain *Homo* (Swartkrans M1 and M2 and Sterkfontein M5) have more than 70% C_4 consumers + mixed feeders. Thus, these data suggest that *Australopithecus africanus* inhabited more "closed," woody environments than *Homo* and its contemporary, *Paranthropus robustus*, which is in close agreement with results from non-isotopic faunal studies (e.g., Vrba, 1985; Reed, 1997).

One distinction between this and most previous faunal studies is that the carbon isotope data suggest that Makapansgat M3 was especially dominated by C_3 consumers, which may suggest a densely wooded environment. Another distinction can be found in Sterkfontein Member 5. Recent excavations have shown that M5 Unit B contains sparse Oldowan tools while Unit C has yielded Acheulean technology, suggesting a more recent age for the latter (Kuman and Clarke, 2000; Clarke and Partridge, 2002).

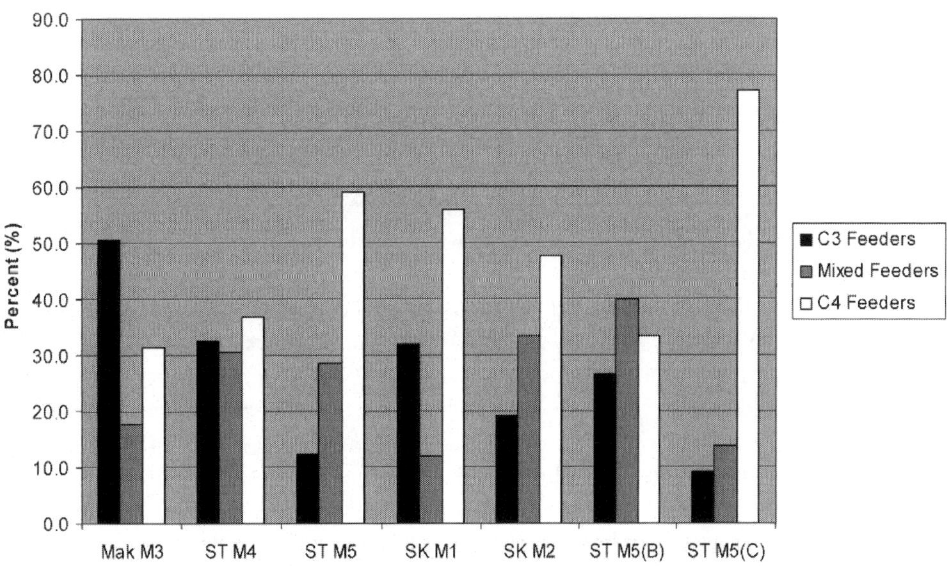

Fig. 16.2 The relative percentages of C_3, C_4, and mixed feeders from a variety of South African early hominin deposits (data from Lee-Thorp et al., 2007). The percentages of C_4 feeders increase sharply during the period of accumulation of Swartkrans M1 and Sterkfontein M5(C), suggesting greater grass cover when *Homo* and *Paranthropus* were present.

And in fact, the carbon isotope data for herbivores in these units differ greatly (Fig. 16.2). Nearly 30% of the Unit B herbivores sampled were C_3 consumers and just over 30% were C_4 consumers; in contrast, fewer than 10% of the Unit C herbivores were C_3 consumers and nearly 80% were C_4 consumers. This suggests that Unit C accumulated when the area was dominated by grassy vegetation, and that a major environmental change occurred between the Oldowan and Acheulean deposits, most probably between 1.8 and 1.6 Ma.

We can also compare continuous, rather than categorical $\delta^{13}C$ data from each Member, which reveals a similar pattern (Fig. 16.3). In aggregate, the herbivores from Makapansgat M3 have the most negative $\delta^{13}C$ values ($\bar{x} = -6.8\%$), followed by Sterkfontein M4 ($\bar{x} = -5.4\%$), Swartkrans M1 ($\bar{x} = -4.4\%$) and Swartkrans M2 ($\bar{x} = -4.4\%$), Sterkfontein M5 Unit B ($\bar{x} = -3.5\%$), and lastly Sterkfontein M5 Unit C ($\bar{x} = -2.1\%$). Thus, once again, the $\delta^{13}C$ values of the *Australopithecus*-bearing members suggest more heavily-wooded environments than those bearing evidence for early *Homo*. Moreover, using both categorical and continuous data, Makapansgat M3 comes out as the most closed environment, with Sterkfontein M5 Unit C appearing to be the grassiest.

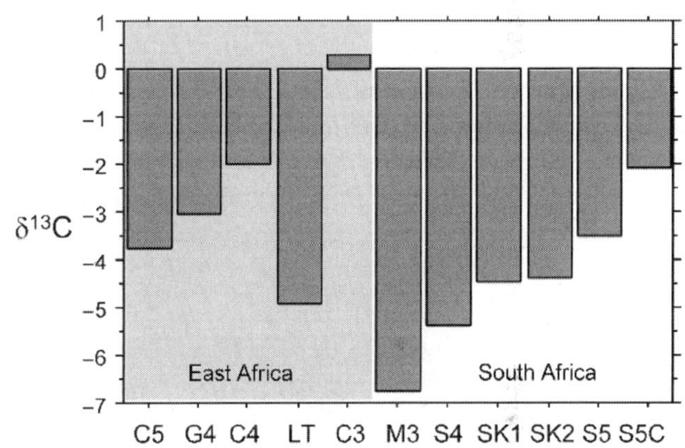

Fig. 16.3 Mean $\delta^{13}C$ values for East (and Central) African and South African herbivore assemblages. The East African deposits have generally more positive $\delta^{13}C$ values than South African sites from similar time periods. This may reflect true environmental differences, but could also be at least partly a reflection of the sampling strategies utilized. C5 is Kossom Bougoudi, Chad (~5.3 Ma), C4 is Kolle, Chad (~5–4 Ma), C3 is Koro Toro, Chad (3.5–3.0 Ma), G4 is Gona, Ethiopia (~4.3 Ma), LT is Laetoli, Tanzania (~3.5 Ma), M3 is Makapansgat M3, South Africa (~3 Ma), ST4 is Sterkfontein Member 4, South Africa (~2.5 Ma), SK1 is Swartkrans Member 1, South Africa (~1.8 Ma), SK2 is Swartkrans Member 2, South Africa (~1.6 Ma), and ST5 is Sterkfontein Member 5, South Africa (~1.8 Ma) (Data from Lee-Thorp, 1989; Sponheimer, 1999; Zazzo et al., 2000; Luyt, 2001; Levin et al., 2004; Kingston and Harrison, 2007).

One potential problem with such isotope-based techniques is that, ideally, one should produce $\delta^{13}C$ values for all herbivore specimens, which is impractical as well as expensive. Thus, one either has to (1) sample a random subset of the fauna preserved in the deposit, or (2) establish mean values for all taxa, and then produce a site mean adjusted for the relative abundance of each taxon (as the site may be dominated by one or a few taxa). The latter will only be possible with well-studied faunal suites in which specimens have been precisely classified to genus or species and for which relative abundance data are available. In contrast, selecting a random sample of herbivores should be quite easy, although it is rarely, if ever, done in practice. Thus, it is difficult to compare our data with published datasets from other sites in Africa (Fig. 16.3), since none of the studies (some of ours included) are explicit with regard to sampling strategy. For instance, the mean $\delta^{13}C$ value for herbivores from Chad suggests that at ~5 Ma (Kossom Bougoudi; $\bar{x} = -3.8‰$), the area was already as open as the areas inhabited by *Homo* in South Africa about 1.8 Ma, and that by ~3 Ma (Koro Toro; $\bar{x} = +0.3‰$) the area was virtually pure grassland (Fig. 16.3; data from Zazzo et al., 2000). This might indicate a real paleoenvironmental difference between the two regions, with Chad having generally more open environments, but it might also partly reflect differences in sampling strategy. Indeed, additional sampling and analysis by Jacques (2007) intimates that C_4 grasses may have been slightly less extensive than the original study suggested.

It is important to note that we are not suggesting that any of the isotopic studies of herbivore enamel in South, Central, or East Africa were flawed, but only that they were not necessarily concerned with determining the eating habitats of the herbivore faunas on the whole. If herbivore carbon isotope data are to be used for paleoenvironmental purposes in the future, however, the results will be more useful when obtained from collections for which the alpha taxonomy and relative abundance studies are well advanced, or from randomly selected specimens.

Another problem, especially at the South African hominin sites, is the unknown length of time in which a fossil assemblage accumulated. Therefore, the assemblages may or may not fairly represent the fauna found in the vicinity of the cave at any given time. Moreover, the relative percentages of taxa found in each deposit may not accurately reflect the living community at any given moment due to collection biases (Brain, 1981; Behrensmeyer, 1991; Lyman, 1994). These difficulties plague all faunal and palynological analyses, however, so are of no special concern for stable carbon isotope studies. And importantly, a recent taphonomic study of the Sterkfontein Valley sites discussed here revealed that while there are taphonomic biases among these assemblages (e.g., more craniodental remains relative to postcranial material in calcified compared to decalcified/uncalcified sediments), there was no evidence that they influenced the taxonomic composition of the faunas (De Ruiter et al., 2008).

Oxygen Isotope Data from South African Early Hominin Sites

Oxygen isotope ratios in herbivore bone and enamel have been used to reconstruct the isotopic composition of environmental waters, which in turn carry climate information (e.g., Longinelli, 1984; Ayliffe and Chivas, 1990; Bryant et al., 1996; Passey et al., 2002; Schoeninger et al., 2003). Oxygen isotope ratio data are regularly produced when analyzing tooth enamel for carbon isotope ratios, and thus studies aimed at providing dietary information using carbon isotopes may also provide paleoclimatic information. Of particular interest here is an idea advanced by Kohn et al. (1996), and recently tested and developed by Levin et al. (2006), that differences in the oxygen isotope compositions of environmentally sensitive (ES) and environmentally insensitive (EI) taxa can be used as paleoaridity indicators. The rationale is that the oxygen isotope compositions of apatite from some mammalian taxa are primarily records of meteoric water (environmental water that originates more directly from precipitation) $\delta^{18}O$ values (EI taxa), while others are more sensitive to differences in aridity as their oxygen isotope compositions track the evaporative enrichment in ^{18}O that occurs in leaves (ES taxa). Thus, in mesic environments there should be little difference between ES and EI taxa as there will be little evaporative enrichment in leaf ^{18}O, but this difference will increase as relative humidity decreases (Fig. 16.4).

This is quite an exciting possibility, and should be more reliable than simple comparisons of $\delta^{18}O$ values between sites, as the latter might be influenced by many factors such as altitude, distance inland, changes in storm tracks, and taxonomic composition. We applied a version of this technique to fossil enamel from Makapansgat, Sterkfontein, and Swartkrans in order to test the hypothesis that *Australopithecus* experienced more humid conditions than early *Homo*. In order to do so we had to determine firstly which taxa are ES and which are EI. This exercise is not straightforward. We can either rely upon the principle of taxonomic uniformitarianism, and assume that fossil relations of modern ES and EI taxa are likely to be ES or EI themselves, or we can rely on the $\delta^{18}O$ values of the taxa themselves, and assume that those with relatively higher $\delta^{18}O$ values are most likely to be ES. Both of these methods produce similar results in the case of Makapansgat Member 3. For instance, taxonomic uniformitarianism suggests that giraffids and perissodactyls should be ES and EI taxa respectively (as in Levin et al., 2006); and thus the former should have higher $\delta^{18}O$ values than the latter, which is indeed the case. However, the difference (ε, see caption of Fig. 16.4) between these taxa in Member 3 is 4.8‰ (ANOVA, F = 63.918; d.f. = 20, $p < 0.0001$; Table 16.1; Fig. 16.3), which is similar to the difference between these two groups, in the hot, arid Turkana region today ($\varepsilon = 5.0‰$, mean annual precipitation (MAP) = 178 mm; data from Levin et al., 2006). If one broadens the analysis to include gazelles and grazing suids, which were found to be ES and EI taxa respectively in Levin et al. (2006), then the difference is 4.9‰ (ANOVA, F = 96.474, d.f. = 33, $p < 0.0001$; Table 16.1; Fig. 16.3). In fact, there is virtually no way to group the oxygen isotope data for Makapansgat (see Sponheimer, 1999) that results in a difference between ES and EI taxa of less than 4‰, which is still considerably greater than the difference found between these groups at Olorgesailie (MAP = 417 mm) and Tsavo

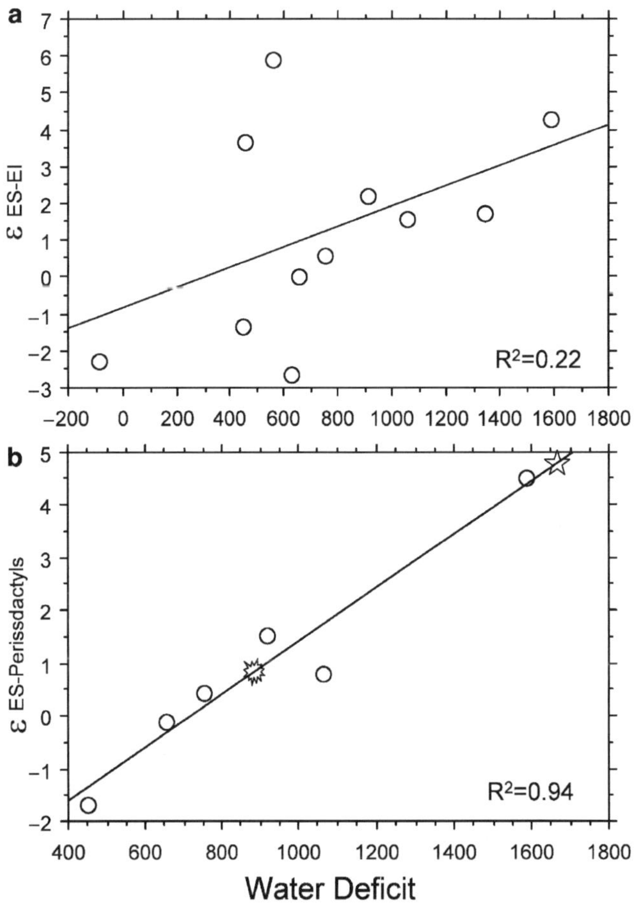

Fig. 16.4 Regression plots showing the differences (ε) between environmentally sensitive (ES) and environmentally insensitive (EI) taxa from a variety of locations with varying water deficits (i.e., the difference between potential evapotranspiration and mean annual precipitation). In general, ε increases in step with aridity. It is worth noting that Δ ($\delta^{13}C_B$–$\delta^{13}C_A$) rather than ε ((1,000 + $\delta^{13}C_B$) / (1,000 + $\delta^{13}C_A$)-1) * 1,000) is often used express such differences as it is more convenient, but as it is not strictly correct we use the latter here (see Cerling and Harris 1999). As can be seen in (a), the relationship is fairly weak when using many taxa simultaneously, but improves greatly when EI taxa are limited to those in the Perissodactyla (b). The star indicates the difference between ES and EI taxa for Makapansgat M3 and the starburst indicates the difference between these groups at Sterkfontein M5 and Swartkrans M2 combined. These data would seem to indicate that early members of the genus *Homo* in South Africa experienced much more humid conditions than *Australopithecus* at Makapansgat.

Table 16.1 Stable isotope data from tooth enamel used to investigate aridity at Makapansgat Limeworks Member 3 (~3 Ma) (Species identifications are from Reed, 1996)

Specimen	Species	$\delta^{13}C$	$\delta^{18}O$	ES
M167	*Ceratotherium simum*	−3.3	−2.3	No
M2088	*Ceratotherium simum*	−2.7	−1.7	No
M8939	*Ceratotherium simum*	−4.3	−2.2	No
M8940	*Ceratotherium simum*	−3.6	−1.6	No
M2108	*Diceros bicornis*	−11.0	−1.1	No
M2109	*Diceros bicornis*	−10.8	−3.5	No
M642/2106	*Diceros bicornis*	−13.1	−5.6	No
M766	*Gazella gracilior*	−12.4	3.9	Yes
M767	*Gazella gracilior*	−10.7	3.1	Yes
M1188	*Gazella vanhoepeni*	−10.6	0.8	Yes
M529	*Gazella vanhoepeni*	−10.3	−0.2	Yes
M7805	*Gazella vanhoepeni*	−12.4	2.3	Yes
M7811	*Gazella vanhoepeni*	−12.8	1.7	Yes
M8823	*Gazella vanhoepeni*	−10.9	3.2	Yes
M9014	*Gazella vanhoepeni*	−11.5	1.9	Yes
M1113	*Giraffa jumae*	−10.8	0.6	Yes
M1798	*Giraffa jumae*	−9.6	1.1	Yes
M2085	*Giraffa jumae*	−10.9	1.6	Yes
M2085	*Giraffa jumae*	−12.6	4.4	Yes
M528	*Giraffa jumae*	−10.5	3.2	Yes
M8853	*Giraffa jumae*	−10.5	1.6	Yes
M936	*Giraffa jumae*	−11.9	5.4	Yes
M938	*Giraffa jumae*	−11.0	2.4	Yes
M193	*Hipparion lybicum*	0.2	−2.4	No
M2476	*Hipparion lybicum*	0.0	−1.5	No
M2480	*Hipparion lybicum*	−1.3	−1.3	No
M2505	*Hipparion lybicum*	−2.0	−3.4	No
MUE1	*Hipparion lybicum*	−0.8	−1.7	No
M2025	*Notochoerus capensis*	−1.0	−2.0	No
M8913	*Notochoerus capensis*	−0.6	−1.2	No
M1826/1890	*Potamochoeroides shawi*	−2.3	−4.9	No
M1859	*Potamochoeroides shawi*	−2.2	−2.4	No
M1886	*Potamochoeroides shawi*	−1.8	−6.0	No
M1876	*Sivatherium marusium*	−11.3	2.5	Yes
M2086	*Sivatherium marusium*	−10.8	1.5	Yes

ES stands for environmentally sensitive as defined in Levin et al. (2006)

(MAP = 549 mm) in Kenya today (Fig. 16.3; data from Levin et al., 2006).

This suggests that Makapansgat at ~3 Ma was much drier than the area is today (600–700 mm), a result which is difficult to reconcile with regional and global paleoclimatic datasets. In addition, the fauna from Makapansgat Member 3 consists of a large number of mesic-associated taxa (e.g., *Cephalophus*, *Redunca*) (Reed 1997). Furthermore, when this paleoaridity indicator was applied to the Sterkfontein Valley fossil sites (a composite of data from similarly aged Sterkfontein M5 and Swartkrans Member 2) using Antilopini and Hippotragini as ES taxa and perissodactyls as EI taxa (as in Levin et al., 2006), it predicts far *wetter* environments (ε = 0.8%, ANOVA, F = 1.999, d.f. = 43, p = 0.1646; Table 16.2; Fig. 16.4) akin to what is expected today in the Masai Mara or Samburu, Kenya (MAP ~700 mm). This is puzzling once again as taxonomic (Vrba, 1985), ecomorphological (Reed, 1997), and carbon isotope (Sponheimer and Lee-Thorp, 2003) studies of the Makapansgat fauna are all consistent with it having had the most mesic (or at least the most heavily wooded) environment of all the South African australopith sites. Either the fauna at these sites tell us very little about environments and climates, or there is a suite of unknown, confounding factors which precludes application of this paleoaridity indicator at these South African sites. We favor the latter interpretation, and thus we fear that we are

Table 16.2 Stable isotope data from tooth enamel used to explore aridity at Sterkfontein and Swartkrans

Specimen	Taxon	Site	Member	$\delta^{13}C$	$\delta^{18}O$	ES
SE 1258	Antilopini	Sterfontein	5	−10.8	−0.5	Yes
SE 1855.1	Antilopini	Sterfontein	5	−12.7	−1.7	Yes
S94-6124	Antilopini	Sterfontein	5	−9.6	−0.7	Yes
BP/3/16974	Antilopini	Sterfontein	5	−9.2	−2.8	Yes
S94-7958	Antilopini	Sterfontein	5	−7.4	−0.7	Yes
S94-7314	Antilopini	Sterfontein	5	0.7	3.7	Yes
S94–6124	Antilopini	Sterfontein	5	−9.6	−0.7	Yes
BP/3/16974	Antilopini	Sterfontein	5	−9.2	−2.8	Yes
SE 1258	Antilopini	Sterfontein	5	−10.8	−0.5	Yes
SE 1855.1	Antilopini	Sterfontein	5	−12.7	−1.7	Yes
S94-7958	Antilopini	Sterfontein	5	−7.4	−0.7	Yes
S94-7314	Antilopini	Sterfontein	5	0.7	3.7	Yes
S94-1787	Equidae	Sterfontein	5	−0.6	−2.9	No
S94-390	Equidae	Sterfontein	5	−2.0	−0.9	No
S94-349	Equidae	Sterfontein	5	−0.9	−2.1	No
S94-1750	Equidae	Sterfontein	5	0.8	−0.7	No
S94-329	Equidae	Sterfontein	5	−4.4	−0.4	No
STS 3006	Equidae	Sterfontein	5	−3.6	−2.5	No
STS 2102	Equidae	Sterfontein	5	−2.9	−1.8	No
STS 1972	Equidae	Sterfontein	5	−4.6	−3.1	No
STS 2313	Equidae	Sterfontein	5	−4.8	−4.4	No
S94-329	Equidae	Sterfontein	5	−4.4	−0.4	No
S94-1787	Equidae	Sterfontein	5	−0.6	−2.9	No
S94-390	Equidae	Sterfontein	5	−2.0	−0.9	No
S94-349	Equidae	Sterfontein	5	−0.9	−2.1	No
S94-1750	Equidae	Sterfontein	5	0.8	−0.7	No
SE 1125.1	Hippotragini	Sterfontein	5	−5.3	−0.6	Yes
SE 1125.1	Hippotragini	Sterfontein	5	−5.3	−0.6	Yes
SKX 811	Antilopini	Swartkrans	2	−12.9	0.6	Yes
SKX 1896	Antilopini	Swartkrans	2	−10.6	1.4	Yes
SKX 2736	Antilopini	Swartkrans	2	−11.5	3.6	Yes
SKX 12067	Antilopini	Swartkrans	2	−2.3	−0.5	Yes
SK 2574	Antilopini	Swartkrans	2	−4.5	−2.4	Yes
SK 6023	Antilopini	Swartkrans	2	−4.3	−2.0	Yes
SKX 5907	Antilopini	Swartkrans	2	−2.9	0.3	Yes
SKX 9385	Antilopini	Swartkrans	2	−3.3	−2.7	Yes
SKX 5962	Antilopini	Swartkrans	2	−4.2	−2.8	Yes
SK 3841	Antilopini	Swartkrans	2	−1.7	−0.9	Yes
SK 5922	Antilopini	Swartkrans	2	−2.4	−4.0	Yes
SKX 12273	Antilopini	Swartkrans	2	−3.8	−3.1	Yes
SK 3160	Equidae	Swartkrans	2	−6.4	0.5	No
SK 2626	Equidae	Swartkrans	2	−2.1	0.0	No
Sk 2626	Equidae	Swartkrans	2	−3.5	−5.0	No
SK 3990	Equidae	Swartkrans	2	0.0	0.4	No
SK 3992	Equidae	Swartkrans	2	−2.8	0.9	No

All of the specimens are believed to be roughly contemporaneous at about 1.8–1.5 Ma. ES stands for environmentally sensitive as defined in Levin et al. (2006).

unable to meaningfully test hypotheses about the relative humidty/aridity experienced by *A. africanus* and *Homo* in South Africa at this time.

So What Have We Learned?

The herbivore $\delta^{13}C$ data discussed herein support the hypothesis that in South Africa early members of the genus *Homo* inhabited more open environments than their predecessors, *A. africanus*. They broadly reinforce the pattern that has been advanced based on other methods for an increase in grassy vegetation between about 2.5 Ma and the first appearance of *Homo* in South Africa at about 1.8 Ma. This is also concordant with what has been observed elsewhere in Africa using taxonomic (Behrensmeyer et al., 1997; Bobe et al., 2002), palynological (Bonnefille, 1995; Dupont and Leroy, 1995), and pedogenic carbonate isotopic approaches (Cerling, 1992; Wynn, 2004). The data presented here intimate, however, that the situation may also be more complex. For one, the isotopic proxies we have looked at suggest Makapansgat M3 was significantly more wooded than the other South African sites including Sterkfontein M4. Thus, there may have been a significant change in the environments experienced by *A. africanus* through time before the advent of *Homo* in the region. They also suggest that the Sterkfontein Valley may have opened dramatically at around the time that Acheulean tools are first found, which could indicate that this toolkit reflects a behavioral modification to new, more open conditions, or may even denote the appearance of a new species of *Homo* in South Africa (and see Grine et al., 1996 for potential differences between eastern and southern African *Homo*).

In contrast, our attempt to examine differences in the aridity experienced by *A. africanus* and *Homo* was either unsuccessful, or suggests that contrary to our hypothesis, *A. africanus* at Makapansgat experienced much drier conditions than *Homo*. We argue that the more likely scenario is that the complexities of oxygen isotope systematics in mammals (and in the hydrology of the local environment) obscured the desired aridity signal. Yet, a microscopic examination of modern and fossil dolomite soils concluded that *A. africanus* from Makapansgat lived in a xeric environment, while *Homo* at Swartkrans lived in a climate comparable to today (Brain, 1958; Brain and Meester, 1964). Although this idea has long been abandoned (e.g., Butzer, 1971), it is intriguing that Trauth et al.'s (2005) study of lake sediments suggests that *H. ergaster* arose during a relatively humid period, which is echoed in a study by Reed and Russak (2009) which found an increase in the abundance of mesic-associated fauna with this taxon in East Africa. This possibility merits further study in South Africa.

Some Final Thoughts on Aridity, Chronology, and the Evolution of *Homo*

The "opening" of African landscapes cannot *a priori* be assumed to be a product of aridification (or *vice versa*). The large-scale emergence of C_4 grasses is often assumed to coincide with increased aridity because of the water-efficiency of C_4 plants (Ehleringer and Monson, 1993), yet temperature and irradiance during the growing season (Teeri and Stowe, 1976) rather than water availability are the chief controls on their distribution. Also, changes in the seasonality of rainfall (Monson et al., 1983), fire regime (van Wilgen et al., 2003), herbivore community ecology (Owen-Smith, 1999), or even atmospheric CO_2 concentration (Cerling et al., 1997) can result in decreased woody vegetation. Moreover, at local and regional scales variations in geology, soils, and topography are determinants of vegetation structure (Andrews and Bamford, 2008), and this could drive landscape evolution, especially in tectonically active regions such as the East African Rift.

Of course, increased aridity could also result in the opening of landscapes, although not necessarily in a straightforward manner. For instance, a recent study showed that just 650 mm of annual rainfall are necessary to establish woody canopy closure – thus a great deal of precipitation is not a prerequisite for heavily wooded environments (Sankaran et al., 2005). Furthermore, when precipitation declines, dense thicket may replace forest (Scholes and Walker, 1993), which is still far from an "open" environment. And while some evidence exists that the "opening" of Africa was accompanied by a decrease in precipitation, the most extensive and continuous record reflects conditions in the Sahara (deMenocal, 1995) and may not have much bearing on what happens elsewhere on the continent, including East and particularly southern Africa. Thus, the posited linkage between decreased woody vegetation and increased aridity in this region needs further investigation.

The chronological accord between climatic events and the first appearance of *Homo* and the later emergence of *H. ergaster* is also problematic. It is often casually argued that while the earliest reasonable evidence for *Homo* is at about 2.3 Ma, and the beginning of northern hemisphere glaciation was about 2.7 Ma, that this is consistent with the idea that climate/environmental change drove this speciation event. For after all, the fossil record is sparse and imperfect. Yet, one also hears that given the first appearance of *H. ergaster/erectus* and the development of the Walker circulation both occur at about 1.8 Ma, the link between climate change and the emergence of this taxon is particularly strong.

But is this the only reasonable way to interpret these data? As noted by Darwin (1859), the fossil record is very patchy indeed, so the argument about the link between climate change and the origin of *Homo* may be defensible. However, whatever happened to the imperfection of the fossil record with regard to the origin of *H. ergaster*? Would it not be equally parsimonious to suggest that the climate/speciation link is in fact poor, since we have very likely underestimated *H. ergaster*'s first appearance given the failings of the fossil record? In short, when the chronological link between speciation/extinction events and climatic change is good we marvel at the concordance, and when it is not we appeal to the imperfection of the fossil record. We cannot have it both ways.

It is also worth noting that because our understanding of the dietary ecology of early *Homo* and its predecessors remains nascent, it is very difficult to meaningfully predict likely outcomes of climatic and environmental changes for these taxa. For instance, given the dietary differences between *Pan troglodytes* and *Papio ursinus* today, would we not expect the two taxa to respond differently to prolonged climate-induced deforestation? The hard data for the diets of *Australopithecus* are increasingly robust (e.g., Scott et al., 2005; Sponheimer et al., 2005; Grine et al., 2006a, b), but remain maddeningly difficult to interpret. Furthermore, our understanding of the diet of early *Homo* is certainly poorer, as direct, non-genetic evidence (e.g., dental microwear, stable isotopes) of the diets of individuals is very scanty (but see van der Merwe et al., 2008; Ungar and Scott, 2009). Stone tools and butchered animal bones almost certainly indicate that early *Homo* ate meat (e.g., Blumenschine, 1987; Semaw et al., 1997, Dominguez-Rodrigo et al., 2005), but they reveal nothing about meat's relative dietary importance. The archeological record is equally consistent with early *Homo* having processed a carcass a day, or a carcass a year. Thus, do we really have a strong basis for believing that early *Homo* and later *Australopithecus* had different diets and hence habitat tolerances? Could the emergence of *Homo* be explained by intrinsic (e.g., social, cognitive, reproductive) rather than extrinsic (e.g., climate, environment, community ecology) factors?

Acknowledgments We thank Richard Leakey, Fred Grine, and John Fleagle for the opportunity to write this chapter. MS also thanks them for the opportunity to participate in the Stony Brook Human Evolution Symposium and Workshop, and the other participants for their stimulating conversation, companionship, and in one case, a fine rendition of Gilbert and Sullivan. MS also thanks the families who took us into their homes at various points during the week, the graduate students and staff who made it all possible and the anonymous reviewers whose comments helped us to revise this chapter.

References

Andrews, P., Bamford, M., 2008. Past and present vegetation ecology of Laetoli, Tanzania. Journal of Human Evolution 54, 78–98.

Ayliffe, L.K., Chivas, A.R., 1990. Oxygen isotope composition of the bone phosphate of Australian kangaroos: potential as a paleoenvironmental recorder. Geochimica et Cosmochimica Acta 54, 2603–2609.

Behrensmeyer, A., Todd, N., Potts, R., McBrinn, G., 1997. Late Pliocene faunal turnover in the Turkana Basin, Kenya and Ethiopia. Science 278, 1589–1594.

Behrensmeyer, A.K., 1991. Terrestrial vertebrate accumulations. In: Allison, P.A., Briggs, D.E.G. (Eds), Taphonomy: Releasing the Data Locked in the Fossil Record. Plenum, New York, pp. 229–335.

Berger, L.R., Lacruz, R., de Ruiter, D.J., 2002. Revised age estimates of *Australopithecus* bearing deposits at Sterkfontein, South Africa. American Journal of Physical Anthropology 119, 192–197.

Blumenschine, R.J., 1987. Characteristics of an early hominid scavenging niche. Current Anthropology 28, 383–407.

Bobe, R., Behrensmeyer, A.K., 2004. The expansion of grassland ecosystems in Africa in relation to mammalian evolution and the origin of the genus *Homo*. Palaeogeography, Palaeoclimatology, Palaeoecology 207, 399–420.

Bobe, R., Behrensmeyer, A.K., Chaplin, R.E., 2002. Faunal change, environmental variability and late Pliocene hominid evolution. Journal of Human Evolution 42, 475–497.

Bonnefille, R., 1995. A reassessment of the Plio-Pleistocene pollen record of East Africa. In: Vrba, E.S., Denton, G.H., Partridge, T.C., Burkle L.H. (Eds), Paleoclimate and Evolution. Yale University Press, New Haven, CT, pp. 299–310.

Brain, C.K., 1958. The Transvaal Ape-Man-Bearing Cave Deposits. Transvaal Museum, Pretoria.

Brain, C.K., 1981. The Hunters or the Hunted? University of Chicago Press, Chicago, IL.

Brain, C.K., 1993. Swartkrans: A Cave's Chronicle of Early Man. Transvaal Museum, Pretoria.

Brain, C.K., Meester, J., 1964. Past climatic changes as biological isolating mechanisms in Southern Africa. Ecological Studies in Southern Africa, 332–340.

Bryant, J.D., Koch, P.L., Froelich, P.N., Showers, W.J., Genna, B.J., 1996. Oxygen isotope partitioning between phosphate and carbonate in mammalian apatite. Geochimica et Cosmochimica Acta 60, 5145–5148.

Butzer, K.W., 1971. Another look at the australopithecine cave breccias of the Transvaal. American Anthropologist 73, 1197–1201.

Cerling, T.E., 1992. Development of grasslands and savannas in East Africa during the Neogene. Palaeogeography, Palaeoclimatology, Palaeoecology 97, 241–247.

Cerling, T.E., Harris, J.M., 1999. Carbon isotope fractionation between diet and bioapatite in ungulate mammals and implications for ecological and paleoecological studies. Oecologia 120, 347–363.

Cerling, T.E., Harris, J.M., Macfadden, B.J., Leakey, M.G., Quade, J., Eisenman, V., Ehleringer, J.R., 1997. Global vegetation change through the Miocene/Pliocene boundary. Nature 389, 153–158.

Clarke, R.J., Partridge, T.C., 2002. On the unrealistic 'revised age estimates' for Sterkfontein. South African Journal of Science 98, 415–419.

Codron, J., Codron, D., Lee-Thorp, J., Sponheimer, M., Bond, W.J., de Ruiter, D., Grant, R., 2005. Taxonomic, anatomical, and spatio-temporal variations in stable carbon and nitrogen isotopic composition of plants from an African savanna. Journal of Archaeological Science 32, 1757–1772.

Cooke, H.B.S., 1978. Faunal evidence for the biotic setting of early African hominids. In: Jolly, C. (Ed), Early Hominids of Africa. Duckworth, London, pp. 267–281.

Darwin, C., 1859. On the Origin of Species. John Murray, London.

deMenocal, P.B., 1995. Plio-Pleistocene African climate. Science 270, 53–59.

de Ruiter, D., Sponheimer, M., Lee-Thorp, J.A., 2008. Indications of habitat association of *Australopithecus robustus* in the Bloubank Valley, South Africa. Journal of Human Evolution 55, 115–130.

Dominguez-Rodrigo, M., Pickering, T.R., Semaw, S., Rogers M.J., 2005. Cutmarked bones from Pliocene archaeological sites, at Gona, Afar, Ethiopia: implications for the function of the world's oldest stone tools. Journal of Human Evolution 48, 109–121.

Dupont, L. M., Leroy, S.A., 1995. Steps toward drier climatic conditions in northwestern Africa during the Upper Pliocene. In: Vrba, E.S., Denton, G.H., Partridge, T.C., Burckle, L.H. (Eds), Paleoclimate and Evolution, with Emphasis on Human Origins. Yale University Press, New Haven, pp. 289–298.

Ehleringer, J. R., Monson, R.K., 1993. Evolutionary and ecological aspects of photosynthetic pathway variation. Annual Review of Ecology and Systematics 24, 411–439.

Feibel, C.S., Brown, F.H., McDougall, I., 1989. Stratigraphic context of fossil hominids from Omo Group deposits: northern Turkana Basin, Kenya and Ethiopia. American Journal of Physical Anthropology 78, 595–622.

Grine, F.E., Jungers, W.L., Schultz, J., 1996. Phenetic affinities among early *Homo* crania from East and South Africa. Journal of Human Evolution 30, 189–225.

Grine, F.E., Ungar, P.S., Teaford, M.F., 2006a. Was the Early Pliocene hominin *'Australopithecus' anamensis* a hard object feeder? South African Journal of Science 102, 301–310.

Grine, F.E., Ungar, P.S., Teaford, M.F., El-Zaatari, S., 2006b. Molar microwear in *Praeanthropus afarensis*: evidence for dietary stasis through time and under diverse paleoecological conditions. Journal of Human Evolution 51, 297–319.

Hill, A., Ward, S., Deino, A., Curtis, G., Drake, R., 1992. Earliest *Homo*. Nature 355, 719–722.

Jacques, L., 2007. Les préférences écologiques (paléorégimes alimentaires, paléohabitats) des grands mammifères herbivores des sites à Hominidés du Miocène supérieur du Nord Tchad. Restitution au moyen de l-anlyse isotopique en carbon et oxygène du carbonate de l'émail dentaire. Ph.D. thesis, Université de Poitiers.

Kimbel, W.H., Walter, R.C., Johanson, D.C., Reed, K.E., Aronson, J.L., Assefa, Z., Marean, C.W., Eck, G.G., Bobe, R., Hovers, E., Rak, Y., Vondra, C., Yemane, T., York, D., Chen, Y., Evensen, N.M., Smith, P.E., 1996. Late Pliocene *Homo* and Oldowan tools from the Hadar Formation (Kada Hadar Member), Ethiopia. Journal of Human Evolution 31, 549–561.

Kingston, J.D., Harrison, T., 2007. Isotopic dietary reconstructions of Pliocene herbivores at Laetoli: implications for early hominin paleoecology. Palaeogeography, Palaeoclimatology and Palaeoecology 243, 272–306.

Kohn, M.J., Schoeninger, M.J., Valley, J.W., 1996. Herbivore tooth oxygen isotope compositions: effects of diet and physiology. Geochimica et Cosmochimica Acta 60, 3889–3896.

Kuman, K., Clarke, R.J., 2000. Stratigraphy, artefact industries and hominid associations for Sterkfontein, Member 5. Journal of Human Evolution 38, 827–847.

Lee-Thorp, J., Sponheimer, M., Luyt, J., 2007. Tracking changing Pliocene environments of the South African hominin sites using stable isotopes in faunal tooth enamel. Journal of Human Evolution 53, 595–601.

Lee-Thorp, J.A., 1989. Stable carbon isotopes in deep time: the diets of fossil fauna and hominids, Ph.D. thesis, University of Cape Town.

Lee-Thorp, J.A., van der Merwe, N.J., 1987. Carbon isotope analysis of fossil bone apatite. South African Journal of Science 83, 712–715.

Levin, N.E., Quade, J., Simpson, S.W., Semaw, S., Rogers, M., 2004. Isotopic evidence for Plio-Pleistocene environmental change at Gona, Ethiopia. Earth Planetary Science Letters 219, 93–110.

Levin, N.E., Cerling, T.E., Passey, B.H., Harris, J.M., Ehleringer, J.R., 2006. A stable isotope aridity index for terrestrial environments. Proceedings of the National Academy of Sciences USA 103, 11201–11205.

Longinelli, A., 1984. Oxygen isotopes in mammal bone phosphate: a new tool for paleohydrological and paleoclimatological research? Geochimica et Cosmochimica Acta 48, 385–390.

Luyt, J., 2001. Revisiting the palaeoenvironments of South African hominid-bearing Plio-Pleistocene sites: new isotopic evidence from Sterkfontein. M.Sc. thesis, University of Cape Town.

Lyman, R.L., 1994. Vertebrate Taphonomy. Cambridge University Press, Cambridge.

Maguire, J.M., 1985. Recent geological, stratigraphic and palaeontological studies at Makapansgat Limeworks. In: Tobias, P.V. (Ed), Hominid Evolution: Past, Present and Future. Alan R. Liss, New York, pp. 151–164.

McFadden, P.L., Brock, A., Partridge, T.C., 1979. Paleomagnetism and the age of the Makapansgat hominid site. Earth Planetary Science Letters 44, 411–415.

Monson, R.K., Littlejohn, R.O., Williams, III G.J., 1983. Photosynthetic adaptation to temperature in four species from the Colorado shortgrass steppe: a physiological model for coexistence. Oecologia 58, 43–51.

Owen-Smith, N., 1999. Ecological links between African savanna environments, climate change, and early hominid evolution. In: Schrenk, F., Bromage, T.G. (Eds), African Biogeography, Climate Change, and Human Evolution. Oxford University Press, New York, pp. 138–149.

Partridge, T.C., 2000. Hominin-bearing cave and tufa deposits. In: Partridge, T.C., Maud, R.R. (Eds), The Cenozoic of Southern Africa, Oxford Monographs on Geology & Geophysics, no. 40, Oxford University Press, New York, pp. 100–133.

Partridge, T.C., Scott, L., Schneider, R.R., 2004. Between Agulhas and Benguela: responses of Southern African climates of the Late Pleistocene to current fluxes, orbital precession and extent of the Circum-Antarctic vortex. In: Battarbee, R.W., Gasse, F., Stickley, C.E. (Eds), Past Climate Variability Through Europe and Africa. Springer, Dordrecht, The Netherlands, pp. 45–68.

Passey, B.H., Cerling, T.E., Perkins, M.E., Voorhies, M.R., Harris, J.M., Tucker, S.T., 2002. Environmental change in the Great Plains: an isotopic record from fossil horses. Journal of Geology 110, 123–140.

Prentice, M.L., Denton, G.H., 1988. The deep-sea oxygen isotope record, the global icesheet system and hominid evolution. In: Grine, F.E. (Ed), Evolutionary History of the "Robust" Australopithecines. Aldine de Gruyter, New York, pp. 383–403.

Quade, J., Cerling, T.E., 1995. Expansion of C_4 grasses in the late Miocene of northern Pakistan: evidence from stable istopes in paleosols. Palaeogeography, Palaeoclimatology, Palaeoecology 115, 91–116.

Reed, K.E., 1996. The Paleoecology of Makapansgat and other African Plio-Pleistocene hominid localities. Ph.D. dissertation. State University of New York at Stony Brook.

Reed, K.E., 1997. Early hominid evolution and ecological change through the African Plio-Pleistocene. Journal of Human Evolution 32, 289–322.

Reed, K.E., Russak, S.M., 2009. Tracking ecological change in relation to the emergence of *Homo* at the Plio-Pleistocene boundary. In: Grine, F.E., Fleagle, J.G., Leakey, R.E. (Eds), The First Humans: Origin and Early Evolution of the Genus *Homo*. Springer, Dordrecht, pp. 159–171.

Sankaran, M., Hanan, N., Scholes, R.J., Ratnam, J., Augustine, D.J., Cade, B.S., Gignoux, J., Higgins, S.I., Le Roux, X., Ludwig, F., 2005. Determinants of woody cover in African savannas. Nature 438, 846–849.

Schoeninger, M.J., Reeser, H., Hallin, K., 2003. Paleoenvironment of *Australopithecus anamensis* at Allia Bay, East Turkana, Kenya: evidence from mammalian herbivore enamel stable isotopes. Journal of Anthropological Archaeology 22, 200–207.

Scholes, R.J., Walker, B.H., 1993. An African Savanna – synthesis of the Nylsvlei Study. Cambridge University Press, Cambridge.

Schrenk F., Bromage, T.G., Betzler, C.G., Ring, U., Juwayeyi, Y.M., 1993. Oldest *Homo* and Pliocene biogeography of the Malawi Rift. Nature 365, 833–836.

Schubert, B., Ungar, P., Sponheimer, M., Reed, K., 2006. Microwear evidence for Plio-Pleistocene bovid diets from Makapansgat Limeworks Cave, South Africa. Palaeogeography, Palaeoclimatology, Palaeoecology 241, 301–319.

Scott, R.S., Ungar, P.S., Bergstrom, T.S., Brown, C.A., Grine, F.E., Teaford, M.F., Walker, A., 2005. Dental microwear texture analysis shows within-species diet variability in fossil hominins. Nature 436, 693–695.

Semaw, S., Renne, P., Harris, J.W.K., Feibel, C.S., Bernor, R.L., Fesseha, N., Mowbray, K., 1997. 2.5-million-year-old stone tools from Gona, Ethiopia. Nature 385, 333–336.

Sikes, N.E., 1994. Early hominid habitat preferences in East Africa: paleosol carbon isotopic evidence. Journal of Human Evolution 27, 25–45.

Smith, B.N., Epstein, S., 1971. Two categories of $^{13}C/^{12}C$ ratios for higher plants. Plant Physiology 47, 380–384.

Spencer, L.M., 1997. Dietary adaptations of Plio-Pleistocene Bovidae: implications for hominid habitat use. Journal of Human Evolution 32, 201–228.

Sponheimer, M., 1999. Isotopic ecology of the Makapansgat Limeworks fauna. Ph.D. dissertation, Rutgers University.

Sponheimer, M., Lee-Thorp. J.A., 2001. The oxygen isotope composition of mammalian enamel carbonate: a case study from Morea Estate, Mpumalanga Province, South Africa. Oecologia 126, 153–157.

Sponheimer, M., Lee-Thorp, J., 2003. Using carbon isotope data of fossil bovid communities for palaeoenvironmental reconstruction. South African Journal of Science 99, 273–275.

Sponheimer, M., Lee-Thorp, J., de Ruiter, D., Codron, D., Codron, J., Baugh, A.T., Thackeray, F., 2005. Hominins, sedges, and termites: new carbon isotope data from the Sterkfontein Valley and Kruger National Park. Journal of Human Evolution 48, 301–312.

Teeri, J.A., Stowe, L.G., 1976. Climatic patterns and the distribution of C_4 grasses in North America. Oecologia (Berlin) 23, 1–12.

Trauth, M.H., Maslin, M.A., Deino, A., Strecker, M.R., 2005. Late Cenozoic moisture history of East Africa. Science 309, 2051–2053.

Ungar, P.S., Scott, R.S., 2009. Dental evidence for diets of early *Homo*. In: Grine, F.E., Fleagle, J.G., Leakey, R.E. (Eds), The First Humans: Origin and Early Evolution of the Genus *Homo*. Springer, Dordrecht, pp. 121–134.

van der Merwe, N.J., Masao, F.T., Bamford, M.K., 2008. Isotopic evidence for contrasting diets of early hominins *Homo habilis* and *Australopithecus boisei* of Tanzania. South African Journal of Science 104, 153–155.

van Wilgen, B.W., Trollope, W.S.W., Biggs, H.C., Potgieter, A.L.F., Brockett, B.H., 2003. Fire as a driver of ecosystem variability. In: du Toit, J.T., Rogers, K.H., Biggs, H.C. (Eds), The Kruger Experience: Ecology and Management of Savanna Heterogeneity. Island Press, Washington, DC, pp. 149–170.

Vogel, J.C., 1978. Isotopic assessment of the dietary habits of ungulates. South African Journal of Science 74, 298–301.

Vrba, E., 1980. The significance of bovid remains as indicators of environment and predation patterns. In: Behernsmeyer, A.K., Hill, A.P. (Ed), Fossils in the Making. University of Chicago Press, Chicago, IL, pp. 247–272.

Vrba, E., 1982. Biostratigraphy and chronology, based particularly on Bovidae, of southern hominid-associated assemblages: Makapansgat, Sterkfontein, Taung, Kromdraai, Swartkrans; also Elandsfontein (Saldanha), Broken Hill (now Kabwe) and Cave of Hearths. Premier Congrés International de Paléontologie Humaine, Nice.

Vrba, E., 1985. Ecological and adaptive changes associated with early hominid evolution. In: Delson, E. (Ed), Ancestors: The Hard Evidence. Alan R. Liss, New York, pp. 63–71.

Wesselman H.B., 1985. Fossil micromammals as indicators of climate change about 2.4 million years ago in the Omo Valley, Ethiopia. South African Journal of Science 81, 260–261.

Wynn, J.G., 2004. Influence of Plio-Pleistocene aridification on human evolution: evidence from paleosols of the Turkana Basin, Kenya. American Journal of Physical Anthropology 123, 106–118.

Zazzo, A., Bocherens, H., Brunet, M., Beauvilain, A., Billiou, D., Mackaye, H.T., Vignaud, P., Mariotti, A., 2000. Herbivore paleodiet and paleoenvironment changes in Chad during the Pliocene using stable isotope ratios in tooth enamel carbonate. Paleobiology 26, 294–309.

Part VI
Summary Perspective on the Workshop

Chapter 17
The First Humans: A Summary Perspective on the Origin and Early Evolution of the Genus *Homo*

Frederick E. Grine and John G. Fleagle

Keywords *Homo habilis* • *Homo rudolfensis* • *Homo erectus* • Africa • Asia • cranium • dentition • skeleton • adaptation • origin • genus • climate • environment • habitat • Pliocene • Pleistocene • diet • archeology • development • paleontology

Introduction

Origin, adaptation and diversity are eternal themes in human evolution. These issues are equally timeless with respect to our own lineage. Human paleontologists continue to grapple with questions surrounding the origin and early evolution of our own genus. How do we identify the earliest members the genus *Homo*? How many species of *Homo* were there in the Pliocene and Pleistocene, and how do they relate to one another? Where and when did they evolve? Other burning issues relate to questions about body size, proportions and the functional adaptations of the locomotor skeleton. When did the human postcranial "Bauplan" evolve, and for what reasons? What behaviors (and what behavioral limitations) can be inferred from the postcranial bones that have been attributed to *Homo habilis* and *Homo erectus*?

Other issues of significance relate to growth, development and life history strategies, and the biological and archeological evidence for diet and behavior in early *Homo*. Additional issues of importance pertain to the environmental and climatic context in which the genus *Homo* evolved. Were there global or pan-African climatic events that relate to the appearance and/or extinction of *Homo* species, and if so, can they be tied to the appearance or disappearance of these species in any meaningful way? Did *Homo* species live in environments that differed from those inhabited by earlier hominins, and can any general trends through time be inferred from paleontological and isotopic evidence?

The announcement in 1964 of the fossil remains of *Homo habilis* from Olduvai Gorge by Louis Leakey, Phillip Tobias, and John Napier was a momentous event with regard to our interpretation of human evolution. We have certainly come a long way in the intervening 4 decades, with new finds and analyses having provided critical information. Such advances have, of course, led to new questions. At the same time, some questions remain stubbornly unresolved, but new technologies and approaches – not to mention the ever present requirement of additional fossils – will hopefully shed light on these in the future.

In the chapters that constitute the present volume, the world's acknowledged experts in their respective fields have contributed state of the art information to address the origin, adaptation and diversity of the genus *Homo*. What do we know, and what knowledge do we still wish for in the quest to understand the evolution of the first humans?

Part I

The three chapters that comprise the first part of the present volume present historical and theoretical perspectives on the interpretation of the paleontological record for early *Homo*.

Richard Leakey (Chapter 1) addresses the question of why we should be concerned with trying to identify where we came from. It is not, he notes, a trivial task to revisit the earliest species of the genus *Homo*. Were there several species, only one of which gave rise to *H. erectus*? Did one or more disappear for reasons that remain unclear? These are important biological and adaptive questions. The issues surrounding the recognition of early *Homo* species and the physical, cultural and behavioral adaptations that may have been related to the acquisition of a diet containing meat are addressed by several of the authors who have contributed succeeding chapters. Leakey argues that that the earliest

F.E. Grine (✉)
Departments of Anthropology and Anatomical Sciences,
Stony Brook University, Stony Brook NY 11794-4364, USA
e-mail: frederick.grine@stonybrook.edu

J.G. Fleagle
Department of Anatomical Sciences,
Stony Brook University Medical Center, Stony Brook,
NY 11794-8081, USA
e-mail: john.fleagle@stonybrook.edu

human story is, indeed, a very complex one, and it is probably safe to say that if one message can taken from this volume, this is it.

Until the early 1960s, it was widely recognized that the geochronologically oldest member of our genus was *Homo erectus*, a species whose remains had been identified in East Asia, Eurasia and Africa. Louis Leakey, Phillip Tobias and John Napier changed all that in 1964, with the announcement of *Homo habilis*. Tobias, one of the founding fathers of *H. habilis*, recounts its early intellectual history in Chapter 2. The initial diagnosis and claims for it were repudiated by a host of formidable figures, and opposition to it continued for over a decade. In light of the contentious early history of *Homo habilis* and its subsequent acceptance, Tobias asks: "Was *H. habilis* ahead of its time?" He argues that it was a "premature discovery" according to Stent's (1972) definition, meaning that the implications of this proposed species could not be connected by simple, logical steps to the prevailing paradigm of the early 1960s. Just as the "prematurity" of Dart's (1925) *Australopithecus africanus* lasted a quarter of a century before being overcome, that of *Homo habilis* took about 15 years before the species gained wide acceptance. Tobias dates the turning of the tide to its recognition by F. Clark Howell (1978) in his compendious chapter in the *Evolution of African Mammals*. He notes that a consensus then seemingly developed that *H. habilis* was a "good species," but that two other proposals developed soon thereafter.

The first entails the possibility that at least two taxa, namely *H. habilis* and *H. rudolfensis*, can be distinguished among the fossils traditionally regarded as representing *Homo habilis sensu lato* (e.g., OH 7, OH 13, OH 16, OH 24 from Olduvai Gorge, KNM-ER 1470 and KNM-ER 1813 from Koobi Fora) regardless of how the hypodigms of these various species are constructed (Lieberman et al., 1988; Wood, 1991, 1992, 1993; Rightmire, 1993; Leakey et al., 2001; Blumenschine et al., 2003). Although the evidence for such a distinction has not been accepted by all workers (e.g., Suwa et al., 1996; Miller, 1991, 2000; Dunsworth and Walker, 2002; Lee and Wolpoff, 2005), including Tobias, the number of *Homo* species represented by the Early Pleistocene fossils in East Africa remains very much an open issue (Baab, 2008).

The second proposal to which Tobias alludes addresses the nature of the genus itself. In Chapter 3, Bernard Wood asks whether the fossils attributed to *Homo habilis* and/or *H. rudolfensis* indeed warrant that generic appellation. Wood methodically addresses the issue first by considering the various proposals that have been made about how a genus should be defined, and then by relating the history of the genus *Homo* from its introduction in 1758 to the addition of the species *H. habilis* some 206 years later. He then reviews the evidence relevant to the nature of *Homo* that has accumulated since 1964, citing newly discovered fossils and new information gleaned from the existing fossil record about the possible adaptive grade and phylogenetic relationships of potential early *Homo* species. In so doing, Wood reviews reconstructions of diet, locomotion and life history, and concludes that the adaptive strategies of *H. habilis* and *H. rudolfensis* were probably closer to that of the type species of *Australopithecus* (*A. africanus*) than to that of *H. sapiens*, the type species of *Homo*. Thus, he maintains the opinion first expressed by Wood and Collard (1999), that *H. habilis* and *H. rudolfensis* should be excluded from the genus *Homo*. Wood's arguments for a major adaptive shift with the appearance of *Homo erectus* received mixed support from the authors of other chapters in this volume.

Lieberman and colleagues (Chapter 8) would likely champion this conclusion insofar as postcranial proportions that relate to endurance running would seem to favor *Homo erectus* as the earliest representative of the genus with a novel physical "Bauplan" and method of food acquisition. Certainly the postcranial remains of *H. habilis*, particularly the sizes and proportions reconstructed from the OH 62 limb bones, differ from those of *H. erectus*. Larson (Chapter 7) also finds that the first major structural reorganization of the shoulder is exhibited in *H. erectus*, although even then it was not completely human-like in its morphology. Unfortunately, we have no meaningful postcranial bones that can be attributed to *Homo rudolfensis*.

On the other hand, Dean and Smith (Chapter 10), point out that the reconstructed life history of *Homo erectus* differs significantly from our own. Was it more "human-like" than *Homo habilis* and *H. rudolfensis*? These questions await new fossils and new analyses of those already in museum collections. The archeological evidence from the Late Pliocene and Early Pleistocene (i.e., between about 2.7 and 1.6 Ma), as discussed by Roche and colleagues (Chapter 12), is perhaps more equivocal on this issue. Their review of the Oldowan suggests that while these artefacts were perhaps not as refined and/or as varied as those of the Acheulean, they certainly do not comprise a static entity. They also note that the zooarcheological record speaks to an apparent rarity of Oldowan hominin carnivory, something that stands in stark contrast to the evidence from some immediate post-Oldowan assemblages.

With reference to the phylogenetic relationships among *H. habilis*, *H. rudolfensis* and other hominin species, Wood observes that most recent analyses find that *Homo* conforms to a monophyletic clade with *H. habilis* and *H. rudolfensis* at or near its base, even if the statistical support for this arrangement is sometimes rather weak. However, as noted by Kimbel in Chapter 4, such statistics do not *falsify* the hypothesis of monophyly. Moreover, the fact that multiple analyses using different data sets result in very similar conclusions about the relationships of these two species would seem to favor it (e.g., Strait et al., 2007; González-José et al., 2008). Kimbel further posits that the most appropriate solution is not to assign *H. habilis* and *H. rudolfensis* to a manifestly paraphyletic taxon such as *Australopithecus*, or to create yet another genus to accommodate them, but rather to keep the basal species of the clade that includes *H. sapiens* within the genus *Homo*.

Part II

The three chapters that constitute the second part of this volume address the craniodental evidence for early *Homo*, tackling taxonomic issues, and questions pertaining to the origin of *Homo* and of *H. erectus* in Africa.

In Chapter 4, William Kimbel observes that as many as three species (*Homo habilis, H. rudolfensis* and *H. erectus*) that bear a closer phylogenetic relationship to *Homo sapiens* than to any known australopith taxon are present in Africa by ca. 1.7–1.8 Ma. The morphology associated with *H. habilis* and *H. rudolfensis* is recorded in a few teeth and jaws in the 2.0–2.5 Ma interval, but the origins of these taxa remain clouded due to an impoverished East African fossil record between 2.5 and 3.0 Ma. Kimbel discusses in detail the half dozen or so specimens that have been variously postulated to represent the earliest members of the genus, thus highlighting the poor nature of the fossil record at about 2.5 Ma.

Kimbel notes that although an adaptive complex comprising encephalization, lithic tool manufacture, and the strategic acquisition of meat protein is commonly implicated in the early evolution of the *Homo* lineage, it is not presently clear whether any of these attributes factored in the foundation of the lineage. Indeed, he questions whether it is even reasonable to expect them to have played such a role. In addressing the issue of the taxonomic boundary of the genus *Homo*, Kimbel notes that redrawing it between *H. erectus* (including *H. ergaster*) and *H. habilis* + *H. rudolfensis* on the argument that the latter were adaptively not substantially different from mid-Pliocene australopiths is confronted by the inherently arbitrary criterion of "adaptive unity." He argues that this, together with evidence for monophyly, suggests that retaining *H. habilis* and *H. rudolfensis* within the genus *Homo* is the most reasonable alternative at the present.

In Chapter 5, Philip Rightmire and David Lordkipanidze discuss the similarities and differences among Early Pleistocene *Homo* crania from East Africa and western Eurasia as evidence bearing on both the origin and the species taxonomy of the genus. They observe that the crania from Dmanisi, which they interpret as representing a single paleodeme, demonstrate substantial morphological variation. This assertion discounts the possibility that the Georgian fossils sample more than one taxon (Gabunia et al., 2002), and emphasizes the single-species interpretation that has gained support from independent studies of the crania that have been published to date (Baab, 2008). Given this frame of reference, the Dmanisi fossils demonstrate that the level and pattern of variation in ancient populations may not always conform to what is expected from studies of modern reference samples. Rightmire and Lordkipanidze observe that this serves to further cloud the taxonomic sorting of the early *Homo* fossils from Olduvai Gorge and Koobi Fora. They note that, if anything, it is now less clear than before how intragroup variation is to be partitioned from differences that denote species boundaries in the fossil record. As such, they are unable to provide resolution to the taxonomic questions that surround the early *Homo* fossils from Africa.

Rightmire and Lordkipanidze argue that while the Dmanisi sample can be referred to *Homo erectus*, it displays some differences from African and East Asian conspecifics, and shares primitive features with specimens attributed to *H. habilis*, such as KNM-ER 1813 and OH 13. They suggest that the Dmanisi hominins therefore had a *H. habilis*-like ancestor, and that this form was possibly the first to disperse from East Africa. Accordingly, *H. erectus* might have originated in western Asia, and only later evolved the larger brain and greater body size that characterize the African and East Asian representatives of the species. While this "Asian Origins" hypothesis differs from the generally accepted view that *H. erectus* evolved in Africa before dispersing into other regions of the Old World, it is not inconsistent with some of the arguments that have been put forward relating to the hominin finds from Liang Bua, Indonesia (Morwood et al., 2005; Argue et al., 2006). Rightmire and Lordkipanidze conclude that apart from the Dmanisi discoveries, there is currently little hard evidence on which to base such an alternative, but William Jungers does just this in his contribution to the present volume. Further fieldwork and additional analyses (including those of the bones attributed to *Homo floresiensis*) might further refine and test these biogeographic hypotheses.

In Chapter 6, Frederick Grine, Heather Smith, Christopher Heesy and Emma Smith examine molar cusp proportions in an attempt to provide at least partial answers to the question of whether the South and East African early *Homo* specimens sample the same taxa, or whether distinct forms (or lineages) might be represented by at least some (if not all) of the South African material (Grine et al., 1993, 1996; Grine, 2001, 2005). Having established the efficacy of cusp proportions to successfully discriminate species samples of living great apes, they use these data to examine the relationships between the South and East African fossils. Some of the South African specimens (those from Sterkfontein Member 5A and Swartkrans Member 1) have affinities with *H. habilis*, whereas others (those from Sterkfontein Member 5C and Swartkrans Member 2) associate most closely with *H. erectus*. These results are consistent with suggestions that the Swartkrans Member 2 and Sterkfontein Member 5C fossils should be attributed to the same species (*H. erectus*) (Kuman and Clarke, 2000), while those from Swartkrans Member 1 and Sterkfontein Member 5A may represent a different one.

At the same time, however, the South African teeth tend to have closer phenetic resemblances among themselves, regardless of derivation, than with East African homologues. This may suggest that some of the fossils from South Africa differ from those in East Africa at the species level. If *Homo habilis* and *Homo erectus* are both known from South and East Africa, or, alternatively, if only *H. erectus* is known from South and East Africa, this might well say something about the adaptive (or at least biogeographic) versatility of the earliest members of our genus.

Part III

Postcranial skeletal morphologies and proportions, and their possible adaptations and evolution form the focus of the three chapters that comprise the third section of this volume.

Susan Larson addresses the structure of the shoulder in and its behavioral significance for early *Homo* in Chapter 7. Her work indicates that *Homo habilis* retained much of the presumed ancestral condition in its shoulder morphology. The fossil evidence, which is rather scanty for shoulder elements that are attributable to *H. habilis* (OH 48 and KNM-ER 3735), and non-existent for *H. rudolfensis*, suggests that the first major structural reorganization occurred in early *Homo erectus*, as represented by KNM-WT 15000. Larson notes that this entailed loss of a cranial orientation to the glenoid fossa, which was probably accompanied by a caudal shift in scapular position. She observes further that these changes would have been constrained by the retention of a relatively short clavicle, resulting in a scapula that shifted caudally, but also somewhat anteriorly. As such, the transition in glenoid fossa orientation was not from cranial to lateral, but rather from cranial to anterior. Larson argues that this would have maintained parasagittal elbow function with a humeral head that displayed only low to modest torsion, probably another retained primitive characteristic. She notes that while this configuration in *H. erectus* would have placed no limitations on manipulatory activities, an abducted upper limb would not have permitted a very large range of posterior motion. Would this have seriously compromised throwing ability in *H. erectus*? Quite probably, she argues, a conclusion that is supported by recent kinematic data (Meyer et al., 2008). This serves to remove a key, human-like element that has been claimed for *Homo erectus* (cf. Bingham, 1999).

Larson points out that the transformation from the configuration displayed by *H. erectus* to something more closely resembling that of modern humans was accomplished largely by elongation of the clavicle. This would serve to spread the shoulders apart, pushing the scapula into a dorsal rather than lateral position on the rib cage, so that the glenoid fossae would come to face laterally. A significant increase in humeral torsion would also be necessary in order to maintain a parasagittally functioning elbow joint with a dorsally positioned scapula. Larson argues that a likely selective factor favoring these changes would be the advantages accrued from a dramatic increase in range of motion at the shoulder. In particular, she notes two potential relationships of relevance.

The first pertains to throwing. As discussed by Hélène Roche and colleagues in Chapter 12, there is no archeological evidence that early *Homo* (*H. habilis* and/or *H. rudolfensis*) or *H. erectus* engaged in projectile throwing. They note that neither the Oldowan nor the early Acheulean evince any evidence for the manufacture and/or use of sophisticated stone-tipped or bone-tipped spears that could have been thrown with lethal accuracy from a distance. Rather, the earliest evidence for throwing spears occurs only in the Middle Pleistocene (Thieme, 1997). Although it has been postulated that Acheulean handaxes served as thrown hunting weapons (O'Brien, 1981), they actually perform very poorly as accurate projectiles (Whittaker and McCall, 2001).

The second potential relationship that Larson discusses pertains to running, which forms the subject of the next chapter by Daniel Lieberman and colleagues. Larson notes that running – whether for speed or endurance – requires shoulder and upper body rotation to counteract the destabilizing torque created by the oppositely moving lower limbs. The relatively narrow shoulders of *H. erectus* suggest that an effective upper body counter-rotation mechanism was not yet an important selective factor in that species. Larson posits that running could well have been an impetus for clavicular elongation to spread the shoulders apart in order to enhance the upper body counter-rotation mechanism at a somewhat later stage of human evolution.

In Chapter 8, Daniel Lieberman, Dennis Bramble, David Raichlen and John Shea address the question of how the genus *Homo* evolved to become an effective carnivore well in advance of the invention of sophisticated projectile technologies. They propose that scavenging and hunting behaviors in *H. erectus* were made possible by a suite of anatomical and physiological features for endurance running. These derived features, many of which play little biomechanical role in walking, make humans the most exceptional endurance runners in the mammalian world, especially in hot conditions. They find that many, but perhaps not all of the anatomical attributes that make humans so good at endurance running were probably present in *H. erectus*. Those that may be discerned from the fossils include enlarged anterior and posterior semicircular canals, a strong nuchal ligament, a substantially enlarged cranial portion of the gluteus maximus muscle, relatively large centra of the (lumbar) vertebrae, a relatively short forearm, relatively long legs, and relatively large hindlimb joints. Others, such as a spring-like plantar arch, have been argued to have been possessed by *Homo habilis* (Susman and Stern, 1982; Susman, 1983), although this particular feature would be relevant to habitual bipedal walking as well. Although, as discussed by Larson, the presence of low, widely separated shoulders in *H. erectus* is highly questionable, the remainder of the features noted by Leiberman and colleagues would have provided members of this species with endurance running capability. Lieberman and colleagues concede that we cannot rule out the possibilities that *H. habilis* had some endurance running capabilities, or that later humans had better performance capability than *H. erectus*.

Endurance running capability would have helped early *Homo* scavenge, but it also would have provided *H. erectus* with the ability to drive large ungulates into hyperthermia by chasing them over long distances in hot, open habitats. Lieberman and colleagues argue that selection for running may have concomitantly served to select against arboreal capabilities in *Homo erectus*. Although new technologies have rendered persistence hunting no longer necessary, they perhaps highlight the key role of endurance athleticism in human evolution (Bramble and Lieberman, 2004).

William Jungers addresses variability and scaling of interlimb proportions in modern humans and their implications for fossil hominins in Chapter 9. In so doing, he discusses the usual cast of characters with reasonably intact and reasonably attributed postcranial remains – *Australopithecus afarensis* (A.L. 288-1), *Homo erectus* (KNM-WT 15000) – and wisely eschews specimens that are manifestly fragmentary and/or for which taxonomic allocation are unreliable. These include the fragmentary OH 62 skeleton that is most reasonably attributed to *H. habilis*, and BOU-VP-12/1, which may represent "*Australopithecus*" *garhi*, but equally possibly any other hominin species that is known at about 2.5 Ma. Jungers, however, brings a new perspective to the discussion from his first-hand experience with the recently described skeleton of *Homo floresiensis* (LB 1).

Jungers demonstrates that the often repeated claims that small-bodied fossil hominins like A.L. 288-1 and LB 1 necessarily have unusual limb proportions simply because they are extrapolated examples of size-required allometries in modern humans are wrong. No small-bodied human remotely approaches the high humerofemoral index found in both of these diminutive fossils.

Jungers speculates that, just as there are other primitive features in the skeleton of *H. floresiensis* (Tocheri et al., 2007, 2008; Larson et al., 2007), the high humerofemoral index exhibited by LB 1, which is a result of relatively short hind limbs rather than long upper extremities, may also be a primitive retention. He further surmises that this would probably rule out *H. erectus* as a likely ancestor in view of the human-like limb proportions seen in KNM-WT 15000. Such a scenario has been forwarded by Morwood et al. (2005) and Argue et al. (2006), who postulated an earlier diffusion from Africa for the ancestry of *H. floresiensis*. However, not all evidence is consistent with the exclusion of *H. erectus* from the potential ancestry of this species (Gordon et al., 2008). While it is easy to doubt a scenario in which *Homo habilis* (or a similar form) emigrated from Africa to serve as a founder for the *H. floresiensis* lineage because of the lack of any good evidence for such a form outside Africa in the Late Pliocene or Early Pleistocene, the fossil record over this period is hardly satisfactory in Africa, let alone Asia. It is perhaps wise to keep a mind open to the possibility – discussed also by Rightmire and Lordkipanidze in Chapter 5 – that *Homo habilis* or another similarly primitive taxon found its way out of Africa prior to the emergence of *H. erectus* in Asia and Africa.

Part IV

Growth, development and life history strategies, and the biological and archeological evidence for diet and behavior in early *Homo* are the subject of the three chapters that comprise the fourth part of this volume.

In Chapter 10, M. Christopher Dean and B. Holly Smith bring together what is known about the body mass, skeletal age and stature of the adolescent *Homo erectus* skeleton from Nariokotome (KNM-WT 15000) and interpret that evidence in light of its dental development. They set out evidence that suggests this individual had a skeletal age at death of 13 years, a dental age of about 8 years, a stature of 160 cm and a body mass of about 50 kg – an unusual combination of physical attributes. Although the aptly nicknamed "strapping youth" was clearly a young adolescent at death, he was nearly the size of a *Homo erectus* adult! As such, he would have attained a greater proportion of adult stature and body mass at a relatively earlier age than in modern humans. Dean and Smith propose that the Nariokotome youth did not experience a long period of slow growth between the end of weaning and the onset of puberty, but rather that he followed a growth curve more typical of chimpanzees. That is, one which lacked the slowdown and later adolescent spurt in linear dimensions which are characteristic of humans today. They point out that the microanatomy of enamel and dentine in other *H. erectus* individuals provides independent corroboration that tooth development in this species was faster than it is in living humans. The discrepancy between chronological age and skeletal age in *H. erectus* is of a degree that leads Dean and Smith to conclude that it should not be assessed by human growth standards. In addition, the flatter neurocranium of

Homo erectus compared with that of modern humans suggests that stature estimates for that species based on *Homo sapiens* regressions would overestimate the height of *H. erectus* (Delson, 1995).

Their new analysis supports earlier conclusions (Smith, 1993; Dean et al., 2001) that *H. erectus* possessed a life history unlike any species living today. They deduce that rapid growth and the lack of a significant slowdown in late childhood may signify that *Homo erectus* individuals would have been independent relatively early on in their juvenile periods. Thus, a *Homo erectus* adolescent of 8–9 years of age may have been much more independent than a modern human adolescent of 12–13 in a traditional society. Working out elements of human life history, as Dean and Smith have done so elegantly, will certainly provide clues to answer other questions that lie behind the appearance and evolution of the human species.

In Chapter 11, Peter Ungar and Robert Scott provide a review of the evidence and analogies that have suggested to many authors a shift in *Homo* toward keystone foods (e.g., meat, marrow, underground storage organs of xeric adapted plants) associated with the spread of C_4 grasslands across parts of Africa in the Late Pliocene. They observe that these models generate valuable hypotheses, some of which can be evaluated using the hominin fossil record, and they do just that by considering the dietary implications of tooth size, shape and wear in specimens that have been attributed to *Homo habilis*, *H. rudolfensis*, and early African *H. erectus*. Dental topographic analysis suggests that the molars of early *Homo* have more occlusal relief than those of their australopith predecessors, which implies increased efficiency for fracturing tougher foods. On the other hand, their analysis of molar microwear textures suggests that early *Homo* individuals did not specialize on particularly tough foods (or hard and brittle items for that matter). By the same token, these fossils evince moderate variation in most microwear texture attributes, and Ungar and Scott suggest that this may hint at possible differences in the mechanical properties of the foods chewed by *H. habilis* and *H. erectus*. Ungar and Scott acknowledge that although the fossil evidence for early *Homo* is meager, the teeth are consistent with adaptations for flexible, versatile subsistence strategies that would have served them well in the variable environments of the African Plio-Pleistocene.

In Chapter 12, Hélène Roche, Robert Blumenschine and John Shea review the lithic and zooarcheological evidence for the behavior of early *Homo* in Africa between roughly 2.6 and 1.7 Ma. They refer to the lithic artefacts from this period as constituting the "Earliest Oldowan." The earliest Oldowan therefore spans from the first evidence for stone-knapping (Semaw et al., 1997, 2003) to the widespread appearance of Acheulean assemblages in Africa and beyond (Klein, 1983; Asfaw et al., 1992; Clark, 1993; Roche, 1995; Kuman and Clarke, 2000; Bar-Yosef and Belfer-Cohen, 2001; Domínguez-Rodrigo et al., 2001; Gabunia et al., 2001; Roche et al., 2003; Lycett and von Cramon-Taubadel, 2008). Roche and colleagues note that over this duration of approximately 1 million years, the earliest Oldowan is not the static entity that it has sometimes been portrayed (Semaw, 2000), a conclusion also reached as a result of independent studies by Braun et al. (2008a). Instead, the evidence indicates a wide variety of strategies for flake production (see also Braun et al., 2008b).

Roche and colleagues observe that although the lithic record appears to be most straightforwardly involved in butchery, more varied patterns of tool use are likely. They observe that Oldowan stone tools are found over such a long temporal stretch, and in so wide a range of contexts that any hypothesis linking their appearance to a one-time-only behavioral shift within a single hominin species is almost certainly wrong.

Roche and colleagues also address the identity of the species (either singular or plural) responsible for the earliest Oldowan stone tools. They consider the arguments of association and evolutionary trends within the *Paranthropus* and *Homo* lineages and make the succinct observation that the extinction of the former leaves not the slightest trace in the pattern of Early Paleolithic industrial variability. While they cannot rule out stone tool production and use by *Paranthropus* (or "*Australopithecus*" *garhi*), they conclude that the principal beneficiaries of knapped-stone tool technology appear to have been early representatives of our own genus. Change and variability within the earliest Oldowan, and also across the Oldowan–Acheulean transition, almost certainly reflect evolutionary trends in the genus *Homo*.

Cut marks and fracture patterns on bones indicate systematic efforts to feed on large mammal carcasses, but Roche and colleagues note that of 16 Oldowan localities with butchery-marked bones, only two (both from Olduvai Gorge) possess substantial proportions of elements showing such marks. They observe that while several factors, including poor preservation, might account for this paucity, the apparent rarity of Oldowan hominin carnivory stands in stark contrast to some immediate post-Oldowan assemblages, where butchery marks are common. Roche and colleagues discuss scavenging as a method of acquiring meat, and note that confrontational scavenging of carcasses from large predators by Oldowan hominins denotes advanced behavioral and technological capabilities. Hunting, like confrontational scavenging, would also yield large quantities of food from complete carcasses and, as such, the two types of carcass acquisition would be largely indistinguishable archeologically. Roche and colleagues conclude that if hunting can be shown to account for the range of prey species at Oldowan sites, this would imply top predator status for these hominins, a postulate that is seemingly incongruent with small body size and simple stone technology. Moreover, the earliest Oldowan and Acheulean assemblages differ markedly in the incidence of butchery marked bones.

Part V

The four chapters that constitute the fifth part of the current volume examine human origins in its environmental and ecological contexts, where these range from global, to pan-African, to individual site contexts.

In Chapter 13, Mark Maslin and Martin Trauth point out that the Late Cenozoic climate of East African is punctuated by episodes of short, alternating periods of extreme wetness and aridity, superimposed on a regime of subdued moisture availability exhibiting a long-term drying trend. The periods of extreme climate variability are characterized by the precession-forced appearance and disappearance of large, deep lakes in the East African Rift Valley, paralleled by wind-driven dust loads reaching the adjacent ocean basins. Over the last 3 million years, these periods of extreme climate variability occur only at the times of major global climatic transitions, such as the intensification of Northern Hemisphere Glaciation (2.7–2.5 Ma), the development of the Walker circulation (1.9–1.7 Ma), and the Mid-Pleistocene Revolution (1.0–0.7 Ma). Maslin and Trauth refer to these periods as 'Pulsed Climate Variability,' and postulate that this may have provided a catalyst for evolutionary change, driving speciation and dispersal events among the hominins (and other mammals) in East Africa. Their hypothesis, like that proposed by Vrba (1985, 1995), is readily testable, since these pulses of climatic variability should correspondence with major evolutionary and/or biogeographic events in the paleontological record. As the authors note, their hypothesis that pulsed climatic variability may have provided a catalyst for evolutionary change is concordant with the Variability Selection Hypothesis put forward by Potts (1998a, b), which posits a link between adaptive change in human evolution to environmental fluctuations throughout the Plio-Pleistocene.

Chapter 14 finds Kaye Reed and Samantha Russak employing mammal community structure to reconstruct habitats on the basis that physical adaptations can be associated with particular habitat physiognomies. They examine fauna from hominin-bearing sites that span the period from 3.18 to 1.6 Ma in East and South Africa to more closely examine the details of the environments associated with the appearance of *Homo habilis/rudolfensis* and of *H. ergaster/erectus*. Reed and Russak note a general trend for the warm, mesic habitats of the earlier Pliocene to give way to more arid, seasonal habitats at varying times, depending on region and site, between about 3.0 and 2.0 Ma. The fossil mammal communities exhibit slight changes in habitat over time; species trend towards occupying more open, but wet environments, as indicated by increases in grazing, aquatic and fresh-grass grazing elements. While this agrees with the larger climatic changes that have been documented for this period of time, their results indicate the importance of examining local variation with respect to global environmental effects.

In Chapter 15, René Bobe and Meave Leakey focus on the question of when *Homo* first appeared in the Omo–Turkana Basin, and on faunal shifts that reflect changing ecological conditions during the Late Pliocene and Early Pleistocene. They argue that the earliest record of a species must be evaluated in terms of its abundance, and the strength of the record prior to its first appearance. Thus, the appearance of the genus *Homo* at about 2.5 – 2.4 Ma represents an upper estimate of the time of its true origination or migration into the basin. Perhaps reflecting the factors discussed by Bobe and Leakey, the earliest lithic artefacts from Ounda Gona, Ethiopia, which date to some 2.6 Ma, predate the earliest well-dated fossil evidence for *Homo* – isolated teeth from Member E of the Shungura Formation and the base of the Kalochoro Member of the Nachukui Formation (Suwa et al., 1996; Prat et al., 2005) and a maxilla from the Kada Hadar Member of the Hadar Formation (Kimbel et al., 1997) – by some 200,000 – 300,000 years.

Their detailed analyses of the fossil mammals, taken in conjunction with other lines of evidence such as paleosol carbonates, indicate that environments in East Africa during the late Pliocene and early Pleistocene were complex and dynamic, and that *Homo* appeared in the Omo–Turkana Basin at a time of moderate, but important increases in the abundance of mammals indicative of grasslands. More pronounced shifts in this direction occurred near the Pliocene–Pleistocene boundary, coincident with the dispersal of *Homo erectus*. Bobe and Leakey argue that the emergence of *Homo* and the spread of *Homo erectus* can be viewed as processes within the larger context of environmental change and mammalian evolution in East Africa.

Matt Sponheimer and Julia Lee-Thorpe in Chapter 16 investigate the environments inhabited by *Homo* in South Africa, using biogeochemical data (stable isotopes) from herbivore tooth enamel to examine the nature of climatic and environmental trends from sites there that date to between about 3 and 1.6 Ma. The carbon isotope data are in broad accord with previous reconstructions from faunal and macrobotanical remains that *Australopithecus africanus* was associated with more woody vegetation than early members of the genus *Homo*. However, their data also suggest that the most marked shift to open, grassy landscapes occurred around the first appearance of *H. erectus*. They also applied an aridity index, based on oxygen isotope data from herbivore enamel (Levin et al., 2006), in an attempt to determine whether these trends in vegetation were associated with aridification, as has often been assumed in the past. This index suggests that *Homo* lived in far moister environments than *Australopithecus*, a finding which is in broad accord with that of Reed and Russak, who found that the fossil mammal species from these (and other sites) tend to occupy more open, but wet environments. The conclusions reached by Sponheimer and Lee-Thorpe, Bobe and Leakey, and Reed and Russak are in agreement with the larger climatic changes that have been documented for this period of time by Maslin and Trauth, but their results indicate the importance of examining local variation with respect to global environmental effects.

Early *Homo* – Conclusions and Questions

Like any volume of this nature, the present one has been perhaps more successful in raising new questions than providing definitive answers to old ones. Nevertheless, there are some broad areas of general agreement that emerged, and these serve both as a summary of our current understanding on the origin and early evolution of the genus *Homo* and as a guide to future research.

The genus *Homo* almost certainly appeared sometime between 3.0 and 2.5 million years ago somewhere in Africa, but there are no specimens of *Homo* yet reported from this relatively poorly sampled time period. Moreover, the earliest fossils that might be attributed to this genus (i.e., those that date to between 2.5 and about 2.0 Ma) are so rare and so anatomically limited as to preclude their clear attribution to any of the named species that are known from later time periods. Certainly these fossils do not permit us to determine much about the nature of the origin of the genus. Although the cladistic relationships of the genus *Homo* are also not clearly resolved, most phylogenetic analyses place its constituent species as the sister taxa to the *Paranthropus* clade. However, *Australopithecus africanus* or perhaps even the problematic *Kenyanthropus platyops* may also assume this role (see Strait et al. [2007] for a detailed review). The strong relationship of the origin of *Homo* to major events of climatic and faunal change suggested earlier by Vrba (e.g., 1985, 1995, 1996) is not broadly supported by either climatic or faunal analyses, as most studies find a series of climatic and faunal changes throughout the period between some 3 and 2 Ma (see also Behrensmeyer, 2006; Frost, 2007). For this period, we desperately require more data from the African fossil record.

Between roughly 2.0 and 1.5 Ma, the fossil record for the genus *Homo* is much better represented, with some excellent material, including cranial and postcranial bones from East and South Africa, as well as the Levant, the Republic of Georgia and East Asia. These fossils have been assigned to four taxa – *Homo habilis*, *H. rudolfensis*, *H. erectus* and *H. ergaster* – and there remains considerable disagreement among individual researchers about the proper allocation of individual specimens to one or other of these taxa. The temporal and biogeographical ranges (and, in fact, the existence) of some of these species are less than clear. These problems, especially as they pertain to the African fossil record, have been particularly highlighted by the specimens from Dmanisi, which in some ways seem to blur the distinction between *H. habilis* and *H. erectus*. Similarly, the primitive features reported for the hominin specimens from Flores that suggest greater similarity to *Australopithecus* or *H. habilis* than to *H. erectus*, further emphasize our poor understanding of the biogeography of early members of the genus *Homo*.

Among the early species of *Homo*, locomotor and dietary adaptations as well as information about development and life history are best known for *Homo erectus* because of the completeness of the Nariokotome boy. Although *Homo erectus* seems to show a number of distinctive skeletal features associated with long distance running, which would have been particularly valuable in hunting or scavenging activities, there are some features of the shoulder, at least, that do not appear to show modifications that would be associated with running. Indeed, in many aspects of upper limb function, *H. erectus* appears to have been quite different from either *Australopithecus* or *Homo sapiens*, and seems not to evince features that are requisites for throwing behavior. Moreover, it is becoming increasingly clear that hominin postcranial evolution was unlikely to have been a linear progression through time (Larson, 2007; McHenry and Brown, 2008), although our understanding of the locomotor adaptations of other early *Homo* species such as *H. habilis* is greatly hampered, or in the case of *H. rudolfensis* precluded, by the lack of reliably associated and relatively complete specimens. Thus, *Homo habilis* has been described alternatively as having possessed hind limb locomotor anatomy that is within the range of modern humans (Susman and Stern, 1982), or having had proportions and presumed locomotor abilities more similar to those of *Australopithecus afarensis* (Johanson et al., 1987).

Indeed, even parts of the type material of *Homo habilis* from Olduvai Gorge (specifically, the OH 7 hand skeleton and the OH 8 foot skeleton) have been argued by some to belong to *Paranthropus boisei* rather than *Homo habilis* (Wood, 1974; Gebo and Schwartz, 2006; Moyà-Solà et al., 2008). However, Susman (2008) has recently offered solid ossific evidence linking the OH 8 foot, the OH 7 hand and the OH 7 mandible to the same juvenile developmental age. He cogently observes that it is much more likely that OH 7 and OH 8 represent the same subadult individual than they do twin adolescents who died at the same time at FLK NN Level 3 and left complementary body parts (Susman, 2008). Accordingly, claims that the OH 8 foot and especially the OH 7 hand represent *Paranthropus boisei* are more unlikely.

Evidence from teeth seems to offer only the most general information about the dietary adaptations in the early species of our genus. As a group, they seem to show morphological evidence suggesting a wide range of dietary adaptations, as well as microwear that is consistent with a diverse diet compared with *Australopithecus*. Comparisons among early *Homo* species are limited by small samples, but suggest a proclivity for *Homo erectus* to have chewed more brittle foods. The Plio-Pleistocene archeological record (if it is all attributable to *Homo*) offers the potential for providing additional insights into diet and many other aspects of the behavior of early members of our genus. While the first appearance of stone tools in African sediments is roughly contemporary with that of fossils attributed to the genus *Homo*, the correlation is not perfect, and there are often multiple hominins, including species of *Paranthropus*, at many sites yielding stone tools. Thus, prior to approximately 1.3 Ma, or the extinction of *Paranthropus*, it is not possible to identify *Homo* unequivocally as the maker or the sole maker and user of stone tools. Nevertheless, this association gains significant support from the fact that the extinction of *Paranthropus* has no impact whatsoever upon the archeological record. All species of *Homo* in the Pliocene and Early Pleistocene (as well as some later species) are found in conjunction with Oldowan tools. Reconstructing their role in the life of early members of our genus is also fraught with assumptions. Stone tools are often associated with butchered bones, and there is strong evidence that these lithics were employed in butchery. Thus, it seems likely that the early evolution of *Homo* was associated with some degree of increased carnivory, although this particular activity appears to have seen a more notable increase later with the Acheulean. The earliest Oldowan tools almost certainly would have been used in many other functions related to the procurement and/or processing of plant foods. In addition to increased levels of carnivory, the dietary breadth and versatility of *Homo* in vegetal consumption would have been increased by the use of stone tools.

Comparative faunal analyses and isotopic studies indicate that the earliest records of *Homo* in the Late Pliocene and Early Pleistocene are associated with increasing evidence of a fluctuating but slow and continuous trend towards more open environments and wetter (rather than drier) conditions than in the preceding Pliocene. Evidence for environmental differences between early *Homo* species is mixed. Some research has found evidence for a more pronounced shift toward grassland environments associated with *Homo erectus*, while other studies have failed to find differences in the environments of these species.

With regard to ontogenetic development, it is clear that *Homo erectus* had a much more rapid pattern of dental and skeletal development than *Homo sapiens* and was more similar to earlier hominins or chimpanzees in the timing of life history parameters. Although there is scant (if any) evidence pertaining to other species of early *Homo*, it is probably reasonable to assume that they were no more modern-human-like than *H. erectus*.

In conclusion, there is evidence among early species attributed to *Homo*, and especially in *Homo erectus*, that by the beginning of the Pleistocene our genus was characterized by lower extremity adaptations for long distance running, a distinctive shoulder mechanism with a forward – rather than laterally-facing shoulder joint, somewhat smaller molar teeth with more occlusal relief than those of earlier hominins with occlusal wear indicative of a diverse diet of tougher foods, and the regular manufacture and use of stone tools that functioned in butchery and probably other foraging activities. Early members of the genus *Homo* are associated with increasingly open, albeit wetter, habitats than earlier hominins. However, these early species of *Homo* retained a rapid pattern of dental and skeletal development similar to that of earlier hominins and chimpanzees.

Unfortunately, at the present time, most of our understanding of the adaptations of early *Homo* is based on the fossil record for *Homo erectus*. Evidence of adaptive differences and/or similarities among the other early species of *Homo* is severely limited by small craniodental samples and especially by the lack of reliably associated postcranial elements. Hence, efforts to identify major adaptive shifts within early *Homo* may be somewhat premature. This problem is certainly something that increased paleontological sampling could – and should – address.

Acknowledgments We thank our colleagues who participated in the Third Stony Brook Human Evolution Symposium and Workshop for providing five days of stimulating discussion, and the manuscripts that comprise the present volume. We are grateful to Eric Delson for his careful reading of our summary chapter and for the numerous suggestions which improved it immeasurably.

References

Argue, D., Donlon, D., Groves, C., Wright, R., 2006. *Homo floresiensis*: microcephalic, pygmoid, *Australopithecus*, or *Homo*? Journal of Human Evolution 51, 360–374.

Asfaw, B., Beyene, Y., Suwa, G., Walter, R.C., White, T.D., WoldeGabriel, G., Yemane, T., 1992. The earliest Acheulean from Konso-Gardula. Nature 360, 732–735.

Baab, K.L., 2008. The taxonomic implications of cranial shape variation in *Homo erectus*. Journal of Human Evolution 54, 827–847.

Bar-Yosef, O., Belfer-Cohen, A., 2001. From Africa to Eurasia - early dispersals. Quaternary International 75, 19–28.

Behrensmeyer, A.K., 2006. Climate change and human evolution. Science 311, 476–478.

Bingham, P.M., 1999. Human uniqueness: a general theory. Quarterly Review of Biology 74, 133–169.

Blumenschine, R.J., Peters, C.R., Masao, F.T., Clarke, R.J., Deino, A.L., Hay, R.L., Swisher, C.C., Stanistreet, I.G., Ashley, G.M., McHenry, L.J., Sikes, N.E., van der Merwe, N.J., Tactikos, J.C., Cushing, A.E.,

Deocampo, D.M., Njau, J.K., Ebert, J.I., 2003. Late Pliocene *Homo* and hominid land use from western Olduvai Gorge, Tanzania. Science 299, 1217–1221.

Bramble, D.M., Lieberman, D.E., 2004. Endurance running and the evolution of *Homo*. Nature 432, 345–352.

Braun, D.R., Plummer, T., Ditchfield, P., Ferraro, J.V., Maina, D., Bishop, L.C., Potts, R., 2008a. Oldowan behavior and raw material transport: perspectives from the Kanjera Formation. Journal of Archaeological Science 35, 2329–2345.

Braun, D.R., Tactikos, J.C., Ferraro, J.V., Arnow, S.L., Harris, J.W.K., 2008b. Oldowan reduction sequences: methodological considerations. Journal of Archaeological Science 35, 2153–2163.

Clark, J.D., 1993. Stone artifact assemblages from Members 1–3, Swartkrans Cave. In: Brain, C.K. (Ed), Swartkrans: A Cave's Chronicle of Early Man. Transvaal Museum Monographs, Pretoria, pp. 167–194.

Dart, R.A., 1925. *Australopithecus africanus*: the man-ape of South Africa. Nature 115, 235–236.

Dean, M.C., Leakey, M.G., Reid, D.J., Schrenk, F., Schwartz, G.T., Stringer, C., Walker, A., 2001. Growth processes in teeth distinguish modern humans from *Homo erectus* and earlier hominins. Nature 414, 628–631.

Delson, E., 1995. The strapping young ancestor. American Scientist 83, 183–185.

Domínguez-Rodrigo, M.J., Serrallonga, J., Juantresserras, L., Alcala, A., luque, L., 2001. Woodworking activities by early humans: a plant residue analysis on Acheulean stone tools from Peninj (Tanzania). Journal of Human Evolution 40, 289–299.

Dunsworth, H., Walker, A.C., 2002. Early genus *Homo*. In: Hartwig, W. (Ed), The Primate Fossil Record. Cambridge University Press, New York, pp. 419–435.

Frost, S.R., 2007. Fossil Cercopithecidae from the Middle Pleistocene Dawaitoli Formation, Middle Awash valley, Afar region, Ethiopia. American Journal of Physical Anthropology 134, 460–471.

Gabunia, L., Antón, S.C., Lordkipanidze, D., Vekua, A., Justus, A., Swisher, C.C., 2001. Dmanisi and dispersal. Evolutionary Anthropology 10, 158–170.

Gabunia, L., Vekua, A.K., de Lumley, M.A., Lordkipanidze, D., 2002. A new species of *Homo* represented by a fossil from the bottom part of the Pleistocene layer at Dmanisi, Georgia. Archaeology, Ethnology and Anthropology of Eurasia 4, 145–152.

Gebo, D.L., Schwartz, G.T., 2006. Foot bones from Omo: implications for hominin evolution. American Journal of Physical Anthropology 129, 499–511.

González-José, R., Escapa, I., Neves, W.A., R. Cúneo, R. and Pucciarelli, H.M., 2008. Cladistic analysis of continuous modularized traits provides phylogenetic signals in *Homo* evolution. Nature 453, 775–778.

Gordon, A.D., Nevell, L., Wood, B.A., 2008. The *Homo floresiensis* cranium (LB1): size, scaling, and early *Homo* affinities. Proceeding of the National Academy of Sciences USA 105, 4650–4655.

Grine, F.E., 2001. Implications of morphological diversity in early *Homo* crania from eastern and southern Africa. In: Tobias, P.V., Raath, M., Moggi-Cecchi, J., Doyle, G. (Eds), Humanity from the African Naissance to the Coming Millennia. Firenza University Press, Firenza, pp. 107–115.

Grine, F.E., 2005. Early *Homo* at Swartkrans, South Africa: a review of the evidence and an evaluation of recently proposed morphs. South African Journal of Science 101, 43–52.

Grine, F.E., Jungers, W.L., Schultz, J., 1996. Phenetic affinities among early *Homo* crania from East and South Africa. Journal of Human Evolution 30, 189–225.

Howell, F.C., 1978. Hominidae. In: Maglio, V.J., Cooke, H.B.S. (Eds), Evolution of African Mammals. Harvard University Press, Cambridge, MA, pp. 154–248.

Johanson, D.C., Masao, F.T., Eck, G., White, T.D., Walter, R.C., Kimbel, W.H., Asfaw, B., Manega, P., Ndessokia, P., Suwa, G., 1987. New partial skeleton of *Homo habilis* from Olduvai Gorge, Tanzania. Nature 327, 205–209.

Kimbel, W.H., Johanson, D.C., Rak, Y., 1997. Systematic assessment of a maxilla of *Homo* from Hadar, Ethiopia. American Journal of Physical Anthropology 103, 235–262.

Klein, R.G., 1983. The Stone Age prehistory of southern Africa. Annual Review of Anthropology 12, 25–48.

Kuman, K., Clarke, R.J., 2000. Stratigraphy, artifact industries and hominid associations for Sterkfontein, Member 5. Journal of Human Evolution 38, 827–847.

Larson, S., 2007. Evolutionary transformation of the hominin shoulder. Evolutionary Anthropology 16, 172–187.

Larson, S., Jungers, W.L., Morwood, M. J., Sutikna, T., Jatmiko, Saptomo, E.W., Due, R.A., Djubiantono, T., 2007. *Homo floresiensis* and the evolution of the hominin shoulder. Journal of Human Evolution 53, 718–731.

Leakey, L.S.B., Tobias, P.V., Napier, J.R., 1964. A new species of the genus *Homo* from Olduvai Gorge. Nature 202, 7–9.

Leakey, M.G., Spoor, F., Brown, F.H., Gathogo, P.N., Kiarie, C., Leakey, L., McDougall, I., 2001. New hominin genus from eastern Africa shows diverse middle Pliocene lineages. Nature 410, 433–440.

Lee, S.H., Wolpoff, M.H., 2005. Habiline variation: a new approach using STET. Theory in Biosciences 124, 25–40.

Levin, N.E., Cerling, T.E., Passey, B.H., Harris, J.M., Ehleringer, J.R., 2006. A stable isotope aridity index for terrestrial environments. Proceedings of the National Academy of Sciences 103, 11201–11205.

Lieberman, D.E., Pilbeam, D.R., Wood, B.A., 1988. A probabilistic approach to the problem of sexual dimorphism in *Homo habilis*: a comparison of KNM-ER 1470 and KNM-ER 1813. Journal of Human Evolution 17, 503–512.

Lycett, S.J., von Cramon-Taubadel, N., 2008. Acheulean variability and hominin dispersals: a model-bound approach. Journal of Archaeological Science 35, 553–562.

McHenry, H.M., Brown, C.C., 2008. Side steps: the erratic pattern of hominin postcranial change through time. Journal of Human Evolution 55, 639–651.

Meyer, K.E., Saether, E.E., Soiney, E.K., Shebeck, M.S., Paddock, K.L., Ludewig, P.M., 2008. Three-dimensional scapular kinematics during the throwing motion. Journal of Applied Biomechanics 24, 24–34.

Miller, J.M.A., 1991. Does brain size variability provide evidence of multiple species in *Homo habilis*? American Journal of Physical Anthropology 84, 385–398.

Miller, J.M.A., 2000. Craniofacial variation in *Homo habilis*: an analysis of the evidence for multiple species. American Journal of Physical Anthropology 112, 103–128.

Morwood, M.J., Brown, P., Jatmiko, Sutikna, T., Wahyu Saptomo, E., Westaway, K.E., Awe Due, R., Roberts, R.G., Maeda, T., Wasisto, S., Djubiantono, T., 2005. Further evidence for small-bodied hominins from the Late Pleistocene of Flores, Indonesia. Nature 437, 1012–1017.

Moyà-Solà, S., Köhler, M., Alba, D.M., Almöcija, S., 2008. Taxonomic attribution of the Olduvai Hominid 7 manual remains and the functional interpretation of hand morphology in robust australopithecines. Folia Primatologica, 79, 215–250.

O'Brien, E.M., 1981. The projectile capabilities of an Acheulean handaxe from Olorgesailie. Current Anthropology 22, 76–79.

Potts, R., 1998a. Environmental hypotheses of hominin evolution. Yearbook of Physical Anthropology 41, 93–136.

Potts, R., 1998b. Variability selection in hominid evolution. Evolutionary Anthropology 7, 81–96.

Prat, S., Brugal, J.-P., Tierceling, J.-J., Barrat, J.-A., Bohn, M., Delagnes, A., Harmand, S., Kimeu, K., Kibunjia, M., Texier, P-J., Roche, H., 2005. First occurrence of early *Homo* in the Nachukui Formation (West Turkana, Kenya) at 2.3–2.4 Myr. Journal of Human Evolution 49, 230–240.

Rightmire, G.P., 1993. Variation among early *Homo* crania from Olduvai Gorge and the Koobi Fora region. American Journal of Physical Anthropology 90, 1–33.
Roche, H., 1995. Les industries de la limite Plio-Pléistocène et du Pléistocène ancien en Afrique. In: Gilbert, J. (Ed), Congresso Internacional de Paleontologia Humana (Orce, September 1995), 3a, Circular: 93. Fundación Caja de, Orce.
Roche, H., Brugal, J.-P., Delagnes, A., Feibel, C., Harmand, S., Kibunjia, M., Prat, S., Texier, P.-J., 2003. Les sites archeologiques plio-pleistocenes de la formation de Nachukui, Ouest-Turkana, Kenya: bilan synthétique 1997–2001. Comptes Rendus Palevol 2, 663–673.
Semaw, S., 2000. The world's oldest stone Artefacts from Gona, Ethiopia: their implications for understanding stone technology and patterns of human evolution between 2.6–1.5 million years ago. Journal of Archaeological Science 27, 1197–1214.
Semaw, S., Renne, P., Harris, J.W.K., Feibel, C.S., Bernor, R.L., Fesseha, N., Mowbray, K., 1997. 2.5-million-year-old stone tools from Gona, Ethiopia. Science 284, 625–629.
Semaw, S., Rogers, M.J., Quade, J., Renne, P.R., Butler, R.F., Dominguez-Rodrigo, M., Stout, D., Hart, W.S., Pickering, T., Simpson, S.W., 2003. 2.6-million-year-old stone tools and associated bones from OGS-6 and OGS-7, Gona, Afar, Ethiopia. Journal of Human Evolution 45, 169–177.
Smith, B.H., 1993. The physiological age of KNM-WT 15000. In: Leakey, R., Walker, A. (Eds), The Nariokotome *Homo erectus* Skeleton. Harvard University Press, Cambridge, MA, pp. 195–220.
Stent, G.S., 1972. Prematurity and uniqueness in scientific discovery. Scientific American 227, 84–93.
Strait, D.S., Grine, F.E., Fleagle, J.G., 2007. Analyzing hominid phylogeny. In: Henke, W., Tattersall, I. (Eds), Handbook of Palaeoanthropology, Volume 3, Phylogeny of Hominids. Springer, New York, pp. 1781–1806.
Susman, R.L., 1983. Evolution of the human foot: evidence from Plio-Pleistocene hominids. Foot and Ankle 3, 365–376.
Susman, R.L., 2008. Evidence bearing on the status of *Homo habilis* at Olduvai Gorge. American Journal of Physical Anthropology 137, 356–361.
Susman, R.L., Stern, J.T., 1982. Functional morphology of *Homo habilis*. Science 217, 931–934.
Suwa, G., White, T.D., Howell, F.C., 1996. Mandibular postcanine dentition from the Shungura Formation, Ethiopia: crown morphology, taxonomic allocations, and Plio-Pleistocene hominid evolution. American Journal of Physical Anthropology 101, 247–282.
Theime, H., 1997. Lower Paleolithic hunting spears from Germany. Nature 385, 807–810.
Tocheri, M., Jungers, W., Larson, S., Orr, C., Sutikna, T., Jatmiko, Saptomo, Due, E.R., Djubiantono, T., Morwood, M., 2007. Morphological affinities of the wrist of *Homo floresiensis*. Abstracts of the 2007 Annual Palaeoanthropology Society Meeting, p. 28. Available at *http://www.paleoanthro.org/meeting.htm*
Tocheri, M.W., Orr, C.M., Jacofsky, M.C., Marzke, M.W., 2008. The evolutionary history of the hominin hand since the last common ancestor of *Pan* and *Homo*. Journal of Anatomy 212, 544–562.
Vrba, E.S., 1985. Environment and evolution - alternative causes of the temporal distribution of evolutionary events. South African Journal of Science 81, 229–236.
Vrba, E.S., 1995. On the connections between paleoclimate and evolution. In: Vrba, E.S., Denton, G.H., Partridge, T.C., Burkle, L.H. (Eds), Paleoclimate and Evolution with Emphasis on Human Origins. Yale University Press, New Haven, CT, pp. 24–45.
Vrba, E.S., 1996. Climate, heterochrony, and human evolution. Journal of Anthropological Research 52, 1–28.
Whittaker, J.C., McCall, G., 2001. Handaxe-hurling hominids: an unlikely story. Current Anthropology 42, 566–572.
Wood, B.A., 1974. Olduvai Bed I post-cranial fossils: a reassessment. Journal of Human Evolution 3, 373–378.
Wood, B.A., 1991. Hominid Cranial Remains. Koobi Fora Research Project, Volume 4. Clarendon Press, Oxford.
Wood, B.A., 1992. Origin and evolution of the genus *Homo*. Nature 355, 783–790.
Wood, B.A., 1993. Early *Homo*: how many species? In: Kimbel, W.H., Martin, L.B. (Eds), Species, Species Concepts, and Primate Evolution. Plenum, New York, pp. 485–522.
Wood, B.A., Collard, M.C., 1999. The human genus. Science 284, 65–71.

Subject Index

Specimen Index

Site/Locality Index

U

W

Printed by Publishers' Graphics LLC